D0386064

Air Quality Analysis for Urban Transportation Planning

MIT Press Series in Transportation Studies
Marvin L. Manheim, editor
Center for Transportation Studies, MIT

1. *The Automobile and the Environment: An International Perspective,* edited by Ralph Gakenheimer, 1978
2. *The Urban Transportation System: Politics and Policy Innovation,* Alan Altshuler with James P. Womack and John R. Pucher, 1979
3. *Planning and Politics: The Metro Toronto Transportation Review,* Juri Pill, 1979
4. *Fundamentals of Transportation Systems Analysis. Volume I: Basic Concepts,* Marvin L. Manheim, 1979
5. *Traffic Flow on Transportation Networks,* Gordon F. Newell, 1980
6. *Miles to Go: European and American Transportation Policies,* James A. Dunn, Jr., 1981
7. *Air Quality Analysis for Urban Transportation Planning,* Joel L. Horowitz, 1982

Air Quality Analysis for Urban Transportation Planning

Joel L. Horowitz

The MIT Press
Cambridge, Massachusetts
London, England

This book was written by the author in his private capacity. No official support or endorsement by the Environmental Protection Agency or any other agency of the federal government is intended or should be inferred.

This book was set in Univers Medium
by Asco Trade Typesetting Ltd., Hong Kong,
and printed and bound by Halliday Lithograph in the United States of America.

Library of Congress Cataloging in Publication Data

Horowitz, Joel.
  Air quality analysis for urban transportation planning.
  (MIT Press series in transportation studies; 7)
  Bibliography: p.
  Includes index.
  1. Urban transportation—Environmental aspects.
2. Air—Pollution. 3. Air quality management.
I. Title. II. Series.
TD195.T7H67 363.7'3921 82-15272
ISBN 0-262-08116-4 AACR2

To Katharine, who has her father back now

Contents

Contents

List of Figures

List of Tables

Series Foreword

The field of transportation has emerged as a recognized profession only in the last fifteen years, although transportation issues have been important throughout history. Today more and more government agencies, universities, researchers, consultants, and private industry groups are becoming truly multimodal in their orientations, and specialists of many different disciplines and professions are working on multidisciplinary approaches to complex transportation issues.

The central role of transportation in our world today and its recent professional status have led The MIT Press and the MIT Center for Transportation Studies to establish The MIT Press Series in Transportation Studies. The series will present works representing the broad spectrum of transportation concerns. Some volumes will report significant new research, while others will give analyses of specific policy, planning, or management issues. Still others will show the interaction between research and policy. Contributions will be drawn from the worldwide transportation community.

Urban air pollution is perceived as a major policy question in many countries and in many urban areas. A central issue facing policymakers, transportation professionals, and citizens is to what extent the design and operation of urban transportation systems may affect air quality. This book, the seventh in the series, will contribute significantly to the illumination of this question by providing a comprehensive scientific and analytical treatment of these relations between urban transportation decisions and air quality.

Marvin L. Manheim

Preface

It has long been recognized that motor vehicles are major contributors to urban air pollution. In addition, it is widely accepted, at least in the United States, that motor vehicles should be equipped with devices to reduce emissions of air pollutants, and new vehicles now sold in this country are designed to meet stringent pollutant emissions standards. Another possible way of reducing air pollution generated by motor vehicles is to design and operate urban transportation systems in ways that tend to reduce emissions. For example, transit service might be improved, carpooling encouraged, and gasoline made more expensive, all with the aim of reducing motor vehicle use and, thereby, emissions. In recent years, many transportation and air quality professionals, advisors to policy-making governmental officials, and members of the public have become involved with the questions whether and to what extent the design and operation of urban transportation systems can or should be changed so as to reduce motor vehicle emissions. This book is directed toward such individuals. Its purpose is to provide a source of basic conceptual, factual, and analytical information on air pollution as it relates to urban transportation.

In contrast to other books on air pollution and its control, this book deals exclusively with topics that are likely to interest and be useful to persons involved in urban transportation planning and decision making. Examples of such topics are the characteristics and harmful effects of transportation-related air pollutants, the effectiveness of transportation measures for controlling air pollution, and methods for predicting concentrations of transportation-related air pollutants. There is little or no discussion of such topics as the characteristics of nontransportation-related air pollutants, industrial air pollution engineering, and air pollution instrumentation. The subjects of motor vehicle emissions standards and emissions control devices have been intensely controversial in the United States during the past ten to fifteen years and have attracted the interest of many observers. However, from the point of view of urban transportation planning, motor vehicle emissions standards and emissions control devices represent fixed parameters, not decision variables. Accordingly, the discussion of these subjects here is relatively brief and is intended mainly as background material rather than as a basis for further analysis.

The orientation of the book is mainly scientific, technical, and analytical. It does not develop policy themes or present policy judgments. Nor does it discuss the many complex institutional (for example, legal, political, and organizational) issues that surround the subject of air quality and urban transportation. The lack of discussion of such issues may strike some readers as a serious omission. However, the institutional issues associated with air quality and those associated with other aspects of urban transportation are closely interwoven and cannot be separated in any meaningful way. Their treatment could easily fill a book by itself and would require a substantial diversion from the subject of air quality analysis.

The book is intended as an information source and reference for individuals working in the area of air quality and urban transportation, as the basis of a one-semester course on this subject, or as a reference source in a course that covers a broader range of transportation or environmental topics. Most of the material in chapters 1–6 can be understood by readers who are comfortable with mathematical notation, ordinary algebra, and the basic concepts involved in mathematical modeling of complex systems. Familiarity with physics, chemistry, calculus, probability theory, and statistics at the level of introductory college courses would be helpful in certain brief parts of the discussion but are not essential. Chapter 7, which deals with dispersion modeling, makes extensive use of calculus and elementary probability theory. Readers who are not comfortable with these subjects may skip chapter 7 without loss of understanding of the material in the other chapters.

The book was written mainly during 1980, and it reflects information that was available then. As the book was entering production, the US Congress was in the early stages of revising the Clean Air Act, which is the main piece of legislation governing air pollution control activity in the United States. The outcome of this revision process cannot be predicted at present, but changes in the US air quality standards (discussed in sections 1.3 and 4.7 of this book) and motor vehicle emissions standards (discussed in section 5.4) are possible results. Even without changes in legislation, the US air quality standards are revised administratively from time to time, and estimates of emissions from motor vehicles and other sources are updated periodically. Indeed, as is noted in table 1.1, two of the air quality standards were in the process of being revised or revoked as this book entered production. None of these actual or potential changes affects the main substantive content of the book. However, readers who require precise and up-to-date information on US air pollution control regulations and practice are advised to consult the most recent government publications on these subjects.

Acknowledgments

Many people have helped to make this book possible. It is a pleasure to acknowledge and thank them for their efforts.

The book evolved from a course on the same subject that I taught in the spring of 1979 while I was a visiting faculty member in the Transportation Systems Division of the Department of Civil Engineering at MIT. I am grateful to the division and the department for giving me the opportunity to teach the course and begin organizing the material that ultimately led to this book.

I have benefited greatly from discussions with Walter F. Dabberdt on the behavior of carbon monoxide near roadways, Kenneth L. Demerjian on atmospheric chemistry, Paul P. Jovanis and Adolph D. May on the interpretation of the outputs of the TRANSYT6C and FREQ models, and Robert G. Lamb on dispersion modeling. George Bonina, Franklin Ching, Elizabeth Deakin, Greig Harvey, Louis E. Keefer, Robert G. Lamb, James L. Repace, and Earl Shirley each read and provided valuable comments on drafts of one or more chapters. I thank all of these people for their help. Naturally, they are absolved of responsibility for any errors that remain in the final product.

My greatest thanks go to my wife, Susan. In addition to providing general moral support and encouragement while the book was being prepared, she served as an editorial and research assistant and typed the early drafts of the manuscript. This book simply could not have been written without her support and assistance.

Barbara Sexton did a superb job of typing the final manuscript.

Air Quality Analysis for
Urban Transportation Planning

1 Introduction

1.1 Objectives

This book is concerned with the relation between urban transportation systems and air pollution. It deals with questions such as:

- What are the transportation-related air pollutants, and why are they harmful?
- How widespread is the problem of transportation-related air pollution?
- What measures are available for reducing transportation-related air pollution?
- What techniques can be used to predict the effectiveness of these measures?

The range of subject areas and intellectual disciplines that affect the answers to these questions is extremely broad. Although urban transportation systems are major sources of urban air pollution, they are by no means the only important sources. Thus, the answers depend on the extent to which transportation systems, as opposed to other sources, are responsible for air pollution. The answers also depend on the characteristics of the physical, chemical, and biological processes that govern the formation and dispersal of air pollutants and that are the sources of the harmful effects or air pollution. In addition, the answers depend on the ways in which potential pollution control measures affect travel behavior and the performance of motor vehicles.

Most people who are involved in the identification, analysis, evaluation, or implementation of transportation options to reduce air pollution are not routinely concerned with all of the subject areas that affect the answers to the foregoing questions. However, as will be seen in the next section, the determination whether implementation of a particular transportation measure would cause or has caused an improvement in air quality, the evaluation of the magnitude of the improvement, and the estimation of the associated benefits and costs all require consideration of chains of connections and interactions that involve most, if not all, of the subject areas. It is the fundamental premise of this book that a basic understanding of these subjects is a prerequisite to effective performance in identifying, analyzing, and evaluating transportation options to improve air quality.

The principal objective of this book is to provide the information needed to achieve such an understanding.

The book deals almost exclusively with what might loosely be called the scientific, technical, and analytic aspects of the relation between urban transportation and air pollution, rather than the institutional (e.g., legal, political, and organizational) aspects. The importance of institutional factors in air pollution control is, of course, enormous. The fact that they are not treated in this book should not be interpreted as suggesting that an understanding of the scientific, technical, and analytic aspects of air pollution control is somehow sufficient to ensure the successful implementation of air pollution control programs. Such a conclusion would be seriously mistaken. However, it is suggested that an understanding of the scientific, technical, and analytic aspects may be a necessary condition for successful implementation of control programs. It is in this spirit that the material in this book is offered.

1.2 A Conceptual Framework

The basic problem of all air pollution control can be stated very simply. Human activities, such as transportation, cause certain noxious materials to be emitted into the air. These materials are harmful in various ways, and it is desirable to identify and implement measures for reducing the emissions and alleviating the harm that they cause. However, controlling air pollution is not as simple as this statement of the problem may make it seem. Human activities and the harmful effects of air pollution are only indirectly related. There are a number of steps and complicating factors that intervene between the human and, in some cases, natural processes that are the basic causes of air pollution and the harmful effects that one would like to alleviate. Consequently, the determination of where, by what means, and by how much emissions should be reduced in order to mitigate the harmful effects of air pollution can be a difficult task.

The relation between pollution sources, pollution control policies, and the harmful effects of pollution is shown schematically in figure 1.1. There are two basic sources of air pollution: human activities and certain natural processes. These activities and processes cause the emission of polluting substances into the air. The emissions correspond physically to flows, that is, the injection of materials into the air at certain rates, and are measured in flow units, such as mass of material emitted per unit of time or per unit of human activity. Examples of these units are grams per second, tons per year, or grams per vehicle kilometer.

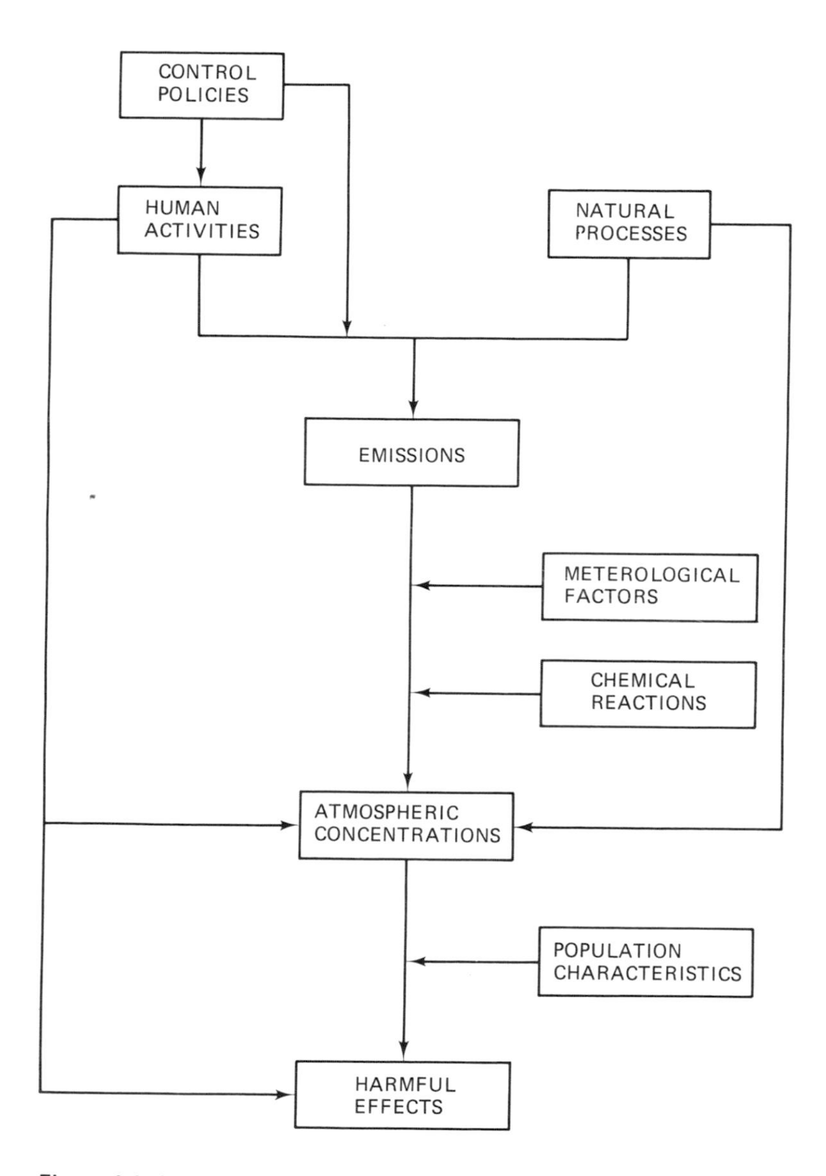

Figure 1.1 A conceptual framework for air quality analysis.

The emissions accumulate in the air to produce concentrations of pollutants, that is, densities that indicate the amount of pollutant present in a unit of air. Concentrations are measured in units such as micrograms of pollutant per cubic meter of air and parts of pollutant per million parts of air.

The relation between emissions and concentrations is influenced by several factors in addition to the gross magnitude of emissions. First, because concentrations represent accumulations of materials, they depend on the locations and times at which emissions occur and, hence, on the locations and times of the human activities and natural processes that produce the emissions. The relation between emissions and concentrations is influenced also by meteorological factors, such as wind speeds and directions. Concentrations tend to be higher, for example, when the wind speed is low and at locations downwind of large emissions sources. Most day-to-day variability in air quality, which can be substantial, is caused by changes in meteorological factors rather than changes in emissions. Finally, the materials emitted into the air by emissions sources may react chemically with each other or with normal atmospheric constituents such as oxygen and water. This can cause the presence in the air of polluting substances that are not emitted by any source, and it can cause the atmospheric concentrations of substances that are emitted in large quantities to be very low.

The direct causes of the harmful effects of air pollution are concentrations rather than emissions. The severity and extent of these harmful effects depend on the magnitudes of the concentrations and on the characteristics of the population that is exposed to the pollution. For example, people with chronic cardiovascular or respiratory diseases are more likely to be harmed by air pollution than are people without these diseases, and all people are more likely to be harmed by high concentrations than by low ones. The extent of harmful effects is influenced also by the human activity system. For example, the extent to which people, crops, and materials are harmed by air pollution depends on their locations and, in the case of people and some materials, movements in polluted areas; this is because pollutant concentrations are virtually never uniform over large areas.

Most pollution control measures affect this conceptual system by means of their influence on either the human activities that cause emissions or the relation between human activities and emissions. For example, a transit service improvement that is intended to induce people to drive less and, therefore, to pollute less is obviously aimed at influencing human activities. On the other hand, the installation of emissions control devices on automobiles is aimed at influencing the relation between human activities and emissions, that is, at reducing the magnitude of emissions per vehicle kilometer or vehicle trip.

It should be clear from figure 1.1 that pollution control measures influence the harmful effects of air pollution only in a very indirect, complex, and, possibly, controversial way. In particular, predicting or measuring the effects of pollution control measures on the severity and extent of the harmful effects of pollution is a multistage process, each stage of which contains numerous potential sources of errors or uncertainties. Moreover, given the complexity of the linkages and the exigencies of public policymaking, numerous opportunities and pressures for oversimplification can arise. For example, many policy studies treat gross emissions as surrogates for concentrations, thereby ignoring the influences of locational and temporal factors, meteorological variables, and chemical reactions. More important, the complexity and current state of understanding of the linkages between pollution control measures and changes in the severity and extent of the harmful effects of pollution are such that it is unrealistic to expect precise or, in some cases, even very accurate measurements or predictions of the effects of pollution control measures. Indeed, as will be seen in later chapters, it sometimes is difficult to predict even whether a particular measure will cause air quality to improve or to worsen.

Because of the complexity of the relations between pollution control measures and their effects on levels of pollution damages, it is necessary to divide air pollution analysis into distinct disciplinary areas. The standard areas, which will be followed in this book (although not in the order presented here), are as follows:

Analysis of Linkages between Public Policies and Human Activities

This deals with problems such as determining what effects transportation system improvements will have on traffic flow characteristics relevant to air quality. This is the area in which planning and transportation agencies are most likely to be involved. It is treated in chapter 6.

Analysis of Linkages between Human Activities and Emissions

Broadly speaking, this is the area dealt with by air pollution engineering. Insofar as it relates to transportation, this segment involves questions such as: Why do motor vehicles cause pollution? What devices are available for reducing motor vehicle emissions? How effective are these devices?

What traffic flow variables affect motor vehicles' emissions rates? What is the relation between these variables and emissions rates? These questions are treated in chapter 5.

Analysis of Linkages between Emissions and Concentrations
This area encompasses air pollution meteorology and atmospheric chemistry. The basic problem in this area is to develop methods for predicting the concentrations that will be caused by specified levels of emissions. This subject is treated qualitatively in chapter 3 and quantitatively in chapter 7.

Analysis of Linkages between Pollutant Concentrations and the Harmful Effects of Air Pollution
This area is concerned with the harm to human health, plants, and materials that air pollution causes. It is also concerned with the quantitative assessment of the benefits of air pollution control. It deals with questions such as: What are the human-health effects of air pollution? What are the quantitative dose-response relations between human, plant, and materials exposures to air pollutants and the harmful effects of these exposures? How much should society be willing to pay to alleviate these harmful effects? These subjects are treated in chapter 4.

These areas of air quality analysis exist in a context that is established by the relative magnitudes of pollutant emissions from transportation and other sources and the extent to which the emissions of nontransportation sources can be controlled. The relative magnitudes of transportation and nontransportation emissions are discussed in chapter 2. The control of emissions from nontransportation sources is not treated in this book. Similarly, institutional factors affecting air pollution control are not treated.

1.3 Air Quality Standards

The linkage between air pollutant concentrations and the harm caused by air pollution is so complex, difficult to study, and poorly understood that for most purposes it is useful to avoid dealing explicitly with this linkage at all. The separation of this linkage from most other aspects of air quality analysis can be achieved by replacing it with a simple surrogate called an air quality standard. An air quality standard is a concentration defined so that if the concentration of a pollutant is below the standard for that pollutant, then for most purposes it can be presumed that the pollutant is not causing harmful effects sufficiently serious to justify corrective measures. Conversely, if the concentration of a particular pollutant exceeds the air quality standard for that pollutant, it can be presumed that significant harmful effects may be occurring and that corrective measures are needed.

Given the surrogates of air quality standards, the operational objective of air pollution control policy becomes achieving these standards. Thus, the complexities of the relations between pollutant concentrations and the harm caused by pollution need not enter the policy analysis process.

In the United States, two types of national air quality standards have been established: primary standards and secondary standards. The primary air quality standards are set to protect human health from the harmful effects of pollution. In other words, if pollutant concentrations exceed the primary air quality standards, it is presumed for most purposes that harm to human health may be occurring and corrective measures are needed. It is presumed that significant harm to human health is not occurring if pollutant concentrations do not exceed the primary standards. The secondary standards are set to protect animals, plants, materials, visibility, soils, water, climate, and things of economic value from the harmful effects of air pollution.[1] The standards that were in effect in the United States during 1980 are shown in table 1.1.

1.4 *Air Pollution Terminology*

Air pollutants are classified as primary or secondary, depending on whether they are emitted directly by man-made or natural sources or are formed in the air by chemical reactions. The primary pollutants are emitted directly, whereas the secondary pollutants are formed by chemical reactions.

Emissions are classified as man-made (or anthropogenic) or natural, depending on whether they originate from man-made or natural sources.

Emissions sources are classified as point, area, or line sources, depending on the magnitudes and geographical distributions of their emissions. A point source is a large, geographically concentrated emitter whose emissions rates are large enough to be significant by themselves, even if no other emissions sources are present. A large, coal-fired electrical power plant is an example of a point source. Transportation sources usually are not treated as point sources. An area source is a collection of small, geographically dispersed emitters that are not significant individually but that are important collectively. Residential and commercial space heaters, considered collectively, and the street system of a city are examples of area sources. A line source is a collection of sources that are small individually and that are distributed roughly uniformly along a line. A

Table 1.1 National air quality standards in effect in the United States during 1980[a]

Pollutant	Averaging time	Primary standard	Secondary standard
Sulfur dioxide	1 yr	80 μg/m³ (0.03 ppm)	
	24 hr	365 μg/m³ (0.14 ppm)	
	3 hr		1,300 μg/m³ (0.5 ppm)
Particulate matter	1 yr[b]	75 μg/m³	
	24 hr	260 μg/m³	150 μg/m³
Nonmethane hydrocarbons[c]	6–9 A.M.	160 μg/m³ (0.24 ppm)	Same as primary standard
Carbon monoxide	1 hr	40 mg/m³ (35 ppm) existing	Same as primary standard
		29 mg/m³ (25 ppm) proposed	
	8 hr	10 mg/m³ (9 ppm)	Same as primary standard
Ozone	1 hr	235 μg/m³ (0.12 ppm)	Same as primary standard
Nitrogen dioxide	1 yr	100 μg/m³ (0.05 ppm)	Same as primary standard
Lead	Calendar quarter	1.5 μg/m³	Same as primary standard

a. Sources: References [1, 2]. The concentrations are specified for a temperature of 25°C and a pressure of 760 mm of mercury. The units used to express the concentrations are explained in section 1.4. The complete specifications of the national air quality standards include descriptions of the air quality measurement techniques and statistical procedures that should be used in comparing pollutant concentrations at a location with the standards. See references [1, 2] for these descriptions.

b. Annual geometric mean.

c. The nonmethane hydrocarbon standard was established as a guide to achieving the ozone standard and is not based on consideration of direct harmful effects of nonmethane hydrocarbons. At the time that the nonmethane hydrocarbon standard was established, it was believed that achieving this standard would insure achievement of the ozone standard. Subsequently, it was found that the ozone and nonmethane hydrocarbon standards could each be achieved without achieving the other. Consequently, the nonmethane hydrocarbon standard is no longer enforced and has ceased to be of significance in air pollution control programs. (In May 1981, the US Environmental Protection Agency proposed revoking this standard.)

heavily traveled section of an urban freeway is an example of a line source.

The distinction between area and line sources is not precise. The same source may be classified either as an area source or a line source, depending on the purpose of the classification. For example, if it is desired to predict the change in the local concentration of a certain pollutant that would be caused by the opening of a new freeway section, it may be convenient to treat the freeway as a line source. However, if it is desired to estimate the contribution of traffic on and in the vicinity of the freeway to concentrations several miles downwind of the freeway, then it may be most convenient to treat the freeway and the nearby surface streets collectively as a single area source.

Most air quality analyses require, as inputs, tabulations of all significant emissions sources and their rates of emission of the pollutants being analyzed. Such tabulations are called emissions inventories. Emissions inventories can be constructed at various levels of detail, depending on their intended use. For example, the entire transportation sector of a city might be considered as a single source category in an emissions inventory, or the transportation sector might be disaggregated into automobiles, gasoline trucks, diesel trucks, and other vehicle types. Similarly, an emissions inventory might specify total annual emissions in a city for each source class, or it might disaggregate each source's emissions according to geographical area and time of day. Table 1.2 shows an example of an emissions inventory for volatile organic compounds in Los Angeles, California.

Emissions inventories are essential inputs to efforts to model the relations between emissions and concentrations of air pollutants. In addition, inventories often permit important qualitative inferences to be made about whether reducing emissions from particular classes of sources is likely to be effective in improving air quality in a region. Emissions inventories will be used for this purpose in chapter 2.

In section 1.2, it was noted that pollutant concentrations can be measured either in mass density units, such as micrograms of pollutant per cubic meter of air ($\mu g/m^3$), or in volumetric units, such as parts of pollutant per million parts of air (ppm). The former units are used for both gaseous pollutants and airborne particles, whereas the latter are used only for gases. Other examples of mass density and volumetric units are milligrams per cubic meter (mg/m^3) and parts per hundred million (pphm), respectively. Measurements of pollutant concentrations normally are

Table 1.2 An emissions inventory for volatile organic compounds in the Los Angeles, California, area in 1972[a]

Source category	Weight emissions		Molar emissions		Average molecular weight
	Tons/day	Weight % of total	10^{-2} ton moles/day	Mole % of total	
Stationary sources: Organic fuels and combustion					
Petroleum production and refining					
Petroleum production	62	2.3	214	5.9	29
Petroleum refining	50	1.9	54	1.5	93
Gasoline marketing					
Underground service station tanks	48	1.8	83	2.3	58
Auto tank filling	104	4.0	141	3.9	74
Fuel combustion	23	0.9	92	2.5	25
Waste burning and fires	41	1.6	124	3.4	33
Stationary sources: Organic chemicals					
Surface coating					
Heat treated	14	0.5	17	0.5	82
Air dried	129	5.0	148	4.1	87
Dry cleaning					
Petroleum-based solvent	16	0.6	13	0.4	126
Synthetic solvent (PCE)	25	1.0	15	0.4	166
Degreasing					
TCE solvent	11	0.4	8	0.2	132
1, 1, 1-T solvent	95	3.6	71	2.0	134
Printing					
Rotogravure	31	1.2	38	1.0	82
Flexigraphic	15	0.6	26	0.7	57

Industrial process sources					
Rubber and plastic manufacturing	42	1.6	58	1.6	73
Pharmaceutical manufacturing	16	0.6	21	0.6	75
Miscellaneous operations	83	3.2	104	2.9	80
Mobile sources					
Gasoline-powered vehicles					
Light duty vehicles					
Exhaust emissions	780	30.0	1,130	31.2	69
Evaporative emissions	481	18.5	529	14.6	91
Heavy duty vehicles					
Exhaust emissions	285	10.9	413	11.4	69
Evaporative emissions	67	2.6	74	2.0	91
Other gasoline-powered equipment					
Exhaust emissions	110	4.2	159	4.4	69
Evaporative emissions	22	0.8	24	0.7	91
Diesel-powered motor vehicles	12	0.5	13	0.4	89
Aircraft					
Jet	20	0.8	17	0.5	121
Piston	22	0.9	39	1.1	56
Total of weighted average	2,604	100	3,625	100	71.9

a. Source: reference [3].

adjusted to refer to a temperature of 25°C (77°F) and a pressure of 760 mm (29.92 in.) of mercury.

Mass density and volumetric units tend to be used interchangeably in studies of gaseous pollutants. Hence, it is useful to be able to convert from one set of units to the other. At a temperature of 25°C and a pressure of 760 mm of mercury, the formula for converting ppm to $\mu g/m^3$ is

$$\mu g/m^3 = (\text{ppm})\,(A)\,(10^6)/24{,}500, \tag{1.1}$$

where A is the molecular weight of the pollutant. For example, ozone (O_3) has a molecular weight of 48. Hence, 0.1225 ppm of ozone corresponds to 240 $\mu g/m^3$.

Instantaneous air pollutant concentrations fluctuate chaotically in response to molecular collisions and random changes in wind currents. These fluctuations are so rapid that they can be neither measured nor predicted. Fortunately, it is unnecessary to be able to measure or predict instantaneous concentrations, as instantaneous concentrations do not cause harmful effects. Rather, harmful effects are caused by average concentrations over longer periods of time. Consequently the concentrations of practical interest are average concentrations over these longer time periods. In other words, if T is the length of the time period of interest, t_1 is the beginning of the period, t_2 is the end of the period, and $C(t)$ is the instantaneous concentration at any time t during the period, the concentration of interest, C, is given by

$$C = (1/T)\int_{t_2}^{t_1} C(t)\,dt. \tag{1.2}$$

T is called the pollutant measurement averaging time. Averaging times of practical interest in air pollution control work range from 1 hr to 1 yr. Pollutant concentrations with these averaging times can be both measured and predicted.

Because pollutant concentrations are not constant in time, the measured or predicted concentration of a pollutant depends on the averaging time. Statements of pollutant concentrations have no meaning unless the corresponding averaging times are specified.

References

1

"National Primary and Secondary Ambient Air Quality Standards," *Code of Federal Regulations*, 40 CFR 50, revised as of 1 July 1980.

2
US Environmental Protection Agency, "Carbon Monoxide: Proposed Revisions to the National Ambient Air Quality Standards," *Federal Register*, 45 FR 55066, 18 August 1980.

3
US Environmental Protection Agency, *Air Quality Criteria for Ozone and Other Photochemical Oxidants*, Report No. EPA-600/8-78-004, Research Triangle Park, NC, April 1978. NTIS Publication No. PB80-124753.*

* References with NTIS publication numbers are available from the National Technical Information Service, 5285 Port Royal Road, Springfield, VA 22161.

2 The Transportation-Related Air Pollutants

2.1 The Principal Pollutants

Human activities cause a vast number of polluting substances to be emitted into the air. The list of pollutants includes common combustion products, such as carbon monoxide, sulfur dioxide, and nitric oxide, and it includes a large array of industrial products, such as arsenic, asbestos, beryllium, and cadmium. Most of these pollutants are products of relatively specialized industrial processes and are causes for concern only in the vicinities of plants that produce them. They are in no sense ubiquitous. However, there are seven air pollutants that occur in high concentrations over such widespread areas that they have become the focus of a nationwide effort to control air pollution: particulate matter, sulfur dioxide, carbon monoxide, hydrocarbons (excluding methane), nitrogen dioxide, ozone, and lead.[1]

Particulate matter, or simply particulates, consists of all solid particles and liquid droplets in the air, except pure water. Particulates encompass a wide variety of chemically distinct substances and include such materials as smoke, fly-ash, aerosols formed in atmospheric chemical reactions, windblown dust, and pollen. The individual particles range in size from roughly 0.01 to 10 μm (1 $\mu m = 10^{-6}$ m). In comparison, visible light has wavelengths ranging from 0.4 to 0.7 μm. The size of a water molecule is roughly 0.0004 μm. The damaging effects of particulate matter depend on the chemical substances involved and on the sizes of the particles. Fine particulates (less than 1 μm in size) can enter the lungs and impair lung function. Acidic and alkaline particulates can cause damage to materials, including accelerated deterioration of paint, masonry, and textiles. Particulates greater than roughly 0.1 μm in size (and, especially, in the size range 0.1–1.0 μm) tend to reduce visibility. With certain exceptions that will be noted in section 2.3, particulate matter is not a major transportation-related air pollutant, and it is not discussed extensively in this book. For additional information on particulate matter as an air pollutant, see reference [1].

Sulfur dioxide (SO_2) is a colorless gas with a pungent, irritating odor. The main source of SO_2 as an air pollutant is combustion of sulfur-containing fuels, such as coal and oil. In the air SO_2 can react chemically with oxygen, water, and particulates to form sulfuric acid and various sulfate salts. The resulting mixture of sulfur oxides and sulfates sometimes is referred to as SO_x. Sulfur oxides are respiratory irritants that are associated with increases in illnesses and deaths from such respiratory diseases as bronchitis and emphysema. Mixtures of sulfur oxides and particulates are especially noxious and have been responsible for some of the notorious "killer smog" episodes of the past (e.g., Donora, Pennsylvania, 1948; London, 1952; New York, 1966). In addition, sulfur oxides, being acidic, are highly corrosive to materials and are important causes of the acid rainfall that has been observed in many parts of the world. SO_2 is not a major transportation-related pollutant. For additional information on SO_2 as an air pollutant, see reference [2].

Carbon monoxide (CO) is a colorless, odorless gas whose principal anthropogenic source is incomplete combustion of organic fuels. It combines with the hemoglobin of the blood to produce carboxyhemoglobin and, thereby, reduces the blood's ability to carry oxygen. At sufficiently high concentrations, CO is fatal to humans. At the concentrations found in urban air, CO is not fatal, but it can aggravate cardiovascular diseases and may impair psychomotor functions. CO is a major transportation-related pollutant. Its harmful effects are described in more detail in chapter 4.

Hydrocarbons (HC) are defined chemically as compounds of carbon and hydrogen. However, in air quality studies, the term "hydrocarbons" often is extended to include a variety of other volatile organic substances, such as aldehydes and alcohols, in addition to true hydrocarbons.[2] Examples of major HC sources are motor vehicle exhaust and evaporation of organic solvents. At the concentrations usually found in urban air, most hydrocarbons are not directly harmful.[3] Their importance as air pollutants arises mainly from their role in atmospheric chemical reactions that produce nitrogen dioxide and ozone, among other compounds, both of which are harmful at or near atmospheric concentrations. One of the most common hydrocarbons, methane, does not participate in these chemical reactions. Accordingly, the term "nonmethane hydrocarbons" (NMHC), meaning all hydrocarbons except methane, frequently is used in air quality studies. Hydrocarbons that participate in the chemical reactions that form nitrogen dioxide and ozone often are called "reactive hydrocarbons" (RHC).

Nitrogen dioxide (NO_2) is a brownish gas with a pungent odor. It is responsible for the brownish color of the sky in many smoggy areas. NO_2 is mainly a secondary pollutant whose presence in the air is caused by oxidation in the air of nitric oxide (NO). NO is formed in high-temperature combustion processes, such as those occurring in fossil-fueled electrical-power plants and in automobile engines. The combination of all nitrogen oxides, consisting mainly of NO and NO_2 but possibly including small quantities of nitrogen trioxide (NO_3), dinitrogen trioxide (N_2O_3), and nitrogen tetroxide (N_2O_4), is referred to as "nitrogen oxides" (NO_x).[4] Nitrogen oxides can react chemically in the air to form nitrous and nitric acid, nitrate salts, and certain organic compounds of nitrogen. Because many of these reaction products are acidic, NO_x is an important contributor to acid rain. NO_2 is a pulmonary irritant, and short exposures to it may increase susceptibility to acute respiratory diseases. In addition, NO_2 is harmful to plants at sufficiently high concentrations, although most plants are not harmed by NO_2 at the concentrations that usually occur in the atmosphere. NO_2 is a major transportation-related pollutant. Its harmful effects are described in more detail in chapter 4.

Ozone (O_3) is a colorless gas with a pungent odor. O_3 has no significant direct emissions sources; it is a secondary pollutant that is formed in the air by photochemical reactions involving HC and NO_x.[5] In addition, large quantities of O_3 are formed naturally in the stratosphere. The reactions of HC and NO_x that form O_3 also oxidize NO to NO_2 and produce small quantities of other inorganic and organic compounds, such as nitric acid and peroxyacetylnitrate (PAN). O_3 and the other products of the photochemical reactions that produce O_3 often are referred to collectively as "photochemical oxidants" or "photochemical smog."[6] Although the photochemical smog mixture contains a large number of compounds that are either known or suspected to be harmful to living organisms or materials, O_3 and NO_x, which are the principal and best-understood smog constituents, are the only ones that currently are the objects of air pollution control efforts. O_3 is a strong pulmonary irritant that causes significant discomfort, clinical symptoms of respiratory illness, and reduced pulmonary function in sensitive individuals. O_3 also may cause increased susceptibility to respiratory infections. In addition, it is toxic to plants and damages many materials. O_3 is a major transporation-related pollutant. Its harmful effects are described in more detail in chapter 4.

Lead (Pb) is a heavy metal that is poisonous to humans. It causes damage to the nervous system and kidneys and inhibits hemoglobin synthesis, among other toxic effects. The main source of atmospheric lead is the combustion in motor vehicles of gasoline that contains lead antiknock compounds. This source of lead can be controlled by substantially reducing or eliminating the use of lead additives in gasoline. The reduction of the lead content of gasoline in the United States has been underway since 1975. By the mid-1980s, it is expected that motor vehicles no longer will be significant sources of lead. Accordingly, the control of transporation-related lead emissions is not discussed further in this book. For additional information on lead as an air pollutant, see reference [3].

2.2 *Natural Sources of the Principal Pollutants*

The principal air pollutants have significant natural as well as anthropogenic sources. Natural sources of particulate matter include soil and rock debris, forest fires, the oceans (as sources of spray and airborn salt), volcanoes, and conversion of naturally produced hydrocarbons and nitrogen- and sulfur-containing compounds into aerosols [4]. Natural sources of sulfur dioxide include volcanoes and the oxidation of hydrogen sulfide produced by the decay of vegetation [5, 6]. CO is produced by the oceans, forest fires, green plants, and oxidation of naturally produced hydrocarbons [6–8]. Natural HC sources include anaerobic decomposition of plants in swamps and marshes, seepage from natural gas and oil fields, and emissions from forest trees. The first two sources mainly produce methane. The third source produces photochemically reactive HC [6, 9]. Nitrogen oxides are produced by soil bacteria [6, 10]. Lead is a natural constituent of the earth's crust and enters the air as windblown soil and rock particles [3]. O_3 has no direct natural emissions sources. However, it can be produced in small quantities by atmospheric chemical reactions involving naturally emitted reactive HC and NO_x. More important, O_3 is formed naturally in the stratosphere through collisions between oxygen atoms and oxygen molecules. Stratospheric O_3 is transported to ground level in small quantities by the normal vertical circulation of the atmosphere. Stratospheric O_3 can be transported to ground level in much larger quantities during a meteorological condition called "tropopause folding" that produces very strong vertical mixing of the atmosphere [9].

Estimates of global rates of production of air pollutants by natural sources are available for particulates, SO_2, CO, HC, and NO_x [4–10]. These estimates are shown in table 2.1, together with estimates of global production rates of the same pollutants by anthropogenic sources. In all cases, natural emissions equal or exceed anthropogenic emissions.

The fact that natural production rates of air pollutants equal or exceed anthropogenic production rates does not necessarily imply that natural sources are more significant contributors to atmospheric concentrations of pollutants than anthropogenic sources are. Natural sources tend to be dispersed, whereas anthropogenic sources tend to be concentrated in cities. Moreover, natural and anthropogenic sources often are in widely

Table 2.1 Global production rates of principal air pollutants[a,b]

Pollutant	Natural production (kg/yr)	Man-made emissions (kg/yr)
Particulate matter	10^{12}	10^{11}
SO_2	10^{11}	10^{11}
CO	10^{12}	10^{11}
Methane	10^{12}	10^{11}
NMHC	10^{11}	10^{10}
NO_x	10^{12}	10^{11}

a. Sources: references [4–10]. All estimates have been rounded to the nearest order of magnitude.
b. Includes secondary production by atmospheric chemical reactions.

separated locations. For example, most natural production of HC occurs in swamps and forests, whereas most anthropogenic production of HC occurs in cities. Therefore, the question whether natural sources' contributions to atmospheric concentrations of pollutants are significant in comparison to anthropogenic contributions needs to be considered separately from the question of the relative magnitudes of emissions from natural and anthropogenic sources.

Natural sources' contributions to air pollutant concentrations, or natural background concentrations, can be estimated by measuring pollutant concentrations in areas that are remote from anthropogenic emissions sources. Table 2.2 shows estimates of natural background concentrations of the principal pollutants, together with the 1980 US air quality standards and examples of peak concentrations in polluted cities. The air quality standards are included in the table as indicators of the minimum concentrations that are likely to be considered excessive under most circumstances. In the case of pollutants for which there are several air quality standards, the standard with the lowest concentration has been used in table 2.2. The examples of peak urban concentrations have been selected arbitrarily from measurements of maximum concentrations in cities for the averaging times shown in the table. Most of these measurements were made during the 1960s, prior to the widespread implementation of emissions control measures. Thus, the examples illustrate the concentrations that urban air pollution control programs might have to deal with if these programs were starting from a condition in which there were no emissions controls.

Table 2.2 Natural background concentrations and examples of peak urban concentrations of air pollutants

Pollutant	Natural background concentration ($\mu g/m^3$)	Air quality standard ($\mu g/m^3$)	Peak urban concentration ($\mu g/m^3$)	References
Particulate matter	10	75 annual geometric mean	160 annual geometric mean	[1]
SO_2	1–4	80 annual average	130 annual average	[2, 11]
CO	50–200	10,000 8-hr average	40,000 8-hr average	[12–14]
NMHC	5–30	160[a] 6–9 A.M. average	1,800 6–9 A.M. average	[9, 16]
NO_2	0.2–10	100 annual average	110 annual average	[6, 13, 17, 18]
O_3	40–100 (normal) 450 or more during tropopause folding	235 1-hr average	900 1-hr average	[9, 19, 20]
Pb	0.0004–0.0012	1.5 quarterly average	3 quarterly average	[3, 21]

a. The NMHC air quality standard was set erroneously and cannot be used as a guide for judging the significance of naturally produced NMHC concentrations. See footnote c to table 1.1.

It can be seen from table 2.2 that with the exception of O_3 concentrations caused by tropopause folding, natural background concentrations of the principal air pollutants are well below both the air quality standards and the example urban concentrations. Thus, with the possible exception of O_3, urban air pollution cannot be attributed to natural sources, despite the large magnitudes of natural emissions. Stratospheric injections of O_3 due to tropopause folding that are large enough to cause high O_3 concentrations at ground level are highly infrequent, occurring only once or twice per year on the average [19, 22]. Moreover, these injections tend to occur during the spring and in association with the passage of cold fronts, whereas high urban O_3 concentrations occur most frequently during the summer and fall and in connection with stagnant weather conditions [9]. Although stratospheric O_3 may occasionally cause high urban O_3 concentrations, the high concentrations that are observed regularly in many cities, like the high concentrations of the other principal pollutants, cannot be attributed to natural sources. Thus, urban air pollution is caused almost entirely by anthropogenic emissions.

2.3 *Anthropogenic Sources of the Principal Pollutants*

Given that anthropogenic emissions are virtually the only cause of urban air pollution, it is necessary to inquire as to the specific sources of these emissions. Table 2.3 shows estimates of total nationwide anthropogenic emissions of the primary principal pollutants in 1975, disaggregated according to source class.[7] Stationary fuel combustion (especially by electrical utilities), metals industries, and mineral products industries are the largest sources of particulate and sulfur oxides emissions. These sources collectively account for over two thirds of nationwide emissions of these pollutants. Transportation sources cause less than 10% of nationwide emissions of either pollutant. Thus, particulates and sulfur oxides are not important transportation-related pollutants in most areas. However, exceptions to this generalization can arise. For example, measurements of the chemical composition of particulate matter in the Los Angeles area have indicated that motor vehicles are responsible for between 20 and 35% of primary airborne particulates in much of that area [25–27]. Moreover, in areas where there are unpaved roads, motor vehicles can generate large quantities of particulate matter in the form of road dust. Finally, motor vehicles contribute to the formation of certain secondary aerosols that result from atmospheric chemical reactions involving hydrocarbons and oxides of nitrogen. These aerosols contribute to reduced visibility in polluted air.

Table 2.3 Nationwide emissions estimates for 1975[a,b]

Source category	Particu-lates	SO_x	NO_x	HC	CO
Transportation	1.1	0.7	8.6	11.3	82.0
Highway vehicles	0.8	0.3	6.4	9.8	73.8
Nonhighway vehicles	0.3	0.4	2.2	1.5	8.2
Stationary fuel combustion	5.0	20.8	11.5	1.4	1.1
Electric utilities	3.7	16.8	6.2	0.1	0.3
Industrial	1.1	2.6	4.5	1.2	0.5
Residential, commercial & institutional	0.2	1.4	0.8	0.1	0.3
Industrial processes	6.5	4.6	0.7	9.2	7.3
Chemicals	0.2	0.3	0.2	2.1	2.2
Petroleum refining	0.1	0.8	0.4	1.0	2.4
Metals	1.4	2.7	0	0.2	1.8
Mineral products	3.7	0.6	0.1	0.1	0
Oil & gas production and marketing	0	0.1	0	2.9	0
Industrial organic solvent use	0	0	0	2.7	0
Other processes	1.1	0.1	0	0.2	0.9
Solid waste	0.5	0	0.1	0.8	2.9
Miscellaneous	0.6	0	0.1	4.2	3.6
Forest wildfires and managed burning	0.4	0	0.1	0.5	3.0
Agricultural burning	0.1	0	0.1	0	0.5
Coal refuse burning	0	0	0	0	0
Structural fires	0.1	0	0	0	0.1
Miscellaneous organic solvent use	0	0	0	3.6	0
Total	13.7	26.1	21.0	26.9	96.9

a. Source: reference [24].
b. Units are 10^6 metric tons per year. A zero indicates emissions of less than 50,000 metric tons per year.

CO, HC, and NO_x clearly are transportation-related pollutants. In 1975 highway vehicles alone accounted for 76% of nationwide CO emissions, 36% of nationwide HC emissions, and 30% of nationwide NO_x emissions. In addition, because O_3 is formed by chemical reactions involving HC and NO_x, motor vehicles were, by implication, major contributors to the occurrence of O_3 concentrations above natural background levels.

In individual cities, the proportions of CO, HC, and NO_x emissions caused by motor vehicles can vary greatly from the nationwide proportions. Because motor vehicles tend to be concentrated in cities, vehicular sources usually, but not always, are responsible for larger proportions of CO, HC, and NO_x emissions in cities than nationwide. Table 2.4 shows estimates of the distributions of CO, HC, and NO_x emissions among

Table 2.4 Distribution of CO, HC, and NO_x emissions among motor vehicles and nonvehicular sources in selected cities during the mid-1970s[a]

		Percentage of emissions caused by	
City	Pollutant	Motor vehicles	Nonvehicular sources
Boston	CO	94	6
	HC	61	39
	NO_x	NA	NA
Baltimore	CO	98	2
	HC	76	24
	NO_x	31	69
Denver	CO	89	11
	HC	82	18
	NO_x	40	60
Los Angeles	CO	91	9
	HC	52	48
	NO_x	53	47
Philadelphia	CO	98	2
	HC	43	57
	NO_x	32	68
Pittsburgh	CO	NA	NA
	HC	55	45
	NO_x	NA	NA
San Francisco	CO	89	11
	HC	46	54
	NO_x	55	45
Washington, DC	CO	100	0
	HC	76	24
	NO_x	74	26

a. Sources: references [28–36]. NA signifies data not available. The details of emissions estimation procedures (e.g., the times of day and seasons to which the estimates apply, whether HC estimates include all HC or only NMHC, and whether emissions from natural sources are included) differ among cities. The estimates presented here give rough indications of the relative importances of vehicular and nonvehicular emissions sources but should not be interpreted in more precise terms without first consulting the cited reference documents for the details of the estimation procedures.

Table 2.5 Proportions of urban HC concentrations due to motor vehicle exhaust during the period 1968–1977[a]

City	Motor vehicle proportion (%)	Year of measurement
New York	25–90	1969
Los Angeles	40–70	1968
Washington, DC	60–100	1976
St. Louis	60–80	1973
Houston	45–65	1977

a. Sources: references [37–40]. The range of motor vehicle proportions within a single city reflects variations in the location and time-of-day of sampling. The Houston estimates have been obtained by applying the method of Lonneman [39] to Houston data [40]. All of the estimates exclude methane.

vehicular and nonvehicular sources in selected cities during the mid-1970s. These estimates were derived from measurements and estimates of the emissions rates of individual sources (see reference [23], for example). Motor vehicles are responsible for 89–100% of the CO emissions, 43–82% of the HC emissions, and 31–74% of the NO_x emissions, depending on the city.

The contribution of motor vehicle exhaust emissions to atmospheric HC concentrations also can be estimated from the chemical composition of atmospheric HC. The estimation procedure is based on the observation that motor vehicle exhaust is the only significant source of acetylene emissions in most places and, therefore, that acetylene can be used as a tracer of motor vehicle exhaust emissions [37]. The estimation procedure is as follows. Let R_{HC} be the ratio of HC to acetylene in motor vehicle exhaust, and let [HC] and $[C_2H_2]$, respectively, denote the concentrations of HC and acetylene in the air. Then the estimated fraction F_{HC} of [HC] that is due to motor vehicle exhaust is

$$F_{HC} = R_{HC}[C_2H_2]/[HC].^{8} \quad (2.1)$$

Table 2.5 shows estimates of the motor vehicle exhaust-related proportions of atmospheric HC concentrations in six cities during the period 1968–1977. These estimates confirm the importance of motor vehicles as sources of HC emissions in urban areas.

The acetylene tracer method also can be used to estimate the proportion of atmospheric CO that is due to motor vehicle exhaust. Let R_{CO} be the ratio of CO to acetylene in motor vehicle exhaust, and let [CO] be the atmospheric CO concentration. Then the fraction F_{CO} of [CO] that is due to motor vehicle exhaust is

$$F_{CO} = R_{CO}[C_2H_2]/[CO]. \quad (2.2)$$

Lonneman et al. have estimated that 80–93% of CO concentrations in downtown Los Angeles are caused by motor vehicle emissions, depending on the time of day and day of the week [37]. Data from the Washington, DC, area indicate that roughly 90% of atmospheric CO in that area is caused by motor vehicle emissions [38]. These estimates are consistent with the results obtained by directly estimating the emissions rates of individual sources (see table 2.4).

2.4 The Extent of Transportation-Related Air Pollution

During the late 1970s, CO concentrations exceeding the CO air quality standards were recorded in 93 metropolitan areas in the United States, O_3 concentrations exceeding the O_3 air quality standard were recorded in 183 metropolitan areas, and NO_2 concentrations exceeding the NO_2 air quality standard were recorded in 6 metropolitan areas. A total of 188 metropolitan areas recorded concentrations exceeding at least one of these air quality standards. In addition, 211 nonurban counties recorded concentrations exceeding one or more of the standards. The areas in which concentrations of CO, O_3, and NO_2 exceeded one or more of the air quality standards for these pollutants contain roughly 75% of the US population and include 88% of the metropolitan areas with populations over 200,000.[9] Thus, the occurrence of concentrations of transportation-related air pollutants exceeding the air quality standards constitutes a widespread problem.[10]

New motor vehicles sold in the United States have been subject to emissions standards of gradually increasing stringency since 1963.[11] These emissions standards have caused significant decreases in average CO, HC, and NO_x emissions per vehicle mile traveled. For example, between 1970 and 1977, average CO emissions per vehicle mile traveled decreased by 20%, average HC emissions per vehicle mile traveled decreased by 30%, and average NO_x emissions per vehicle mile traveled decreased by 10% [41]. However, total emissions of CO, HC, and NO_x and atmospheric concentrations of CO, O_3, and NO_2 have not decreased as much as have emissions per vehicle mile traveled owing to growth in motor vehicle travel and increases in HC and NO_x emissions from nontransportation sources. Figures 2.1–2.9 illustrate the behavior of CO, O_3, and NO_2 concentrations in the United States during the period 1970–1977. The figures suggest that CO concentrations decreased over

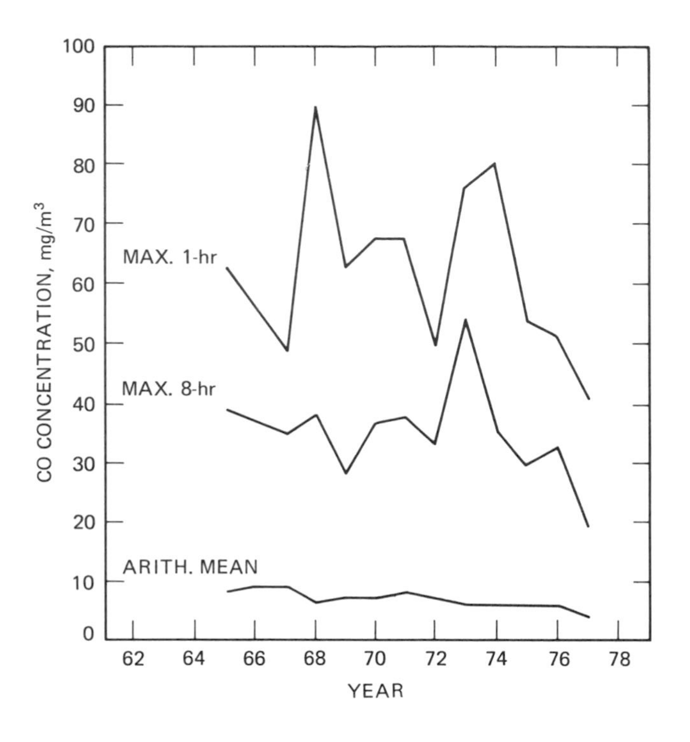

Figure 2.1 Annual variations of CO concentrations at a monitoring site in Denver, Colorado. (Source: reference [8])

Table 2.6 Trends in CO, O_3, and NO_2 concentrations at nationwide monitoring sites during 1970–1976[a]

Pollutant	Number of sites	Number of sites reporting		
		Increase	Decrease	No change
CO	202	40	152	10
O_3	174	77	70	27
NO_2	276	164	96	16

a. Source: reference [42].

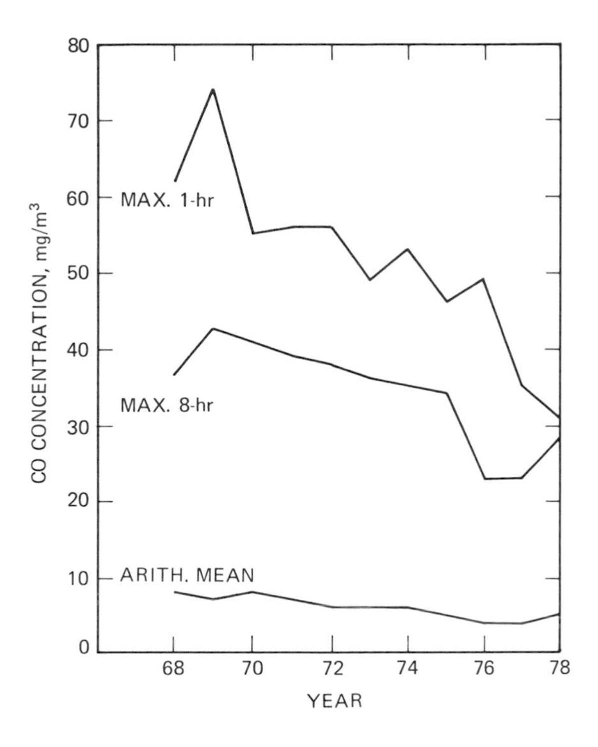

Figure 2.2 Annual variations of CO concentrations at a monitoring site in Los Angeles, California. (Source: reference [8])

the period, O_3 concentrations decreased in California (where HC emissions control requirements are stronger than they are elsewhere) but increased slightly in the rest of the country, and NO_2 concentrations did not change. Table 2.6 summarizes changes in CO, O_3, and NO_2 concentrations during the period 1970–1976 at monitoring sites throughout the United States. These nationwide statistics indicate that CO concentrations decreased over the period, O_3 concentrations did not change significantly, and NO_2 concentrations increased. The trends shown in figures 2.1–2.9 and table 2.6 suggest that concentrations of CO, O_3, and NO_2 that exceed the air quality standards are likely to present a persistent as well as a widespread problem.

The persistence of concentrations of CO, O_3, and NO_2 exceeding the air quality standards is further indicated by projections of future concentrations of these pollutants. During the period 1970–1990, emissions

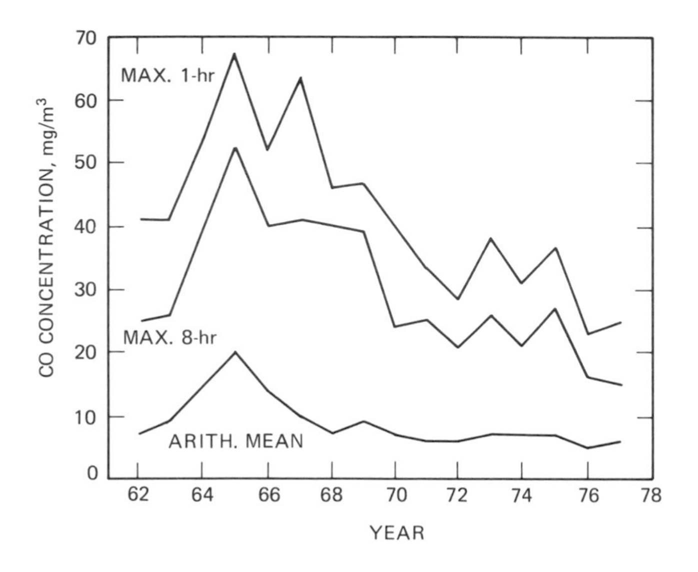

Figure 2.3 Annual variations of CO concentrations at a monitoring site in Chicago, Illinois. (Source: reference [8])

controls for both vehicular and nonvehicular sources are expected to cause nationwide average reductions in CO, O_3, and NO_2 concentrations of roughly 60%, 40%, and 10%, respectively [28]. However, these emissions reductions will not enable the air quality standards for CO, O_3, and NO_2 to be achieved everywhere in the United States. The US Environmental Protection Agency, Department of Transportation, and Federal Energy Administration (now the Department of Energy) have estimated that in 1990 roughly 5 to more than 39 metropolitan areas may continue to experience CO, O_3, or NO_2 concentrations that exceed the air quality standards for these pollutants, depending on the types of emissions control measures that are implemented in each area and their effects on air quality [28, 43].[12]

The emissions standards for new motor vehicles are causing emissions from vehicular sources to decrease more rapidly than emissions from nonvehicular sources are decreasing. Accordingly, the proportions of CO, HC, and NO_x emissions that are attributable to motor vehicles also are decreasing. Nonetheless, motor vehicles are likely to remain important sources of these pollutants for the foreseeable future. Table 2.7 shows projections of the distributions of CO, HC, and NO_x emissions among

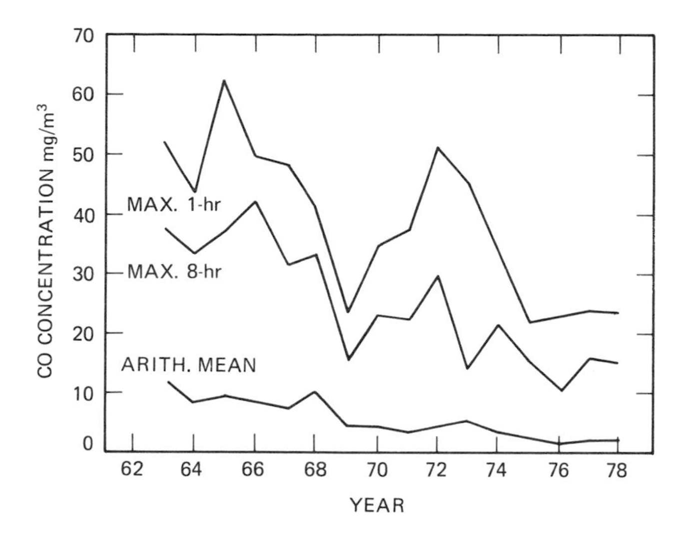

Figure 2.4 Annual variations of CO concentrations at a monitoring site in Philadelphia, Pennsylvania. (Source: reference [8])

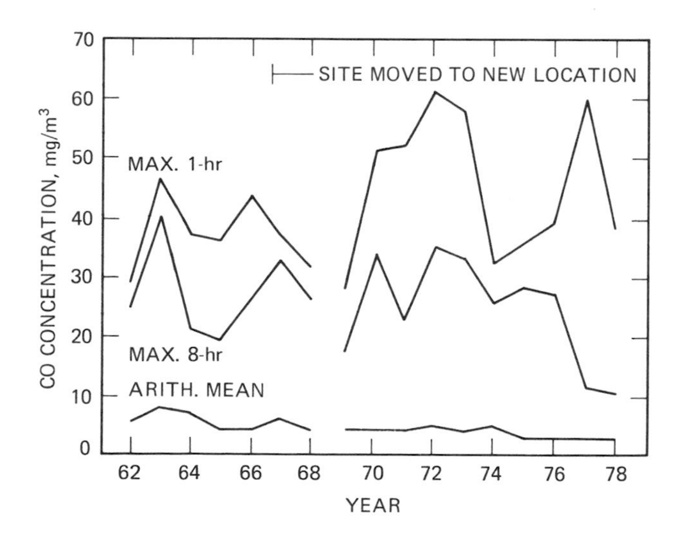

Figure 2.5 Annual variations of CO concentrations at monitoring sites in Washington, DC. (Source: reference [8])

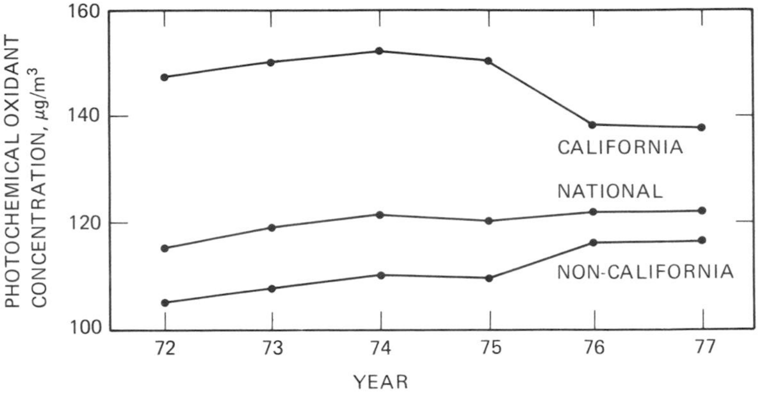

Figure 2.6 Annual variations in mean 90th percentiles of 1-hr average O_3 concentrations measured during the second and third quarters of the year. (Source: reference [24])

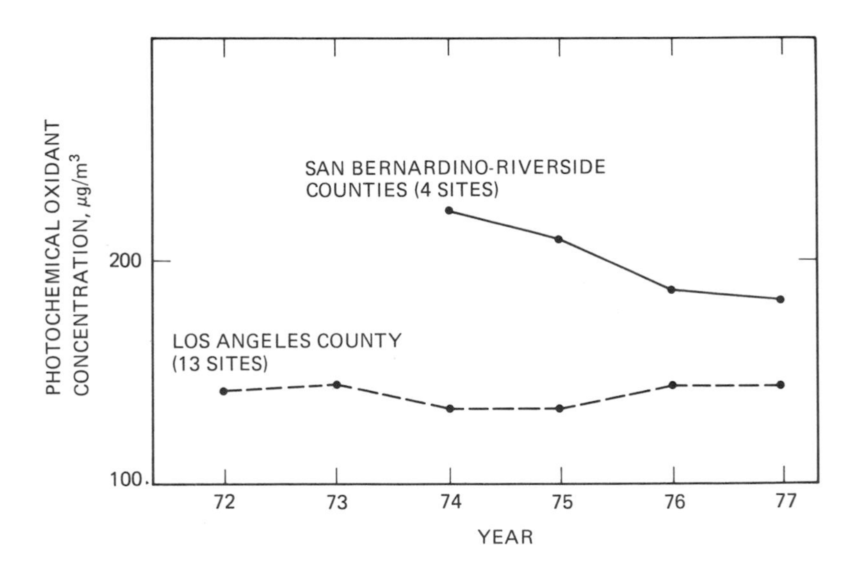

Figure 2.7 Annual variations in the mean 90th percentiles of 1-hr average O_3 concentrations measured throughout the year at monitoring sites in the Los Angeles, California, area. (Source: reference [24])

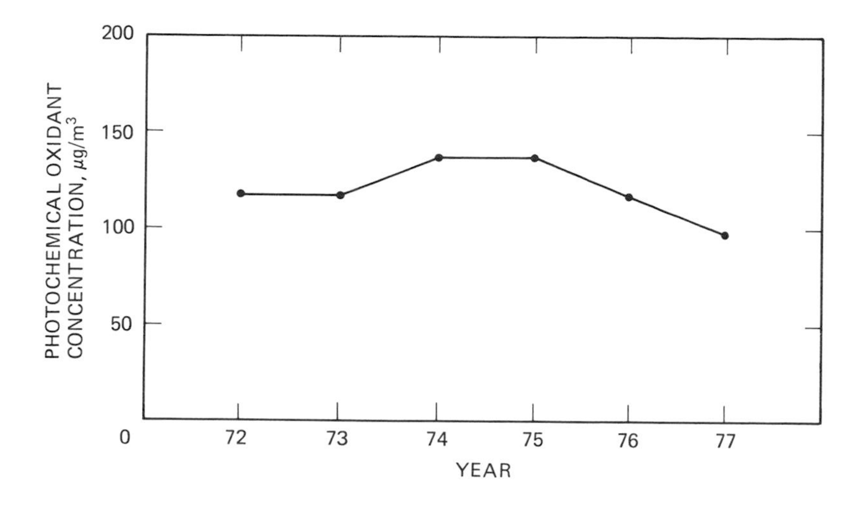

Figure 2.8 Annual variations in the mean of daily maximum 1-hr average O_3 concentrations measured during April through October at six sites in the San Francisco, California, area. (Source: reference [24])

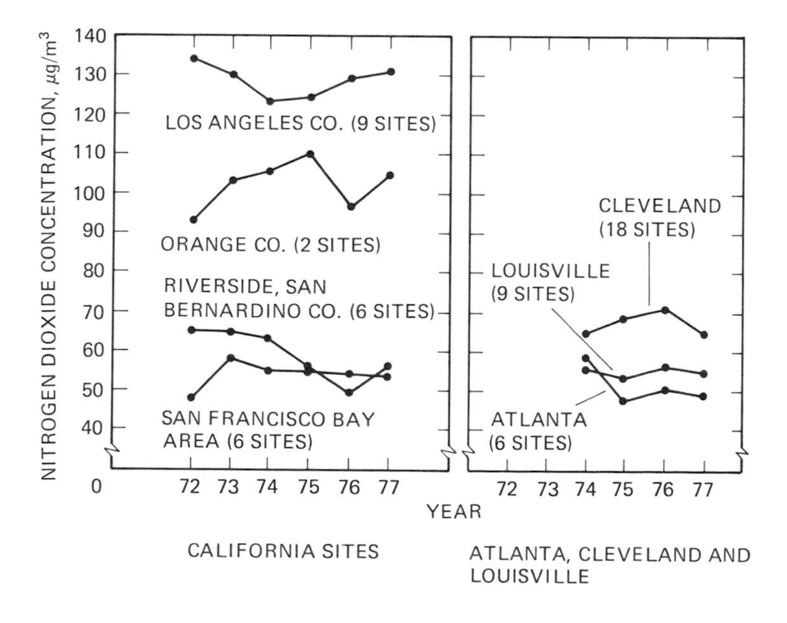

Figure 2.9 Variations in mean annual average NO_2 concentrations. (Source: reference [24])

Table 2.7 Projected distributions of CO, HC, and NO_x emissions among motor vehicles and nonvehicular sources in selected cities in 1987[a]

City	Pollutant	Percentage of emissions caused by	
		Motor vehicles	Nonvehicular sources
Boston	CO	88	12
	HC	34	66
	NO_x	NA	NA
Baltimore	CO	NA	NA
	HC	56	44
	NO_x	NA	NA
Denver[b]	CO	85	15
	HC	78	22
	NO_x	46	54
Los Angeles	CO	78	22
	HC	34	66
	NO_x	41	59
Pittsburgh	CO	NA	NA
	HC	39	61
	NO_x	NA	NA
San Francisco[b]	CO	87	13
	HC	27	73
	NO_x	37	63
Washington, DC	CO	100	0
	HC	54	46
	NO_x	56	44

a. Sources: references [30–33, 35, 36]. NA signifies data not available. The qualifications stated in footnote a of table 2.4 also apply to this table.
b. Emissions distribution in 1985. The San Francisco projections do not incorporate certain relaxations in the stringency of the new motor vehicle emissions standards that occurred in 1977 and, therefore, may underestimate the proportions of emissions that will be caused by motor vehicles in 1985.

vehicular and nonvehicular sources in selected cities in 1987. These projections, which include the effects of the new motor vehicle emissions standards and of emissions control requirements that existed for non-vehicular sources during 1975–1977, indicate the emissions distributions that would be expected to occur if no further emissions control requirements beyond those existing in 1977 were developed.[13] In 1987 motor vehicles are projected to account for 78–100% of CO emissions, 27–78% of HC emissions, and 37–56% of NO_x emissions, depending on the city. Moreover, automobiles are expected to cause roughly 50–75% of vehicular CO, HC, and NO_x emissions in cities in 1987, depending on the pollutant and the city [31, 35, 36]. Thus, motor vehicles generally and automobiles in particular are likely to remain major contributors to future urban air pollution, despite the increasing stringency of the new motor vehicle emissions standards.

In summary, concentrations of CO, O_3, and NO_2 that exceed the air quality standards for these pollutants occur in most large US metropolitan areas and are likely to continue to occur in many of these areas for the foreseeable future. Although the proportions of CO, HC, and NO_x emissions caused by motor vehicles will decrease during the 1980s, owing to the effects of the emissions standards for new motor vehicles, and the importance of controlling emissions from nonvehicular sources will increase, motor vehicles will continue to be substantial sources of CO, HC, and NO_x emissions in cities. Accordingly, it is worth while to consider what measures in addition to the new motor vehicle emissions standards may be available for reducing motor vehicle emissions in cities. These measures are discussed in section 5.5 and chapter 6. However, before doing this, it is necessary to discuss several other aspects of the transportation-related air pollution problem that affect the design, effectiveness, and desirability of implementing measures to control motor vehicle emissions. These aspects include the relations between pollutant concentrations and factors such as emissions rates, source proximity, and meteorological conditions; the types of harmful effects that are caused by transportation-related air pollution; and certain vehicular engineering and operational factors (including the means by which emissions from new vehicles are controlled) that affect motor vehicle emissions rates and the extent to which these emissions rates can be reduced.

References

1
US Department of Health, Education, and Welfare, *Air Quality Criteria for Particulate Matter,* Report No. AP-49, Washington, DC, 1969. NTIS Publication No. PB 190251.

2
US Department of Health, Education, and Welfare, *Air Quality Criteria for Sulfur Oxides,* Report No. AP-50, Washington, DC, 1969.

3
US Environmental Protection Agency, *Air Quality Criteria for Lead,* Report No. EPA-600/8-77-017, Washington, DC, December 1977. NTIS Publication No. PB 280411.

4
National Research Council, *Airborne Particles,* Report No. EPA-611/1-77-053, US Environmental Protection Agency, Washington, DC, November 1977. NTIS Publication No. PB 276723.

5
National Research Council, *Hydrogen Sulfide*, Report No. EPA-600/1-78-018, US Environmental Protection Agency, Washington, DC, February 1978. NTIS Publication No. PB 278576.

6
Rasmussen, K. H., Taheri, M., and Kable, R. L., *Sources and Natural Removal Processes for Some Atmospheric Pollutants*, Report No. EPA-650/4-74-032, US Environmental Protection Agency, Washington, DC, June 1974. NTIS Publication No. PB 237168.

7
National Research Council, *Carbon Monoxide*, Report No. EPA 600/1-77-034, US Environmental Protection Agency, Washington, DC, September 1977. NTIS Publication No. PB 274965.

8
US Environmental Protection Agency, *Air Quality Criteria for Carbon Monoxide*, Report No. EPA-600/8-79-022, Washington, DC, October 1979.

9
US Environmental Protection Agency, *Air Quality Criteria for Ozone and Other Photochemical Oxidants*, Report No. EPA-600/8-78-004, Research Triangle Park, NC, April 1978. NTIS Publication No. PB80-124753.

10
National Research Council, *Nitrogen Oxides*, Report No. EPA-600/1-77-013, US Environmental Protection Agency, Washington, DC, February 1977. NTIS Publication No. PB 264872.

11
Georgii, H. W., "Contribution to the Atmospheric Sulfur Budget," *Journal of Geophysical Research*, Vol. 75, pp. 2365–2371, 1970.

12
Jaffe, L. S., "Carbon Monoxide in the Biosphere: Sources, Distribution and Concentrations," *Journal of Geophysical Research*, Vol 78, pp. 5293–5305, 1973.

13
Robinson, E., and Robbins, R. C., *Abundance and Fate of Gaseous Atmospheric Pollutants*, Final Report of Project PR-6755, prepared for the American Petroleum Institute by the Stanford Research Institute, Menlo Park, CA, February 1968.

14
Seiler, W., "The Cycle of Atmospheric CO," *Tellus*, Vol. 26, pp. 116–135, 1974.

15
Wilkness, P., Lamontagne, R., Larson, R., Swinnerton, R., and Dickson, C., "Atmospheric Trace Gases in the Southern Hemisphere," *Nature: Physical Science*, Vol. 245, pp. 45–47, 1973.

16
Westberg, H., "Review and Analysis," in P. E. Coffey and H. Westberg, eds., *International Conference on Oxidants, 1976—Analysis of Evidence and Viewpoints,*

Part IV, Report No. EPA-600/3-77-116, US Environmental Protection Agency, Research Triangle Park, NC, October 1977, NTIS Publication No. PB 277462.

17
Noxon, J. F., "Nitrogen Dioxide in the Stratosphere and Troposphere Measured by Ground-Based Absorption Spectroscopy," *Science*, Vol. 189, pp. 547–549, 15 August 1975.

18
Noxon, J. F., "Tropospheric NO_2," *Journal of Geophysical Research*, Vol. 83, pp. 3051–3057, 20 June 1978.

19
Mohnen, V. A., "Review and Analysis," in V. A. Mohnen and E. R. Reiter, eds., *International Conference on Oxidants, 1976—Analysis of Evidence and Viewpoints,* Part III, Report No. EPA 600/3-77-115, US Environmental Protection Agency, Research Triangle Park, NC, December 1977. NTIS Publication No. 279010.

20
Lamb, R. G., "A Case Study of Stratospheric Ozone Affecting Ground-Level Oxidant Concentrations," *Journal of Applied Meteorology*, Vol. 16, pp. 780–794, 1977.

21
Akland, G. G., *Air Quality Data for Metals, 1970–1974, from the National Air Surveillance Network*, Report No. EPA-600/4-76-041, US Environmental Protection Agency, Washington, DC, August 1976. NTIS Publication No. PB 260905.

22
Reiter, E. R., "Review and Analysis," in V. A. Mohnen and E. R. Reiter, eds., *International Conference on Oxidants 1976–Analysis of Evidence and Viewpoints,* Part III, Report No. EPA 600/3-77-115, US Environmental Protection Agency, Research Triangle Park, NC, December 1977. NTIS Publication No. PB 279010.

23
US Environmental Protection Agency, *Compilation of Air Pollutant Emission Factors,* 3rd ed., Report No. AP-42. NTIS Publication Nos. PB 284487, PB 284488, PB 288905, PB 295614.

24
US Environmental Protection Agency, *National Air Quality Monitoring and Emissions Trends Report, 1977*, Report No. EPA 450/2-78-052, Research Triangle Park, NC, December 1978.

25
Gartrell, G., Jr., and Friedlander, S. K., "Relating Particulate Pollution to Sources: The 1972 California Aerosol Characterization Study," *Atmospheric Environment*, Vol. 9, pp. 279–299, 1975.

26
Heisler, S. L., Friedlander, S. K., and Husar, R. B., "The Relationship of Smog Aerosol Size and Chemical Element Distributions to Source Characteristics," *Atmospheric Environment*, Vol. 7, pp. 633–649, 1973.

27
Friedlander, S. K., "Chemical Element Balances and Identification of Air Pollution Sources," *Environmental Science and Technology*, Vol. 7, pp. 235–240, March 1973.

28
US Department of Transportation, US Environmental Protection Agency, and US Federal Energy Administration, *An Analysis of Alternative Motor Vehicle Emissions Standards*, Washington, DC, May 1977.

29
Anderson, G. E., Hayes, S. R., Hillyer, M. J., Killus, J. P., and Mundkur, P. V., *Air Quality in the Denver Metropolitan Region*, 1974–2000, Report No. EPA 909/1-77-002, US Environmental Protection Agency, Region VII, Denver, May 1977. NTIS Publication No. PB 271894.

30
Central Transportation Planning Staff, *Transportation Element of the State Implementation Plan for the Boston Region*, Boston, December 1978.

31
Regional Planning Council, *Transportation Control Plan*, Baltimore, September 1978.

32
Southern California Association of Governments, *Air Quality Management Plan*, Los Angeles, January 1979.

33
Southwestern Pennsylvania Regional Planning Commission, *Transportation Component of the State Air Quality Implementation Plan for the Southwestern Pennsylvania Region*, Pittsburgh, January 1979.

34
Delaware Valley Regional Planning Commission, *Transportation Element for Southeastern Pennsylvania of the 1979 State Implementation Plan*, Philadelphia, April 1979.

35
Association of Bay Area Governments, Bay Area Air Pollution Control District, and Metropolitan Transportation Commission, *Air Quality Maintenance Plan*, report prepared for public review and comment, San Francisco, December 1977.

36
Metropolitan Washington Council of Governments, *Washington Metropolitan Air Quality Plan*, draft for public review, Washington, DC, September 1978.

37
Lonneman, W. A., Kopczynski, S. L., Darley, P. E., and Sutterfield, F. D., "Hydrocarbon Composition of Urban Air Pollution," *Environmental Science and Technology*, Vol. 8, pp. 229–236, 1974.

38
Record, F. A., *Characterization of the Washington, D.C. Oxidant Problem*, Report No. EPA 450/3-77-054, US Environmental Protection Agency, Research Triangle Park, NC, September 1977. NTIS Publication No. PB 275458.

39
Lonneman, W. A., "Ozone and Hydrocarbon Measurements in Recent Oxidant Transport Studies," in B. Dimitriades, ed., *International Conference on Photochemical Oxidant Pollution and Its Control: Proceedings*, Vol. 1, Report No. EPA 600/3-77-001a, US Environmental Protection Agency, Research Triangle Park, NC, January 1977, pp. 211–223. NTIS Publication No. PB 263232.

40
Houston Chamber of Commerce, *Houston Area Oxidants Study: Program Summary*, Houston, July 1979.

41
US Environmental Protection Agency, *Mobile Source Emission Factors*, Report No. EPA 400/9-78-005, Washington, DC, March 1978. NTIS Publication No. PB 295672.

42
US Environmental Protection Agency, *National Air Quality and Emissions Trends Report, 1976*, Report No. EPA 450/1-77-002, Research Triangle Park, NC, December 1977. NTIS Publication No. PB 279007.

43
US Environmental Protection Agency, *Cost and Economic Impact Assessment of the National Ambient Air Quality Standard for Ozone*, Research Triangle Park, NC, January 1979.

3 Qualitative Characteristics of Pollutant Concentrations

3.1 Introduction

The design and evaluation of measures to control transportation-related air pollution requires an understanding of the ways in which pollutant concentrations depend on emissions rates, on the proximity of emissions sources to the locations where concentrations are measured, and on certain nonemissions-related factors, such as meteorological conditions and topography. For example, it is necessary to be able to answer questions such as:

- Are high pollutant concentrations likely to be found only in the vicinities of large emissions sources, or can high concentrations also occur far away from emissions sources?
- To reduce excessive pollutant concentrations at a particular location, is it sufficient to reduce emissions only in the immediate vicinity of that location, or must emissions be reduced over a more widespread area?
- Are changes in pollutant concentrations likely to be proportional to changes in emissions rates, other factors remaining constant, or is the relation between emissions and concentrations more complex?
- How are pollutant concentrations affected by meteorological conditions and topography?
- Are there particular times of day or seasons of the year when pollutant concentrations are especially likely to be high?

The purpose of this chapter is to describe important qualitative characteristics of the relations between concentrations of CO, O_3, and NO_2—the principal transportation-related pollutants that are directly harmful—and other factors. Quantitative methods for describing these relations are discussed in chapter 7.

3.2 Carbon Monoxide

The concentration of CO at any particular point is determined mainly by the rate of CO emissions from nearby CO sources; by certain meteorological variables, such as the wind direction, wind speed, and the degree of atmospheric turbulence in the vicinity of the point; and by topographical features near the point. CO concentrations tend to be high on and near congested roadways and at other locations where traffic densities are high, and to decrease rapidly as distance from the roadway increases. In addition, the CO concentration at a point tends to follow the diurnal variation in the traffic volume in the vicinity of the point, being highest during the hours of peak traffic. However, the influence of nearby traffic on CO concentrations is moderated by the meteorological variables. These variables can cause large fluctuations in the CO concentration at a point, even if emissions in the vicinity of the point do not change, and they tend to impose systematic seasonal variations on CO concentrations. CO concentrations near a roadway are influenced also by local topographical features, such as the presence of tall buildings near the roadway, and by whether the roadway is depressed below ground level, at ground level, or elevated. As a result of the influences of traffic flows, meteorological variables, and topography, CO concentrations tend to exhibit great spatial and temporal variability. The CO concentration at one location and time may be very different from the CO concentration at another relatively nearby location or to the CO concentration at the same location but at a different time.

The CO concentration near a roadway is closely related to the traffic volume on the roadway. This relation is illustrated in figure 3.1, which shows measured CO concentrations and traffic volumes on the streets of a large city. The measurements shown in the figure were made at several different locations and times. The considerable scatter in the data reflects

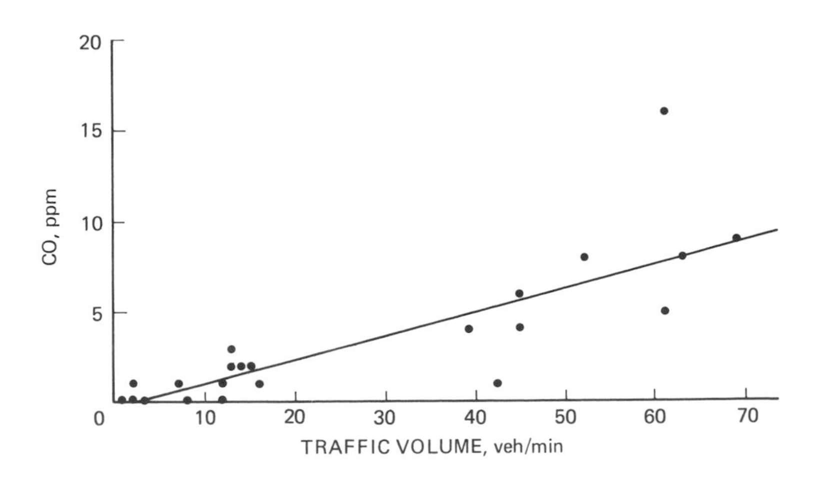

Figure 3.1 The relation between CO concentrations and traffic volumes on the streets of a large city. (Source: reference [1])

variations in traffic flow conditions (e.g., the speed of traffic flow), meteorological conditions, topography, and the geometrical relations between the roadways and the CO sensors at the different locations and times. However, there is a clear and roughly linear association between CO concentrations and traffic volumes, with high CO concentrations occurring in connection with high traffic volumes. Figure 3.2 shows hourly variations in CO concentrations and traffic volumes at a single location in New York City. As in figure 3.1, there is a clear association between high CO concentrations and high traffic volumes. Hourly average CO concentrations and traffic volumes on roadways in Detroit, New York City, and Los Angeles are shown in figure 3.3. The association between high CO concentrations and high traffic volumes is again clear.

The CO concentration at a point is also sensitive to wind direction and speed, atmospheric stability, and the mixing height. The wind direction affects the concentration because the concentration of a pollutant is higher downwind of an emissions source than upwind of the source. Increases in wind speeds tend to reduce the amount of time that any given air parcel spends in the vicinity of an emissions source, thereby reducing the pollutant concentration in that parcel. Accordingly, increases

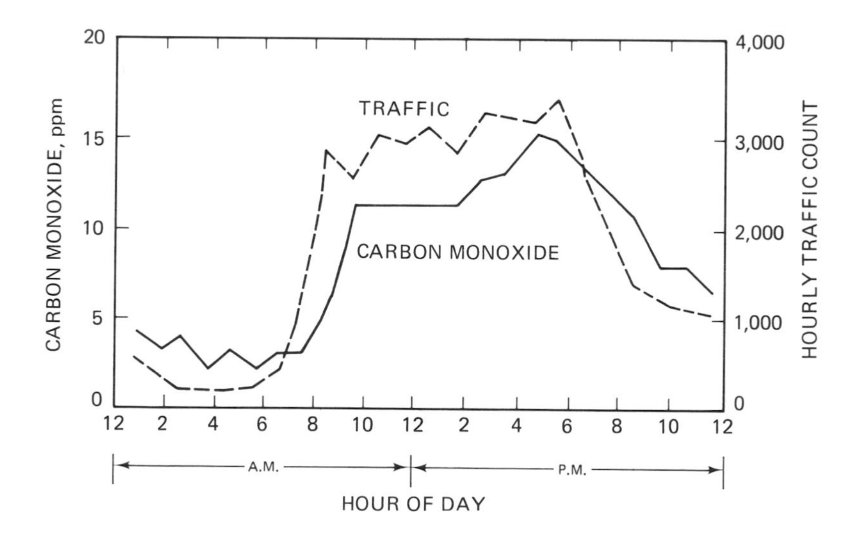

Figure 3.2 Hourly average CO concentrations and traffic counts in midtown Manhattan, New York. (Source: reference [2]. Reprinted with permission, copyright 1968 by the American Association for the Advancement of Science)

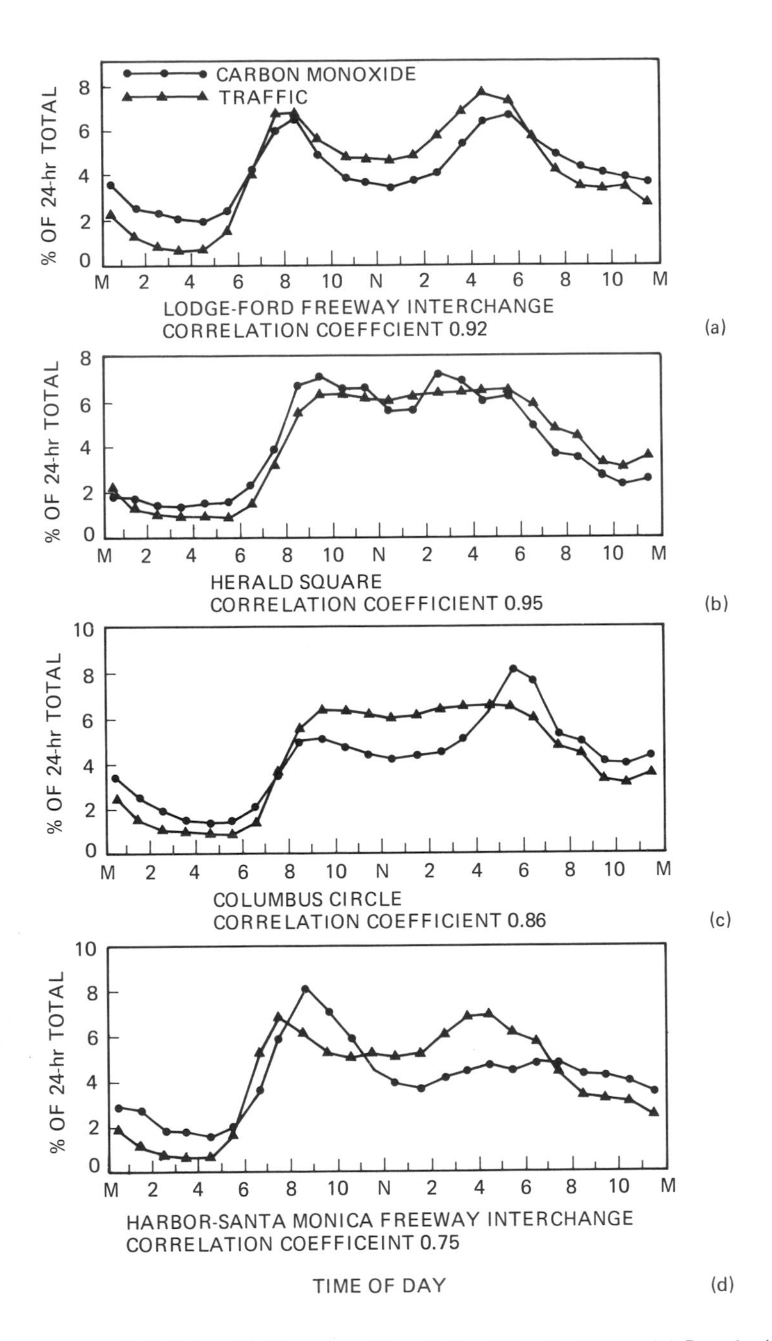

Figure 3.3 CO and traffic versus time of day at four locations: (*a*) Detroit; (*b*) New York; (*c*) New York; (*d*) Los Angeles. (Reprinted with permission from reference [3]: Colucci, J.M., and Begeman, C.R., "Carbon Monoxide in Detroit, New York and Los Angeles," *Environmental Science and Technology*, Vol. 3, pp. 41–46, January 1969, Copyright 1969, American Chemical Society)

in wind speeds tend to decrease air pollutant concentrations.[1] Increasing atmospheric stability is equivalent to reducing atmospheric turbulence. Reduced turbulence causes the rate at which pollutants are mixed into the air and diluted to be reduced. Hence, pollutant concentrations in the vicinity of emissions sources tend to be higher when the air is stable than when it is turbulent. The mixing height refers to the vertical distance above the ground over which pollutants can be mixed and diluted. Reducing the mixing height reduces the effective volume of air that is available for diluting pollutants and, hence, tends to increase pollutant concentrations.

A common cause of low mixing heights is a meteorological condition called a "temperature inversion" or "inversion layer." Normally, the temperature of the air decreases with increasing altitude above the ground. In a temperature inversion, the air temperature increases with increasing altitude. The altitude above ground at which the temperature starts to increase is called the "inversion base" (see figure 3.4). Atmospheric mixing across the inversion base cannot take place. If air that originated below the inversion layer rose through the inversion base, it would be cooler than the surrounding air and would sink back below the inversion base. Conversely, if air that originated in the inversion layer sank below the inversion base, it would be warmer than the surrounding air and would rise again. Thus, a temperature inversion limits the mixing height. Ground-based temperature inversions often occur in the early morning because the ground, which tends to lose heat rapidly at night by

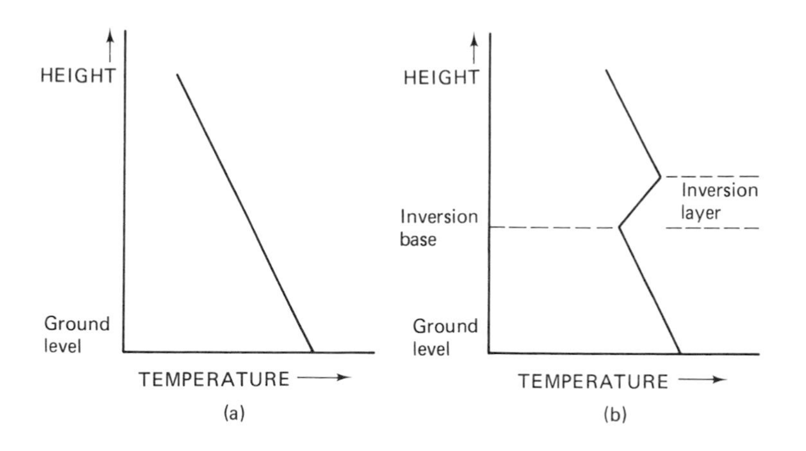

Figure 3.4 Air temperature as a function of distance above the ground (*a*) under normal conditions and (*b*) with a temperature inversion.

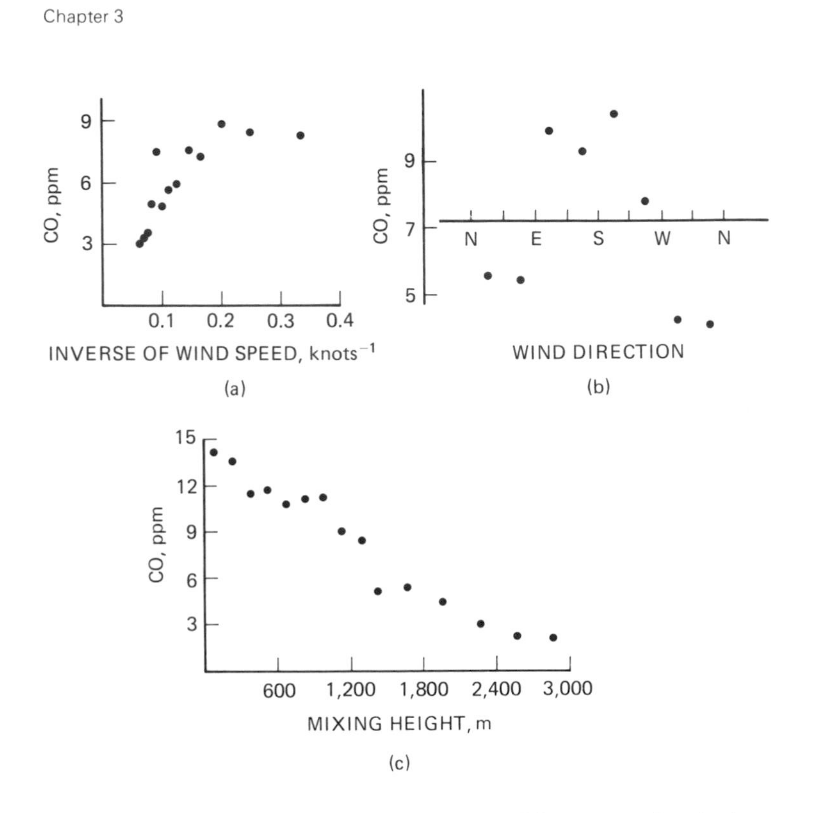

Figure 3.5 Effects of meteorological variables on CO concentrations in Jersey City, New Jersey. Each point represents the average of all observations made with the indicated value of the variable on the horizontal axis: (*a*) CO vs. the inverse of wind speed, summer, 7–8 A.M.; (*b*) CO vs. wind direction, winter, 7–8 A.M.; (*c*) CO vs. mixing height (meters), summer, 4–5 P.M. (Source: reference [4]. Reprinted with permission)

radiation, cools the air close to it. Elevated inversions occur frequently in high-pressure weather systems, as these systems tend to cause a gradual sinking, compression, and, therefore, warming of the upper layers of the atmosphere. Other conditions that cause temperature inversions are the passage of warm and cold fronts, and winds that cause warm air to flow over a cold surface or cold air.

Figure 3.5 illustrates the effects of wind direction, wind speed, and mixing height on CO concentrations at a site in Jersey City, New Jersey. Changes in wind direction are associated with concentration changes of roughly a factor of two, changes in wind speed are associated with concentration changes of roughly a factor of three, and changes in mixing height are associated with concentration changes of roughly a factor of five. Because wind direction, wind speed, and mixing height often vary together over time—for example, increases in mixing height often are associated with increases in wind speed—the plots in figure 3.5 do not necessarily indicate how CO concentrations would change if only one meteorological variable were changed at a time. However, the plots do illustrate the important influence of meteorological variables on CO concentrations.

Meteorological variables moderate the relation between CO concentrations and traffic volumes, often causing CO concentrations to vary in ways that are not attributable to variations in traffic volumes.[2] For example, figure 3.3 shows that there is a distinct afternoon traffic peak at the Harbor-Santa Monica Freeway interchange in Los Angeles but that there is not a corresponding CO peak. The lack of a CO peak is due to the relatively high wind speeds that normally occur in Los Angeles in the late afternoon (figure 3.6).

Systematic seasonal variations in meteorological conditions tend to impose systematic seasonal variations on CO concentrations. Figures 3.7–3.9 illustrate the seasonal patterns of CO concentrations. The highest CO concentrations tend to occur in winter. Atmospheric conditions tend to be more stable and wind speeds tend to be lower in winter than in other seasons, thus causing reduced dispersion of CO emissions in winter and increased CO concentrations. In addition, motor vehicles' CO emissions tend to be higher when the air temperature is low than when the air temperature is high.[3] Thus, CO emissions tend to be higher in winter than in other seasons.

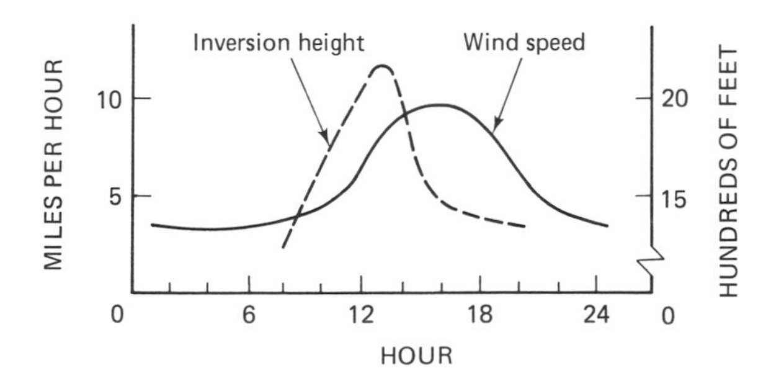

Figure 3.6 Diurnal variation in wind speed and inversion height in downtown Los Angeles, California. (Reprinted with permission from reference [5])

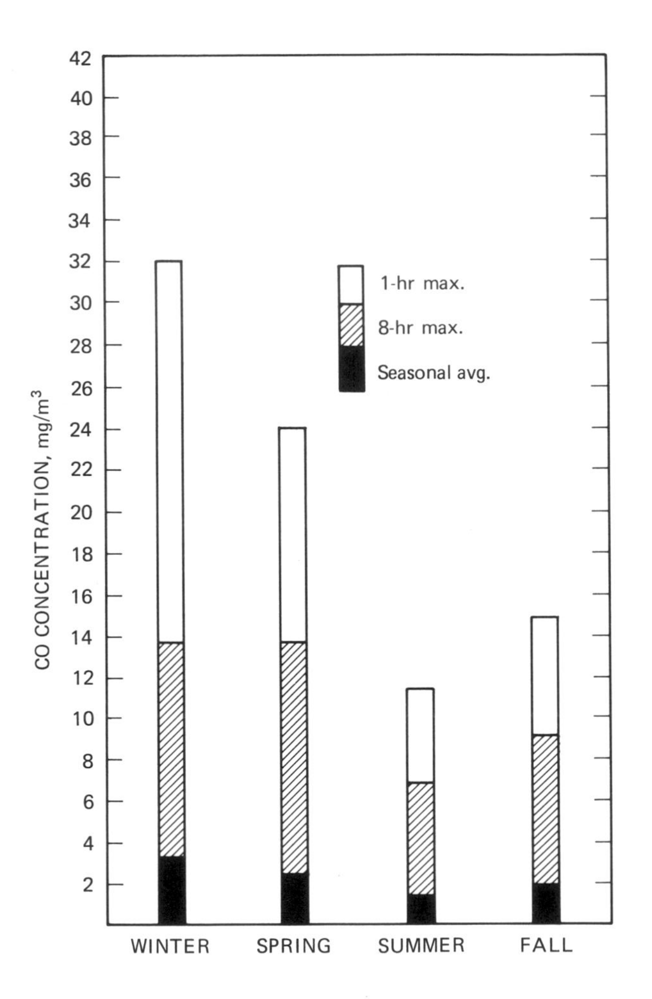

Figure 3.7 Seasonal variations of CO concentrations at a monitoring site in Baltimore, Maryland. (Source: reference [6])

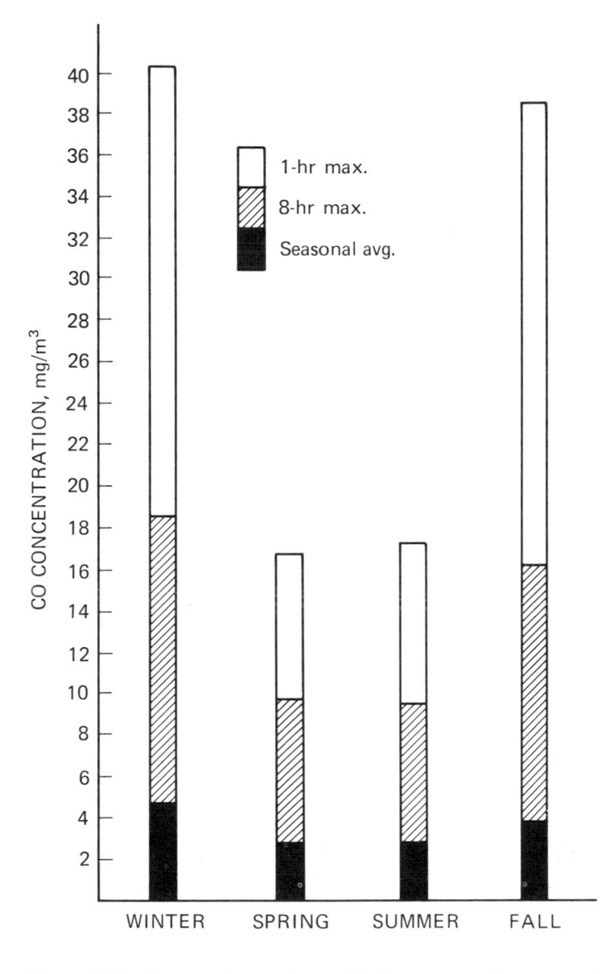

Figure 3.8 Seasonal variations of CO concentrations at a monitoring site in Denver, Colorado. (Source: reference [6])

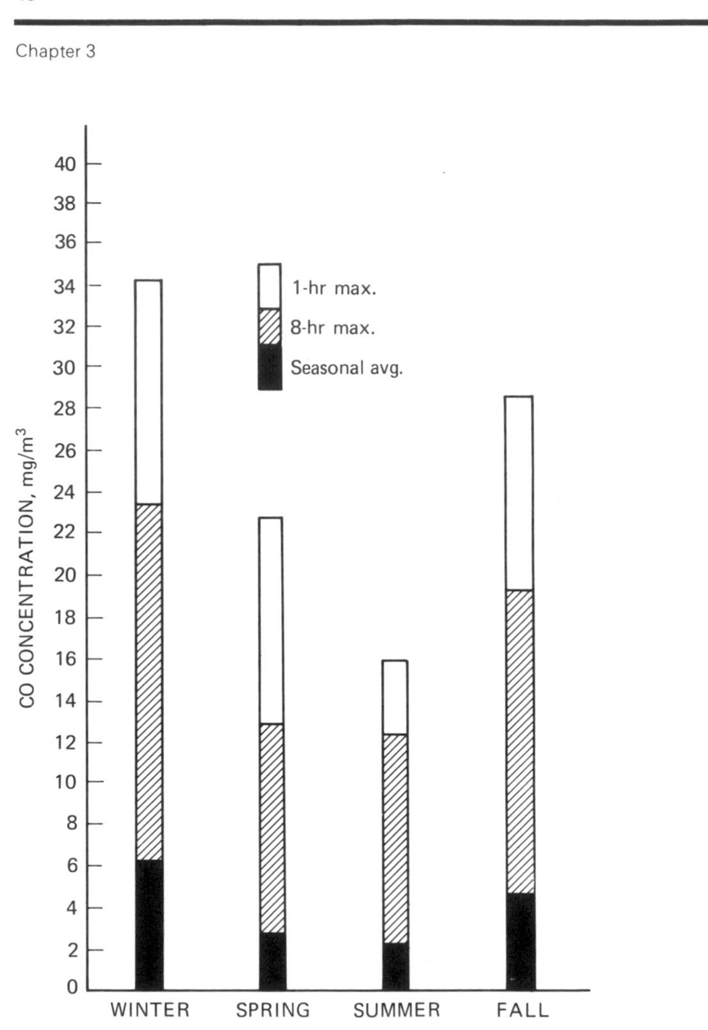

Figure 3.9 Seasonal variations of CO concentrations at a monitoring site in Los Angeles, California. (Source: reference [6])

The close association between CO concentrations, traffic volumes, and meteorological variables makes it possible to develop statistical models that express the CO concentrations at specified locations in terms of traffic and meteorological variables. These models can be very useful in clarifying and illustrating the relations between CO concentrations and the traffic and meteorological variables, as well as in forecasting CO concentrations. A good example of such a model has been developed by G. C. Tiao and S. C. Hillmer to describe 1-hr average CO concentrations at a monitoring site near the San Diego Freeway in Los Angeles [7, 8].[4] This model will now be discussed in some detail, as it illustrates several important characteristics of the behavior of CO concentrations near heavily traveled roadways.

The structure of the Tiao and Hillmer model is as follows. Let CO_t be the average CO concentration at the San Diego Freeway site during hour t, TD_t be the average traffic density on the freeway segment passing the site during hour t, and $WS_\perp$ be the component of the wind speed that is perpendicular to the freeway during hour t. The Tiao and Hillmer model then expresses CO_t as the following function of TD_t and $WS_\perp$:

$$CO_t = a + k(TD_t)\exp[-b(WS_\perp - W_0)^2], \tag{3.1}$$

where a, K, b, and W_0 are constant parameters whose values must be estimated from data. Table 3.1 shows nonlinear regression estimates of the parameter values for several different seasons and days of the week. Figures 3.10–3.12 compare the average observed CO concentrations at different times of day with the CO concentrations predicted by the Tiao and Hillmer model. In all cases, there is good agreement between the observed and predicted concentrations.

In equation (3.1), CO_t increases with increasing values of $WS_\perp$ when $WS_\perp$ is less than W_0 (roughly 3 mph—see table 3.1). This seemingly counter-intuitive dependence of the CO concentration on the wind speed is due mainly to the effects of freeway traffic on the flow of air and dispersion of pollutants near the freeway.[5] Close to the freeway (the monitoring site is only 25 ft from the edge of the roadway) and at low wind speeds, the dispersion of CO is strongly influenced by the turbulent wakes of the moving vehicles and by the tendency of the warm exhaust gases to rise. Under these conditions, a small increase in $WS_\perp$ tends to reduce the degree of dispersion that occurs in the wakes before the exhaust gases reach the monitor and, also, to reduce the distance that the gases rise before reaching the monitor. Consequently, the CO concentration at the monitor tends to increase. In addition, the vehicles on the freeway disrupt the smooth flow of air across the freeway and create a downward

Table 3.1 Parameters of the Tiao and Hillmer CO model[a]

June 1975–October 1975; weekdays

Parameter	Estimate	Standard error
a	1.90	0.14
b	0.035	0.005
W_0	3.27	0.13
K	0.0241	0.00096

June 1975–October 1975; Sundays

Parameter	Estimate	Standard error
a	1.79	0.23
b	0.013	0.011
W_0	2.54	1.11
K	0.0190	0.0016

December 1974–April 1975; weekdays

Parameter	Estimate	Standard error
a	1.42	0.17
b	0.029	0.006
W_0	2.89	0.26
K	0.228	0.0010

a. Reprinted with permission from reference [7]: Tiao, G. C., and Hillmer, S. C., "Statistical Models for Ambient Concentrations of Carbon Monoxide, Lead, and Sulfate Based on the LACS Data," *Environmental Science and Technology*, Vol. 12, pp. 820–828, July 1978, Copyright 1978, American Chemical Society. The model is specified as in equation (2.3). The parameters are estimated for the following units: CO is in ppm; TD is in vehicles per mile; and $WS_{\perp}$ is in mi/hr.

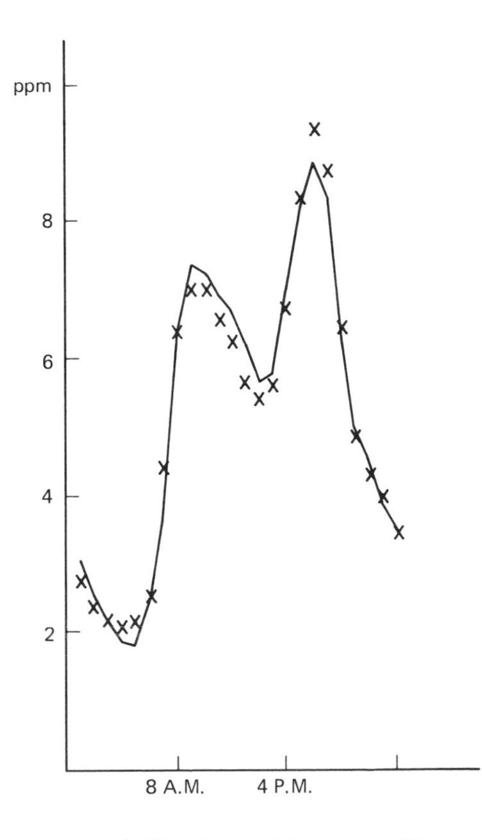

Figure 3.10 Plot of mean 1-hr average CO concentrations observed at the San Diego Freeway monitoring site on summer weekdays and the concentrations predicted by the Tiao and Hillmer model. Key: ×, predicted; –, observed. (Reprinted with permission from reference [7]: Tiao, G.C., and Hillmer, S.C., "Statistical Models for Ambient Concentrations of Carbon Monoxide, Lead, and Sulfate Based on the LACS Data," *Environmental Science and Technology*, Vol. 12, pp. 820–828, July 1978, Copyright 1978, American Chemical Society)

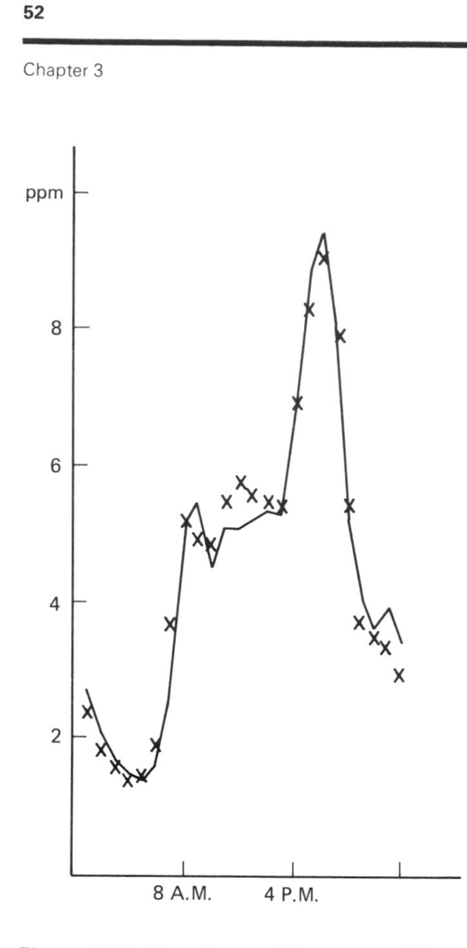

Figure 3.11 Plot of mean 1-hr average CO concentrations observed at the San Diego Freeway monitoring site on winter weekdays and the concentrations predicted by the Tiao and Hillmer model. Key: ×, predicted; —, observed. (Reprinted with permission from reference [7]: Tiao, G.C., and Hillmer, S.C., "Statistical Models for Ambient Concentrations of Carbon Monoxide, Lead, and Sulfate Based on the LACS Data," *Environmental Science and Technology*, Vol. 12, pp. 820–828, July 1978, Copyright 1978, American Chemical Society)

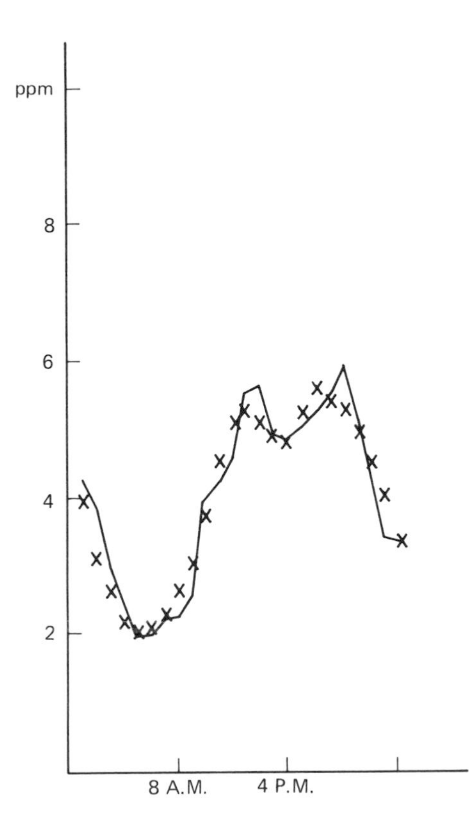

Figure 3.12 Plot of mean 1-hr average CO concentrations observed at the San Diego Freeway monitoring site on summer weekends and the concentrations predicted by the Tiao and Hillmer model. Key: ×, predicted; —, observed. (Reprinted with permission from reference [7]: Tiao, G.C., and Hillmer, S.C., "Statistical Models for Ambient Concentrations of Carbon Monoxide, Lead, and Sulfate Based on the LACS Data," *Environmental Science and Technology,* Vol. 12, pp. 820–828, July 1978, Copyright 1978, American Chemical Society)

movement of air on the downwind side of the freeway. This further enhances the ground-level CO concentration near the freeway. At higher values of $WS_{\perp}$, the effects of wind speed dominate the effects of the turbulent wakes, the buoyancy of the exhaust gases, and the downwash of air on the downwind side of the freeway. Hence, CO concentrations are inversely related to $WS_{\perp}$, as normally is expected.

The term a in equation (3.1) measures the background CO concentration at the San Diego Freeway monitoring site. This background concentration is due to CO emissions from all upwind CO sources except the freeway. The estimates of the value of a shown in table 3.1 indicate that the background CO concentration is less than 2 ppm. In contrast, the total CO concentration, including both background CO and the contribution from traffic on the freeway, can exceed 8 ppm. This illustrates the importance of the freeway traffic as a source of CO at the monitoring site.

The Tiao and Hillmer model is based on the hypothesis that under constant meterological conditions the San Diego Freeway's contribution to the measured CO concentration at the monitoring site—that is, the measured concentration minus the background concentration—is proportional to the total rate of CO emissions from all of the vehicles on the freeway segment that passes the monitoring site. Thus, the term K in equation (3.1) is proportional to the average CO emission rate of these vehicles. The good performance of equation (3.1) in predicting CO concentrations suggests that the hypothesis of proportionality between CO emissions and concentrations, other things being equal, is correct. Additional empirical support for the hypothesis is provided by statistical models relating CO concentrations to traffic flows in downtown Los Angeles [5] and in several cities in New Jersey [4], and by the relation between CO concentrations and traffic volumes shown in figure 3.1. In all of these cases, CO concentrations are linearly related to traffic densities or traffic volumes, which is the form of relation that is to be expected if the hypothesis of proportionality between CO concentrations (excluding background CO) and CO emissions is true. The proportionality between CO emissions on a roadway and the roadway's contribution to CO concentrations at nearby locations will be used in the discussion of CO control measures in chapter 6.

The estimated value of K in equation (3.1) is considerably lower on weekends than on weekdays (see table 3.1). Since K is proportional to the average CO emission rate of vehicles on the freeway, this indicates that the average CO emission rate of these vehicles is lower on weekends than it is on weekdays. The weekend-weekday difference in the average CO

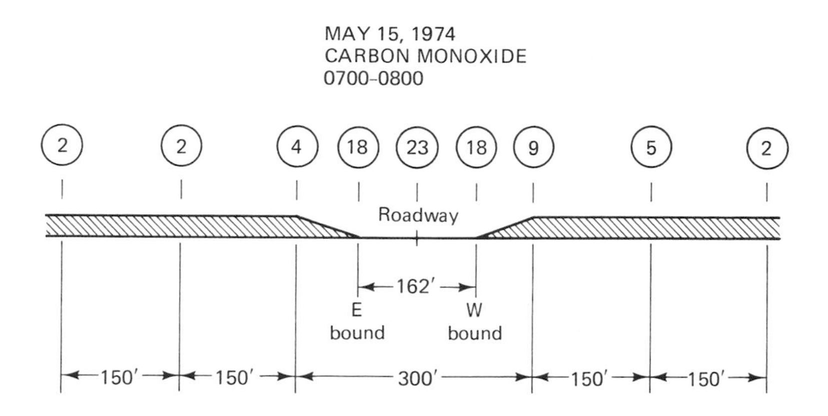

Figure 3.13 One-hour average CO concentrations at points along a line that crosses the Santa Monica Freeway. Circled numbers are 1-hr average CO concentrations (ppm) 4–5 ft above ground surface. May 15, 1974, 7–8 A.M. (Source: reference [10])

emission rate is caused by weekend-weekday differences in traffic flow conditions. Average traffic speeds on the freeway tend to be higher on weekends than they are on weekdays [7, 8], and vehicles' CO emissions rates tend to decrease as their average speeds of travel increase.[6]

The Tiao and Hillmer model illustrates the dependence of CO concentrations near a roadway on traffic flow, emissions, and meteorological variables. CO concentrations are also very sensitive to location. The Tiao and Hillmer model does not illustrate this characteristic of CO concentrations as the model is based on data obtained at a single location. However, as will now be discussed, CO concentrations often vary greatly over small distances.

Figure 3.13 illustrates one example of the high spatial variability of CO concentrations. The figure shows 1-hr average CO concentrations measured along a line that crosses the Santa Monica Freeway in Los Angeles. The CO concentration changes by more than a factor of 10 over distances of 300–450 ft on each side of the freeway center line.

The spatial variation of CO concentrations on the surface streets of a city is illustrated in figure 3.14. This figure is based on 39 measurements of CO concentrations at each of five sites within a one-square-block area in San Jose, California [11]. The quantity α in the figure is the ratio of the average of the CO concentrations measured at the indicated monitoring

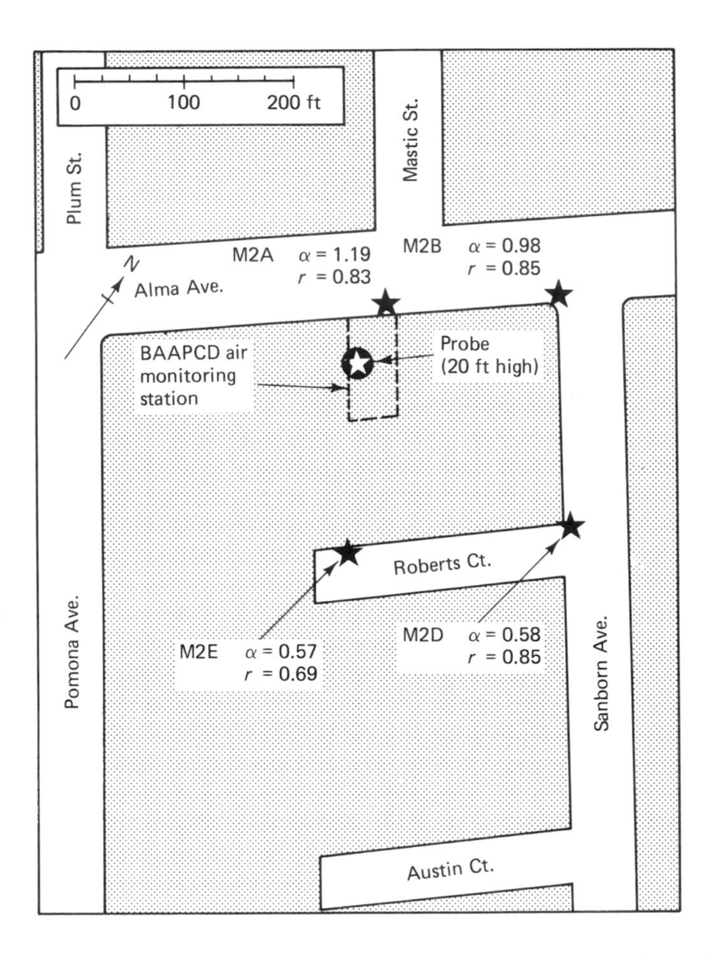

Figure 3.14 Relations between CO concentrations at five sampling points within a one-square-block area in San Jose, California. (Reprinted with permission from reference [11])

site to the average of the concentrations measured at the Bay Area Air Pollution Control District (BAAPCD) monitoring station. The quantity r in the figure is the correlation coefficient between the CO concentrations at the indicated monitoring site and the concentrations at the BAAPCD monitoring station. The concentrations were measured with 5-min averaging times. Alma Avenue in the figure is a four-lane roadway whose traffic volume was roughly 1,300 vehicles/hr at the times that CO monitoring took place. Roberts Court is a residential cul-de-sac that carries very little traffic. The average CO concentration measured at site M2E on Roberts Court, which is 200 ft from the BAAPCD station, is only 57% of the average concentration at the station.

In further CO monitoring carried out over a 1.5-mi-long route in and around downtown San Jose, CO concentrations were found to vary by roughly a factor of six [11]. Similar results have been obtained in other cities. Changes in CO concentrations of a factor of 3.4 over the length of a city block have been observed in Dayton, Ohio [12]. Eight-hour average CO concentrations have been observed to vary by a factor of 4 among different locations in the central business district of Seattle [13].

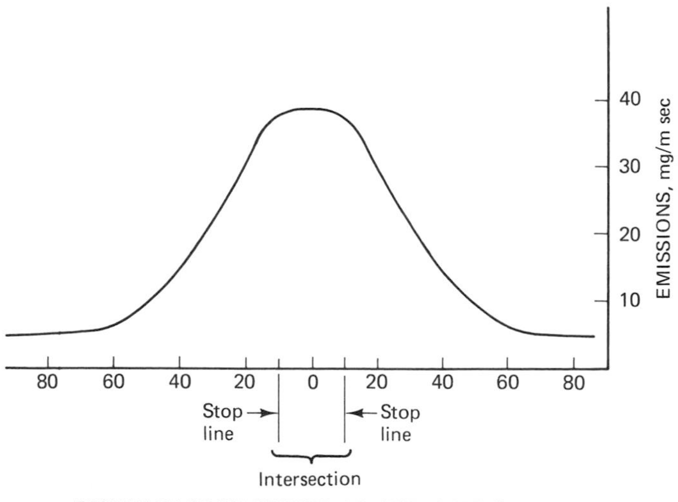

Figure 3.15 CO emissions in the vicinity of a signalized intersection. Source: See section 5.7.

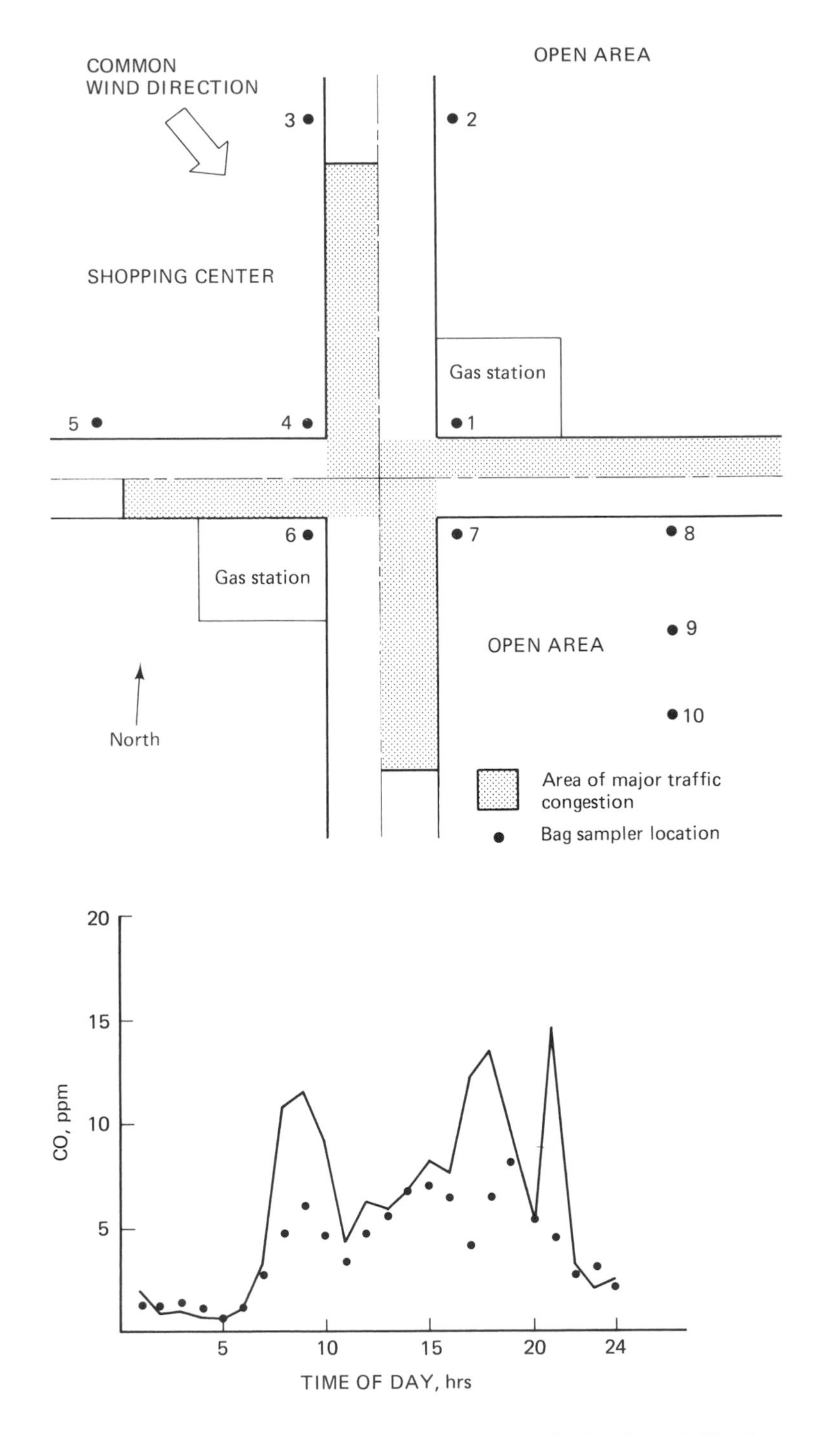

Figure 3.16 CO concentrations near an intersection in San Jose, California: (*a*) locations of sampling points; (*b*) 1-hr average CO concentrations at different times of day (———, average of sites 1, 4, 6, 7; . . . . , average of sites 2, 3, 5, 8, 9, 10). (Source: reference [14])

Another source of spatial variation in CO concentrations is spatial variation in the rate or density of CO emissions along a roadway. CO emissions tend to be higher in the vicinity of points of congestion, such as signalized intersections, than at locations where traffic flows freely. Figure 3.15 shows an example of the variation of the density of CO emissions on a roadway as a function of the distance from a signalized intersection. The emissions density is nearly 10 times as large in the intersection as it is 80 m away from the intersection. Because CO emissions tend to peak at intersections, CO concentrations also tend to peak at intersections. Figure 3.16 shows an example of CO concentrations in the vicinity of an intersection in San Jose. The concentrations at the intersection tend to be larger than those away from the intersection. Moreover, the concentrations at the intersection exhibit strong peaks associated with the morning and afternoon traffic peaks and with the closing of the stores in a nearby shopping center. The peak concentrations away from the intersection are only about half as large as the peak concentrations at the intersection.

Local topographical features, such as tall buildings along a street, also can contribute to large variations in CO concentrations over short distances. When the wind direction is within about 60° of the cross-street direction on a roadway lined with tall buildings, a helical air circulation can be

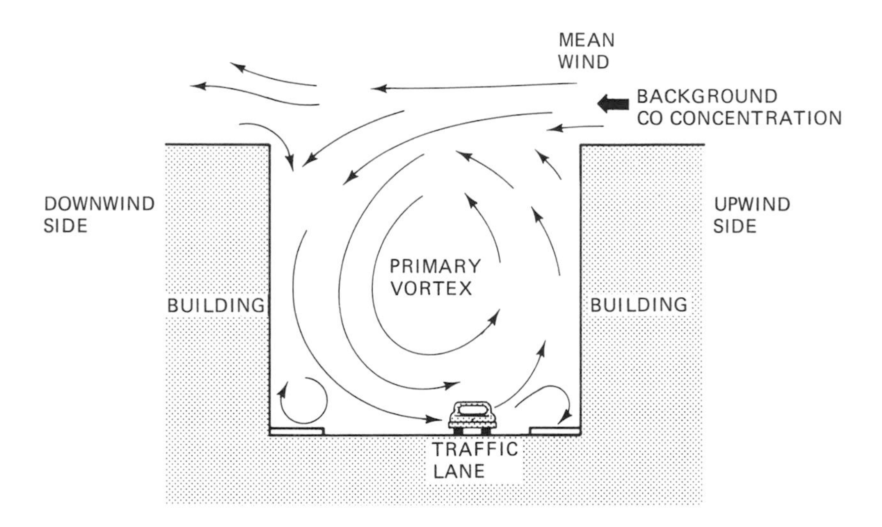

Figure 3.17 Schematic diagram of cross-street circulation between buildings in a street canyon. (Adapted with permission from reference [15])

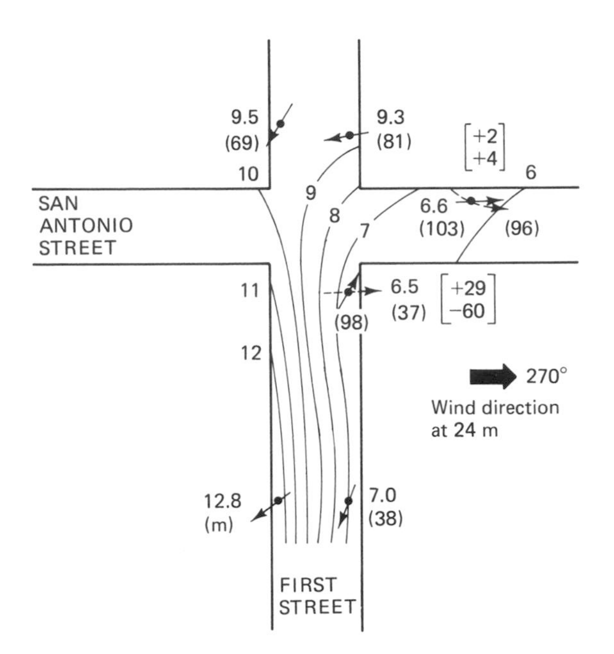

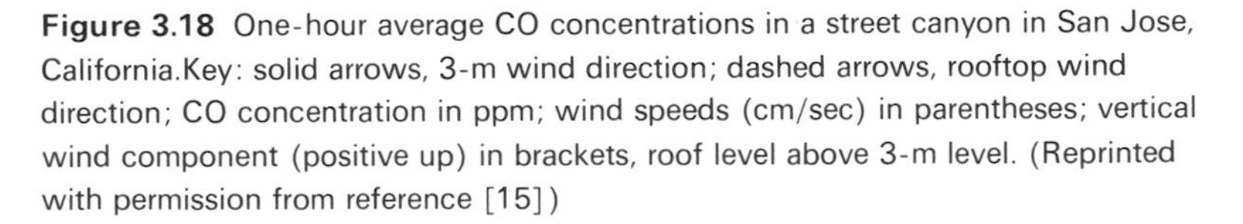
Figure 3.18 One-hour average CO concentrations in a street canyon in San Jose, California.Key: solid arrows, 3-m wind direction; dashed arrows, rooftop wind direction; CO concentration in ppm; wind speeds (cm/sec) in parentheses; vertical wind component (positive up) in brackets, roof level above 3-m level. (Reprinted with permission from reference [15])

created that causes CO concentrations to be considerably higher on the upwind side of the street than they are on the downwind side [15, 16]. This air circulation is illustrated in figure 3.17. Figure 3.18 shows an example of the resulting gradient of 1-hr average CO concentrations on a street in San Jose. The buildings along First Street in the figure are 3–5 stories high. The CO concentration is roughly twice as high on the upwind side of the street as on the downwind side. In Toronto, CO concentrations on the upwind side of a street canyon that are three times as large as downwind-side concentrations have been observed [17].

Because of the high degree of spatial variability of CO concentrations, the concentrations measured at fixed monitoring stations do not necessarily indicate the concentrations to which individuals are exposed. Figure 3.19 compares CO concentrations measured at the BAAPCD monitoring station in San Jose with the concentrations experienced by a pedestrian in

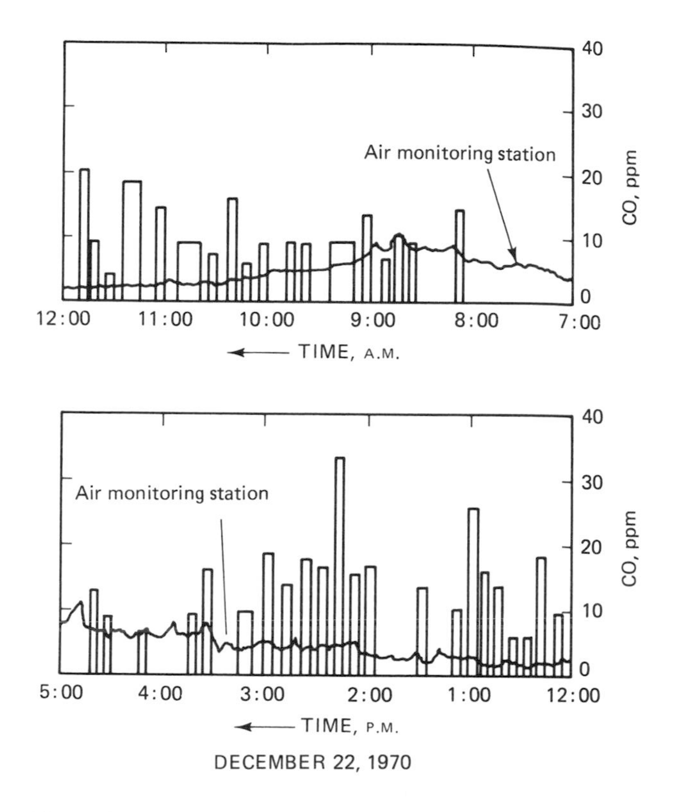

Figure 3.19 CO concentrations measured by walking in downtown San Jose, California, and at the BAAPCD monitoring station. (Reprinted with permission from reference [11])

downtown San Jose. The average concentration to which the pedestrian was exposed is roughly 60% higher than the average concentration measured at the BAAPCD station [11]. A study of the concentrations encountered by 63 persons working in the Boston area showed that these persons encountered maximum 1-hr average CO concentrations that were 40% larger than the 1-hr average concentrations measured at the fixed monitoring stations operated in downtown Boston by the Massachusetts Bureau of Air Quality Control. The 8-hr average concentrations encountered by the same individuals were 35% less than the 8-hr average concentrations measured at the downtown monitoring stations [18].

The spatial variability of CO concentrations has some important implications for the design of programs to reduce excessive CO concentrations. The first implication concerns the problem of characterizing the magnitudes and extent of excessive CO concentrations in a city. Because CO measurements made at fixed site monitors indicate the magnitudes of CO concentrations only in the immediate vicinities of the monitoring sites and do not necessarily represent the concentrations that occur at other locations or to which individuals may be exposed, the characterization of CO concentrations in a city requires either a dense network of fixed monitoring stations or intensive use of mobile monitoring equipment. Although quantitative criteria for determining whether networks of CO monitors are sufficiently dense do not yet exist, it seems clear that the networks in most cities are highly inadequate. For example, in 1977 and 1978, CO concentration data were available at only 12 sites in the city of Philadelphia [19], 9 sites in the city of Washington, DC [20], 6 sites in the Boston metropolitan area [21], and 2 sites in the Pittsburgh area [22]. Thus the geographical extent of excessive CO concentrations in most cities is largely a matter of conjecture.

A second implication of the spatial variability of CO concentrations concerns the degree of spatial detail that is needed in transportation and traffic analyses performed in support of CO control programs. Frequently the smallest geographical units that are resolved in urban transportation analyses are traffic zones that may have surface areas of several square miles, and the links and nodes of networks in which several streets and intersections may be represented by a single link or node. These analyses are capable of indicating the effects of transportation policy measures only on spatially averaged CO concentrations. The analyses cannot resolve the CO peaks that can occur near heavily traveled roadways, intersections or other points of congestion and in street canyons. To resolve CO peaks and to estimate the effects of transportation policy measures on these peaks, it is necessary to carry out transportation studies with a level of spatial detail that enables the traffic conditions that cause CO peaks to be resolved. Unfortunately, as will be discussed in chapter 6, the state of the art of transportation systems analysis is such that studies with the necessary spatial detail can be carried out only with great difficulty, if at all.

Finally, the spatial characteristics of CO enable transportation measures with highly localized impacts to be effective in reducing localized CO concentrations, although these measures are not effective in reducing the concentrations of more uniformly dispersed pollutants, such as ozone. High CO concentrations at a point tend to be caused mainly by CO emissions in the immediate vicinity of the point, rather than by emissions farther away. Thus, CO concentrations at a point usually can be controlled by controlling nearby emissions, even if emissions farther away do not change. For example, it will be seen in chapter 6 that the creation of a vehicle-free zone can cause a substantial reduction in the concentration of CO within the area covered by the zone. Similarly, changes in the timing of the signals at an intersection can substantially reduce CO concentrations in the vicinity of the intersection, provided that traffic volumes do not change. However, neither of these measures significantly affects O_3 concentrations.

3.3 *Ozone*

The most important factor affecting the behavior of O_3 as an air pollutant is that O_3 has no significant anthropogenic emissions sources. It is a secondary pollutant whose presence in the lower atmosphere in excessive concentrations is due mainly to atmospheric chemical reactions involving HC and NO_x emitted by anthropogenic sources. As a result, the characteristics of O_3 are very different from those of CO, which is a primary pollutant. Whereas CO concentrations at a point tend to be linearly related to CO emissions near the point, other things being equal, the relation between O_3 concentrations and HC and NO_x emissions is highly nonlinear. Changes in HC or NO_x emissions rarely produce equal percentage changes in O_3 concentrations; O_3 concentrations often are lower near large emissions sources than they are farther away; and under certain conditions, reductions in HC or NO_x emissions can cause O_3 concentrations to increase. In addition, spatial variations in O_3 concentrations tend to be much more gradual than spatial variations in CO concentrations are. If the concentration of O_3 is excessive at a particular monitoring site, then the O_3 concentration is likely to be excessive over a widespread area—sometimes encompassing tens or hundreds of thousands of square miles—surrounding that site.

Like CO concentrations, O_3 concentrations are sensitive to a variety of meteorological variables. The meteorological variables that affect O_3 concentrations include wind direction and speed, atmospheric stability, mixing height, temperature, and the intensity of solar radiation.

To understand the behavior of O_3, it is necessary to have an elementary understanding of the chemical processes through which O_3 is formed. O_3 formation begins with the photolysis of NO_2 by sunlight to form NO and an oxygen atom:

$$NO_2 \xrightarrow{h\nu} NO + O. \qquad (3.2)$$

The oxygen atom formed by reaction (3.2) combines quickly with atmospheric molecular oxygen to form O_3:

$$O + O_2 \rightarrow O_3. \quad (3.3)$$

However, reactions (3.2) and (3.3), by themselves, cannot cause the accumulation of high concentrations of O_3 because NO and O_3, which are formed in equal quantities by reactions (3.2) and (3.3), cannot coexist in high concentrations. Both are consumed rapidly by the reaction

$$NO + O_3 \rightarrow NO_2 + O_2. \quad (3.4)$$

Hydrocarbons enable O_3 to accumulate by providing a chemical pathway whereby NO can be oxidized to NO_2 without consuming O_3. The oxidation of NO via the hydrocarbon pathway removes the NO formed in reaction (3.2) sufficiently rapidly to prevent reaction (3.4) from consuming a substantial proportion of the O_3 produced in reaction (3.3).

Many complex sequences of chemical reactions are involved in the oxidation of NO to NO_2 in the presence of HC. These reaction sequences enable a single HC molecule to cause the oxidation of several NO molecules. The sequences are initiated by substances such as atomic oxygen, O_3, and free hydroxyl radicals (OH), which react with HC to form oxygenated free radicals.[7] The oxygenated free radicals then react with NO to form NO_2 and other products. Some of these products react further to form more free radicals that can oxidize NO to NO_2. The reaction sequences eventually terminate through the formation of relatively stable products, such as PAN and other nitrates.

The most important reaction sequences leading to NO oxidation are initiated by free OH radicals. These free radicals can be formed, among other ways, by photolysis of nitrous acid (HNO_2) and aldehydes. HNO_2 is produced by reactions involving NO, NO_2, and atmospheric water vapor, and aldehydes are present in motor vehicle exhaust, among other sources. The following is an example of a reaction sequence initiated by free OH radicals:[8]

$$HC + OH \overset{O_2}{\rightarrow} \text{oxygenated organic free radicals } (RO_2); \quad (3.5)$$

$$RO_2 + NO \rightarrow NO_2 + \text{other oxygenated organic free radicals (RO)}; \quad (3.6)$$

$$RO \overset{O_2}{\rightarrow} RO_2 + \text{other products}. \quad (3.7)$$

Because RO_2 radicals are regenerated in reaction (3.7) and, in addition, because some of the other products of this reaction can oxidize NO to NO_2, one HC molecule is able to cause the oxidation of several NO molecules. However, reaction (3.7) produces less RO_2 than reaction (3.5) does, and other reactions cause NO_x to be converted to stable products. Hence, the reaction sequence (3.5) to (3.7) eventually stops if the quantities of HC and NO_x that are available are fixed.

One consequence of the complex process through which O_3 is formed is that the relation between precursor (HC and NO_x) emissions and O_3 concentrations is nonlinear. This is illustrated in figure 3.20, which shows the maximum 1-hr average O_3 concentration that occurs in a simple photochemical system as a function of the HC and NO_x concentrations that are present in the system before photochemical reactions begin. No HC or NO_x is injected into the system after the reactions begin. Therefore, the initial HC and NO_x concentrations, respectively, are proportional to emissions of HC and NO_x into the system. Moreover, given percentage changes in the initial HC and NO_x concentrations are equivalent to equal percentage changes in emissions. Although this system is highly simplified relative to a real urban atmosphere and cannot be extrapolated quantitatively to a real atmosphere, the simple system and real atmospheres

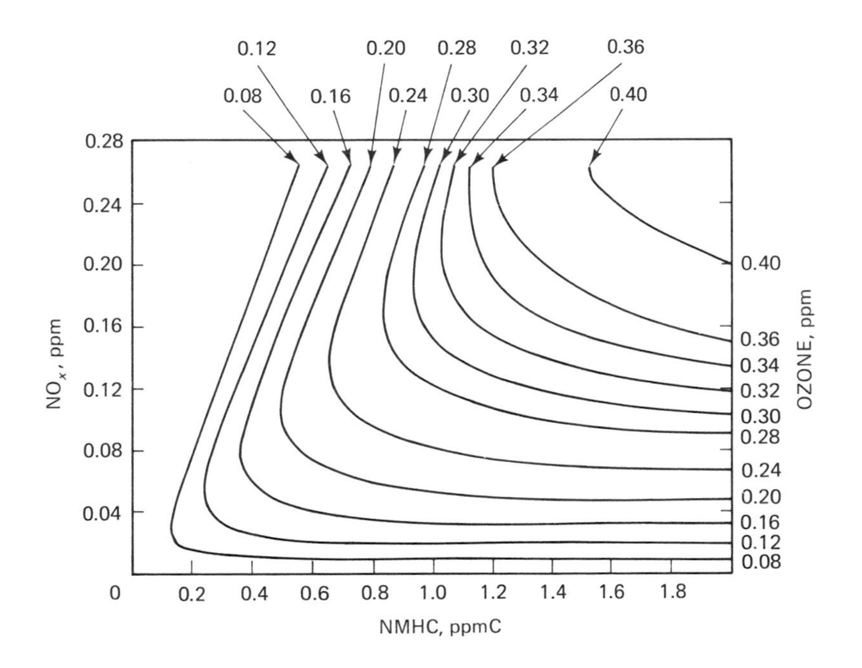

Figure 3.20 Minimum 1-hr average O_3 concentration as a function of the initial HC and NO_x concentrations in a simple photochemical system. (Source: reference [24])

have many qualitative similarities.[9] In particular, the behavior of the simple system illustrates two important general characteristics of the relation between O_3 concentrations and precursor emissions. First, reductions in precursor emissions do not necessarily cause reductions in maximum O_3 concentrations. Referring to figure 3.20, when the ratio of the initial HC concentration to the initial NO_x concentration is relatively low (less than approximately 5), reducing the initial NO_x concentration while keeping the initial HC concentration constant tends to increase the maximum O_3 concentration in the system. When the initial HC-to-NO_x ratio is relatively high (greater than approximately 20), reducing the initial HC concentration while keeping the initial NO_x concentration constant has little or no effect on the maximum O_3 concentration.

The second important characteristic of O_3-precursor relations that figure 3.20 illustrates is the lack of proportionality between changes in O_3 concentrations and changes in precursor emissions. Table 3.2, which is derived from figure 3.20, gives examples of the percentage reductions in

Table 3.2 Examples of percentage reductions in maximum O_3 concentrations caused by various percentage reductions in initial HC and NO_x concentrations[a]

Reduction in HC (%)	Reduction in NO_x (%)	Reduction in maximum O_3 (%)
0	0	0
25	0	7
50	0	19
75	0	60
0	25	12
25	25	17
50	25	26
75	25	50
0	50	26
25	50	30
50	50	35
75	50	50
0	75	49
25	75	49
50	75	52
75	75	60

a. Derived from figure 3.20. The percentage reductions are from base concentrations of 1.4 ppm C, 0.117 ppm, and 0.30 ppm for HC, NO_x, and O_3, respectively.

the maximum O_3 concentration in the simple photochemical system that are caused by various percentage reductions in the initial HC and NO_x concentrations. The O_3 reductions caused by HC reductions of 0, 25, 50 and 75% are 0–49%, 7–49%, 19–52%, and 50–60%, respectively, depending on the degree of NO reduction. In every tabulated case the percentage reduction in the O_3 concentration is less than the percentage reduction in the concentration of at least one of the precursors.[10]

In practical air pollution control planning, it is often assumed that O_3 concentrations and HC emissions are linearly related when meteorological conditions are held constant. Thus,

$$[O_3] = [O_3]_b + kE_{HC}, \tag{3.8}$$

where $[O_3]$ is the measured O_3 concentration, $[O_3]_b$ is the background O_3 concentration, E_{HC} is the rate of HC emissions, and k is a proportionality constant whose value depends on meteorological variables. Equation (3.8) is called the "linear rollback" or "proportional" model. Although such a model can be useful when it is applied to a relatively inert pollutant such as CO [compare equations (3.1) and (3.8), for example], the application of linear rollback to a reactive pollutant such as O_3 cannot be justified on either theoretical or empirical grounds. In general, the changes in O_3 concentrations that the linear rollback model predicts are wrong, as can be seen from the fact that in the simple photochemical system of figure 3.20, in which $[O_3]_b$ is zero, the O_3 concentration is neither directly proportional to, nor uniquely determined by, the initial HC concentration. The only virtue of the linear rollback model for O_3, and the only reason for the continuing use of this model in practical air quality planning, is its simplicity.

The chemical reactions that produce O_3 also produce a variety of less familiar oxidants—such as peroxyacetylnitrate (PAN), hydrogen peroxide, and nitrous and nitric acids—and aldehydes.[11] It is usually assumed that the resulting mixture of oxidants and aldehydes consists mainly of O_3, although few efforts have been made to measure the relative concentrations of O_3 and the other constituents of the mixture. PAN and oxidant measurements made in Los Angeles, St. Louis, Missouri, and Hoboken, New Jersey, indicate that the ratio of the concentration of PAN (in ppm) to the concentration of oxidant (also in ppm) varies between 0.01 and 0.36 (25).[12] Formaldehyde-to-O_3 concentration ratios (with the concentrations expressed in ppm) in the range of 0.1–0.2 have been observed in northern New Jersey [26]. Hydrogen peroxide-to-O_3 concentration ratios (with the concentrations again expressed in ppm) ranging from 0.1 to 0.3 have been observed during the hours of peak O_3 concentrations in the Los Angeles area [27, 28]. Also in the Los Angeles area, the nitric acid-to-O_3 concentration ratio (with the concentrations in ppm) has been observed to be 0.07 [29]. These results are based on relatively small numbers of measurements and, because the measurements were made at different times and places, are not comparable. Nonetheless, the results suggest that even if PAN, formaldehyde, hydrogen peroxide, nitric acid, and, possibly, other aldehydes and non-O_3 oxidants are minor contributors to urban photochemical smog when considered individually, collectively they may represent a substantial proportion of the smog mixture. However, this is only a suggestion; the available data are too few and diverse to permit a stronger conclusion to be reached.

There also have been attempts to measure O_3 and oxidant concentrations simultaneously at the same locations [25]. However, the imprecision of existing instrumentation for measuring oxidant concentrations makes it impossible to reach conclusions about O_3-to-oxidant concentration ratios from these measurements. In general, there have been too few measurements of atmospheric concentrations of non-O_3 oxidants and aldehydes to permit conclusions to be drawn regarding the importance of these compounds as contributors to urban photochemical smog.

Because reaction (3.4) prevents O_3 and NO from coexisting in large concentrations, O_3 concentrations tend to be low in the vicinity of large NO sources, such as heavily traveled roadways, even if O_3 concentrations are high elsewhere. This effect is illustrated schematically in figure 3.21. In the figure, the O_3 concentration is lower and the NO_2 concentration is higher on the downwind side of the roadway than on the

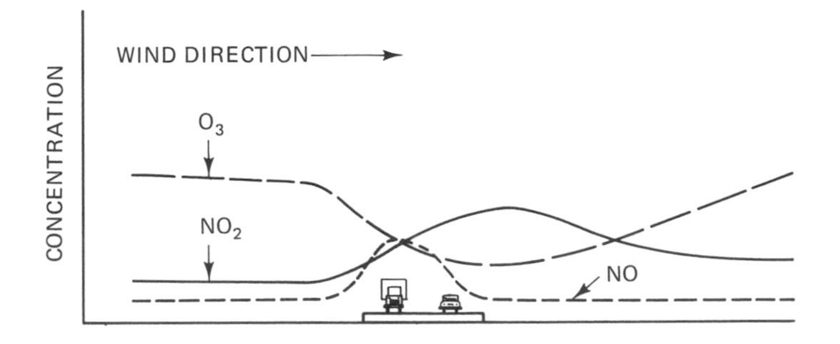

Figure 3.21 Depletion of O_3 and enhancement of NO_2 downwind of a roadway where NO is emitted. (Source: reference [30])

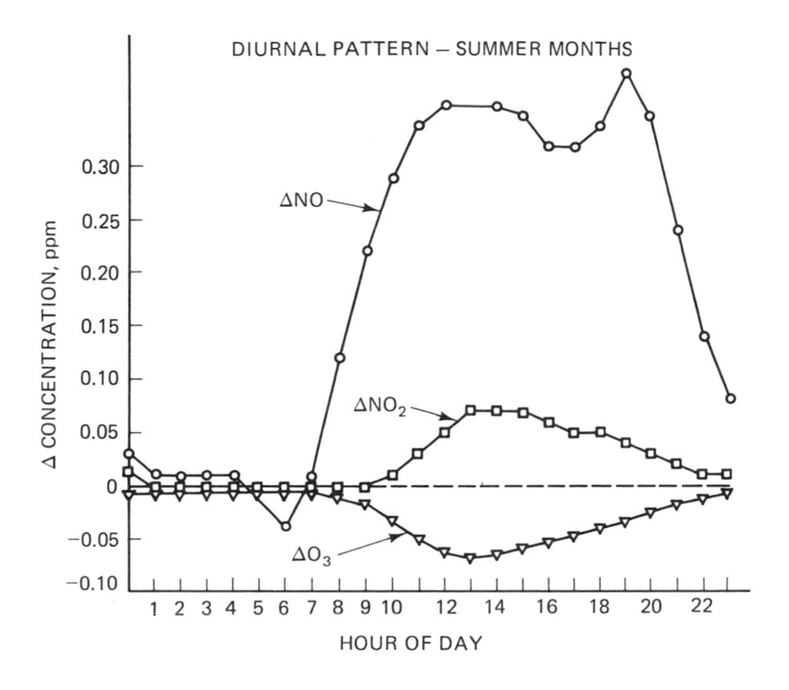

Figure 3.22 Differences between the concentrations of NO, NO_2, and O_3 measured on the northeast and southwest sides of the San Diego Freeway. Δ signifies concentration on the northeast side of the freeway minus the concentration on the southwest side. (Source: reference [31])

upwind side. Figure 3.22 shows diurnal variations in the average differences between O_3, NO, and NO_2 concentrations measured at two sites on opposite sides of the San Diego Freeway in Los Angeles. The wind patterns in Los Angeles and the orientation of the freeway are such that the northeast side tends to be downwind of the roadway and the southwest side tends to be upwind of the roadway. Thus, the cross-freeway concentration differences correspond roughly to differences between downwind and upwind concentrations. The downwind- (i.e., northeast-) side NO concentration greatly exceeds the upwind- (i.e., southwest-) side concentration during most of the day, indicating that the freeway is a major NO source. The differences between the downwind and upwind NO_2 and O_3 concentrations are roughly equal in magnitude and opposite in sign, indicating that most of the O_3 that is carried onto the freeway by the wind is consumed by reaction (3.4).[13]

The formation of O_3 takes place over time periods ranging from several hours to several days in duration, depending on meteorological conditions. During this time the air is being mixed and moved by winds. This mixing and movement causes emissions from large numbers of HC and NO_x sources that are distributed over a widespread area to be stirred into the same polluted air mass. As a result, high O_3 concentrations rarely can be attributed to individual HC and NO_x emissions sources. Rather, high O_3 concentrations are caused by the combined precursor emissions of all of the sources over which the air mass passes during the reaction period. In addition, the atmospheric mixing that occurs during the formation of O_3 tends to cause spatial variations in O_3 concentrations to be very gradual. Except for the abrupt reductions in O_3 concentrations that occur near large NO sources, it usually is the case that if the O_3 concentration is high at a particular monitoring site, it also is high over a widespread area surrounding that site.

The chemical reactions that form O_3 and the need for sunlight to drive these reactions impose characteristic diurnal variations on urban O_3, NO, and NO_2 concentrations. An example of these variations is shown in

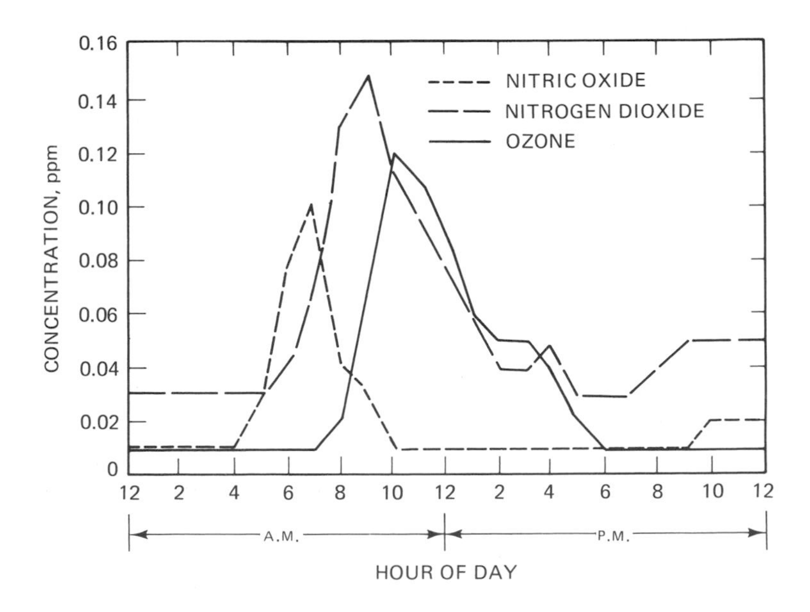

Figure 3.23 Diurnal variation in NO, NO_2, and O_3 concentrations in Los Angeles, California, 19 July 1965. (Source: reference [32])

figure 3.23, which is based on observations of O_3, NO, and NO_2 concentrations in Los Angeles. There is a sharp morning NO peak associated with the morning traffic peak and other activities. This NO is converted to NO_2 by processes such as reactions (3.5)–(3.7) over a period of 2–3 hr in the morning. O_3 accumulation does not begin until the NO conversion is nearly complete, owing to reaction (3.4). After O_3 begins to form, its concentration increases rapidly and, typically, reaches a peak during the late-morning or early-afternoon hours. The O_3 concentration then starts to decrease. There is a small NO_2 peak associated with oxidation of the NO emissions that occur during the afternoon traffic peak.

The afternoon NO_2 peak is considerably smaller than the morning peak because the late-afternoon sunlight is not sufficiently intense to drive the photochemical reactions that produce the morning peak. Moreover, the wind speed in Los Angeles tends to be higher in the afternoon than in the morning (see figure 3.6). O_3 is not produced in the late afternoon or evening due to insufficient sunlight. Late in the evening, as the winds diminish, NO_2 begins to accumulate again. This accumulation provides the NO_2 needed to initiate reaction (3.2) the following day.

The tendency of O_3 concentrations to peak during the late-morning or early-afternoon hours has been observed in many cities [32]. Considering the roughly 3-hr time period needed to convert the NO emissions from the morning traffic peak into NO_2 and the need for this conversion to be nearly complete before O_3 accumulation can begin, a late-morning or early-afternoon O_3 peak is what one would expect to observe if the O_3 were caused only by HC and NO_x emissions during the morning peak traffic period. Accordingly, the diurnal pattern of urban O_3 often has been interpreted to mean that only early-morning HC and NO_x emissions contribute significantly to peak O_3 concentrations. The now inoperative US air quality standard for NMHC, which applies to the 6–9 A.M. average NMHC concentration, reflects this interpretation. However, it now appears that this interpretation is incorrect and that the concentration of O_3 formed in a given volume of air depends on the total mass of precursors emitted into that volume, not just on early-morning precursor emissions [33].

The early-afternoon peaking of O_3 in many cities is most likely due to meteorological factors. For example, the increase in the wind speed that occurs in the afternoon in the Los Angeles area causes polluted air in Los Angeles to be pushed eastward and to be replaced with cleaner air, thus causing the O_3 concentration in Los Angeles to decrease. However, as the polluted air mass moves eastward, it receives new emissions of pre-

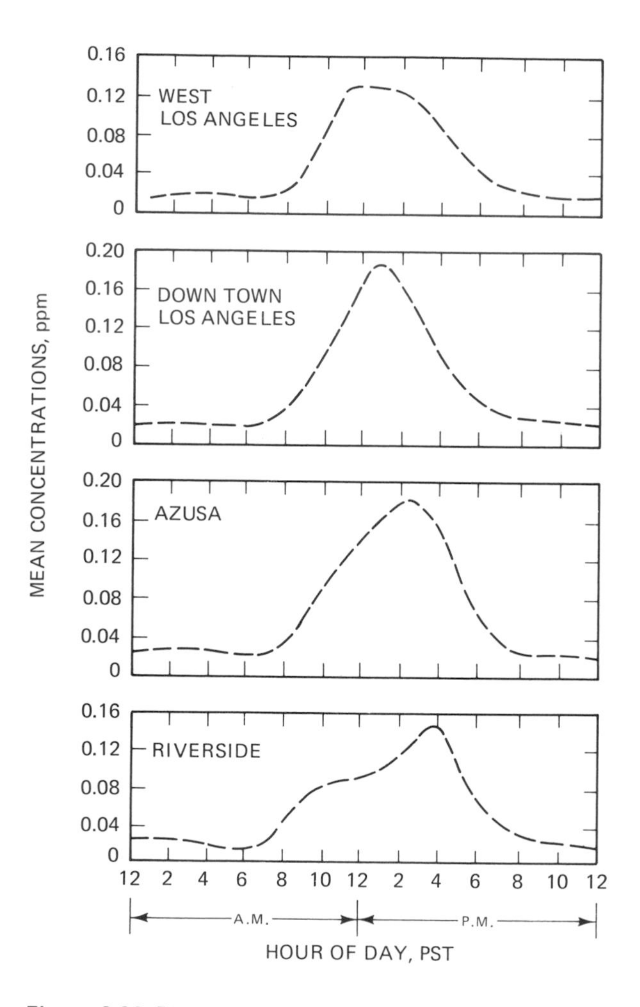

Figure 3.24 Diurnal variation of mean 1-hr average oxidant concentrations at four sites in the Los Angeles, California, area, October 1965. (Source: reference [32])

cursors, and the chemical reactions that produce O_3 continue. The O_3 concentrations in air mass do not decrease and may increase, and the high O_3 concentrations that occurred in Los Angeles in the late morning or early afternoon (and, often, higher concentrations) occur in downwind communities later in the day. This is illustrated in figure 3.24, which shows an example of diurnal variations in 1-hr average oxidant concentrations at four locations in the Los Angeles area. The West Los Angeles monitoring station is approximately 10 mi west of the downtown Los Angeles station. The Azusa and Riverside monitoring stations are approximately 20 and 50 mi east of downtown Los Angeles, respectively. Although the peak oxidant concentrations at the four stations are similar in magnitude, the peak at West Los Angeles occurs about 1 hr earlier than the peak at downtown Los Angeles, whereas the peaks at Azusa and Riverside, respectively, occur roughly 2 and 4 hr later than the downtown Los Angeles peak.

In addition to moving polluted air masses from place to place, the wind and other meteorological factors affect the concentration of O_3 that is generated by a given rate and geographical distribution of precursor emissions. Increases in wind speed, atmospheric turbulence, and the mixing height tend to increase the volume of air into which precursors can be dispersed and, also, to increase the rate of precursor dispersal. Consequently, O_3 concentrations tend to decrease. The effect of wind speed and mixing height on the concentration of O_3 is illustrated in table 3.3, which shows means of 1-day average O_3 concentrations in downtown Los Angeles for various wind speeds and mixing heights. The O_3 concentration decreases monotonically as the wind speed and mixing height increase.

Table 3.3 Means of daily averages of O_3 concentrations (pphm) for various wind speeds and maximum mixing heights in downtown Los Angeles, California: June–October 1967–1970[a]

Wind speed (mph) 6–12 A.M.	Maximum mixing height (ft)			
	0–2,000	2,100–2,500	2,600–3,500	≥3,600
0–4	5.7	5.6	4.3	3.8
4.1–5	4.8	4.5	4.1	3.1
≥5.1	4.3	3.9	3.0	2.2

a. Reprinted with permission from reference [34].

O_3 concentrations also are affected by the intensity of solar radiation and the air temperature. Solar intensity affects O_3 concentrations because of the need for sunlight to drive photochemical reactions such as reaction (3.2). Hence, O_3 concentrations tend to increase with increasing solar intensity. The rates of many of the chemical reactions involved in O_3 formation are sensitive to temperature. Although the temperature sensitivities of individual reactions are not yet fully understood, the net effect of temperature changes on the complete system of O_3-generating reactions is to cause the rate of O_3 formation to increase when the temperature increases. This is illustrated in table 3.4, which shows the O_3 concentrations that were produced when three laboratory mixtures containing the same initial concentrations of HC and NO_x w^re irradiated for 6 hr at different temperatures. Increasing the temperature from 16 to 38.2°C caused the concentration of O_3 formed in 2 hr to increase by a factor of 6.5 and caused the quantity of O_3 formed in 6 hr to increase by a factor of 3.2.

Seasonal variations in meteorological conditions, particularly in solar intensity and temperature, impose systematic seasonal variations in O_3 concentrations. These are illustrated in figure 3.25. O_3 concentrations tend to be higher from midspring to midfall than they are during the rest of the year.

The atmospheric mixing and movement that take place while O_3 is being formed can cause O_3 concentrations to be high at large distances from the sources of the precursors that generated the O_3. In addition, O_3 can be trapped in aboveground strata of the atmosphere and transported for long distances, again leading to high O_3 concentrations far away from the sources of precursors [36, 37]. For example, photochemical air pollution

Table 3.4 Effect of temperature changes on O_3 formation in a laboratory mixture of HC and NO_x[a]

Temperature (°C)	Initial concentration (ppm)			O_3 concentration (ppm)	
	HC	NO	NO_2	After 2 hr	After 6 hr
16.0	0.51	0.53	0.06	0.021	0.130
30.0	0.55	0.52	0.06	0.020	0.270
38.2	0.55	0.55	0.08	0.137	0.410

a. Reprinted in part with permission from reference [35]: Carter, W. P. L., Winer, A. M., Darnall, K. M., and Pitts, J. N., Jr., "Smog Chamber Studies of Temperature Effects in Photochemical Smog," *Environmental Science and Technology*, Vol. 13, pp. 1,094–1,100, September 1979, Copyright 1979, American Chemical Society.

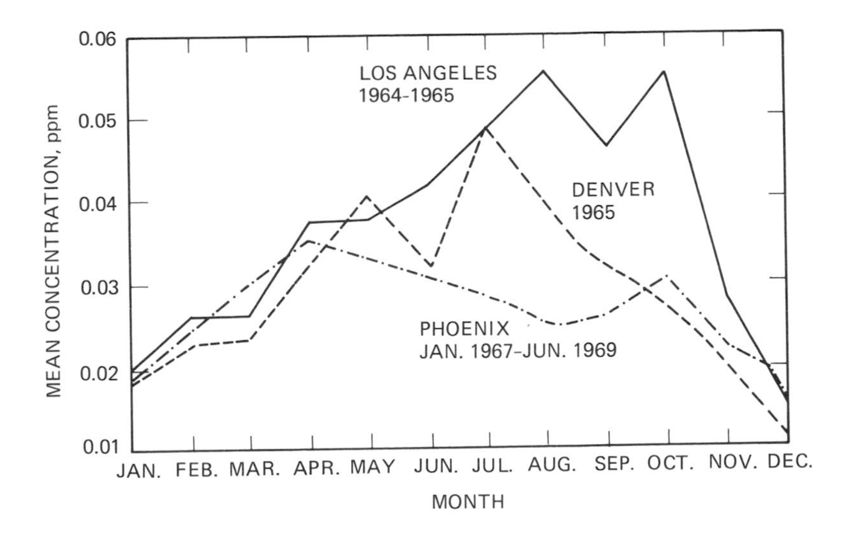

Figure 3.25 Monthly variation of mean hourly oxidant concentrations in three cities. (Source: reference [32])

resulting from precursor emissions in the New York City area is a major cause of high O_3 concentrations in Connecticut and has been observed to be transported as far from New York City as northeastern Massachusetts [38, 39]. Precursor emissions in St. Louis have been found to produce O_3 concentrations that exceed the US air quality standard for O_3 100–150 mi downwind of the city [36, 40]. Transport of photochemical pollution over distances of at least 50 mi has been observed in the Los Angeles area [41, 42]. In addition, O_3 and its precursors can be entrained in the gradual rotation and movement of high-pressure weather systems and, thereby, distributed over areas of tens or hundreds of thousands of square miles [36, 43, 44]. This phenomenon, which is particularly common in the United States east of the Mississippi River, can cause the occurrence of high O_3 concentrations in rural areas that are seemingly well removed from significant sources of precursors. For example, O_3 concentrations exceeding 300 $\mu g/m^3$ have been observed at such relatively isolated locations as McHenry, Maryland; Kane and Dubois, Pennsylvania; and Wilmington, Coshocton, and Wooster, Ohio [45, 46]. During summertime occurrences of stagnating high-pressure weather systems, an area of the United States that may include as many as 20 states can be covered with a blanket of O_3 that is generated by precursor emissions from widely distributed sources within the area.[14]

Because of the O_3 transport phenomenon, the air entering large cities can contain relatively high concentrations of O_3. For example, O_3 concentrations ranging from 120 to 260 $\mu g/m^3$ have been observed in the air entering the urban corridor from Washington, DC, to Boston [43, 47]. Moreover, the highest O_3 concentrations associated with the precursor emissions from a city may occur 30–50 mi downwind of the city. For example, the highest O_3 concentrations in the Los Angeles area tend to occur between Azusa and Riverside which, as has already been noted, are 20 and 50 mi downwind of Los Angeles, respectively. Similarly, the highest O_3 concentrations in the New York City plume often occur in Connecticut, and the Philadelphia plume often produces high O_3 concentrations in rural New Jersey [38, 39, 48].

The spatial characteristics of O_3 concentrations cause the design requirements for O_3-monitoring networks to be different from the requirements for CO networks. Because spatial variations in O_3 concentrations tend to be gradual, dense networks are not needed. For example, to characterize O_3 concentrations in a metropolitan area, it may be sufficient to monitor O_3 and precursor concentrations at several locations distributed 20–50 mi downwind of the city and at several locations within and upwind of the city. At present, most large cities have sufficiently many O_3 monitors to characterize O_3 concentrations within the cities and their nearby suburbs. However, many cities do not have upwind and downwind monitors, and most have inadequate monitors for precursor concentrations. Thus, many of the data needed to develop and evaluate O_3 control programs are unavailable.

The spatial characteristics of O_3 also affect the design of programs to reduce O_3 concentrations. Because winds and weather systems mix and distribute O_3 and its precursors over large areas, achieving significant reductions in urban and downwind O_3 concentrations requires achieving reductions in precursor emissions that are significant on at least a metropolitan regional scale. In densely populated areas, such as the eastern United States, it may be necessary to achieve reductions in precursor emissions that are significant on a multistate regional scale. Therefore, to be effective in contributing to reduced O_3 concentrations, urban transportation measures must be oriented toward reducing emissions of precursors from the transportation system as a whole. Measures that may be appropriate for this purpose include improvements in regionwide transit service (particularly for travel to, from, or within the suburbs and for nonwork purposes), regionwide carpooling programs, and pricing incentives for the use of high-occupancy modes. Isolated measures whose effects are highly localized, such as creation of a vehicle-free zone in the central business district of a city, intersection improvements, and transit improvements in a few radial corridors, are unlikely to have significant effects on O_3 concentrations.[15]

3.4 *Nitrogen Dioxide*

Anthropogenic NO_x emissions consist of a mixture of NO and NO_2 in which NO is by far the dominant constituent, accounting for 90–95% of the total. Although the remaining 5–10% of NO_x emissions consist of NO_2, primary NO_2 emissions are not the main cause of high atmospheric NO_2 concentrations. Rather, high NO_2 concentrations are caused by the oxidation of NO to NO_2 through the system of chemical reactions that produce O_3 [see reactions (3.4) and (3.6), for example].

The atmospheric mixing that occurs during the formation of NO_2 through the photochemical reaction process tends to cause spatial variations in NO_2 concentrations to be gradual away from large NO sources. However, reaction (3.4) can cause sharp peaks in NO_2 concentrations to occur near NO sources. For example, the NO_2 concentration on the downwind side of the San Diego Freeway in Los Angeles can exceed the NO_2 concentration on the upwind side by a factor of 4 or more [31].

Like O_3 concentrations, NO_2 concentrations tend to decrease with increasing wind speed and mixing height and to increase with increasing temperature [49]. However, the dependence of NO_2 concentrations on the intensity of solar radiation is somewhat complicated. Although sunlight is needed to drive the photochemical reactions that generate NO_2, sunlight also destroys NO_2 by photolysis [see reaction (3.2)]. Thus, increases in solar intensity can either increase or decrease NO_2 concentrations, depending on the existing solar intensity level and the stage in the photochemical reaction process at which the increase in intensity occurs [50, 51].

Despite the complexity of the chemical reactions that produce NO_2, NO_2 concentrations within an air mass appear to be roughly proportional to NO_x emissions into the mass. This is illustrated in figures 3.26 and 3.27, which show, respectively, the peak and time-averaged NO_2 concentrations that were formed when various mixtures of HC and NO_x were irradiated by sunlight for 6 hr in a laboratory environment.[16] The experimental setup was such that no HC or NO_x was injected into the system after the photochemical reactions began. Hence, the initial concentrations of HC and NO_x in the laboratory system are proportional to the emissions of HC and NO_x into the system. The scatter in figure 3.26 is caused by variations

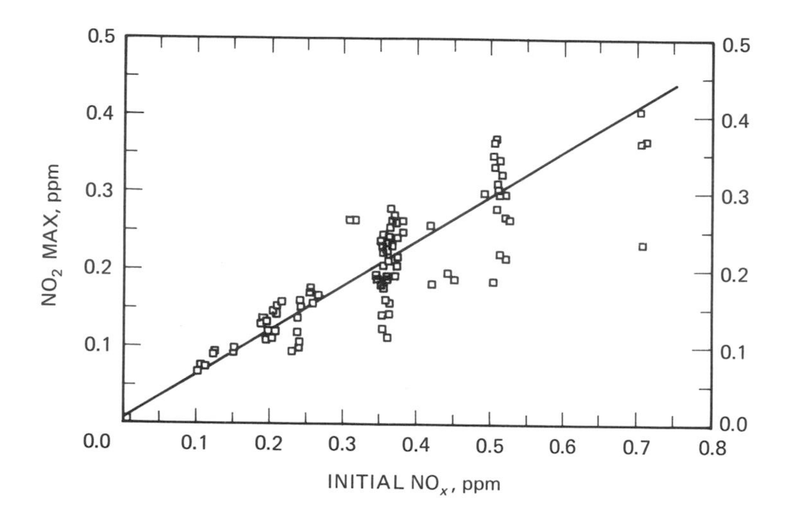

Figure 3.26 Maximum NO_2 concentration in a laboratory photochemical system as a function of the initial NO_x concentration. (Source: reference [51])

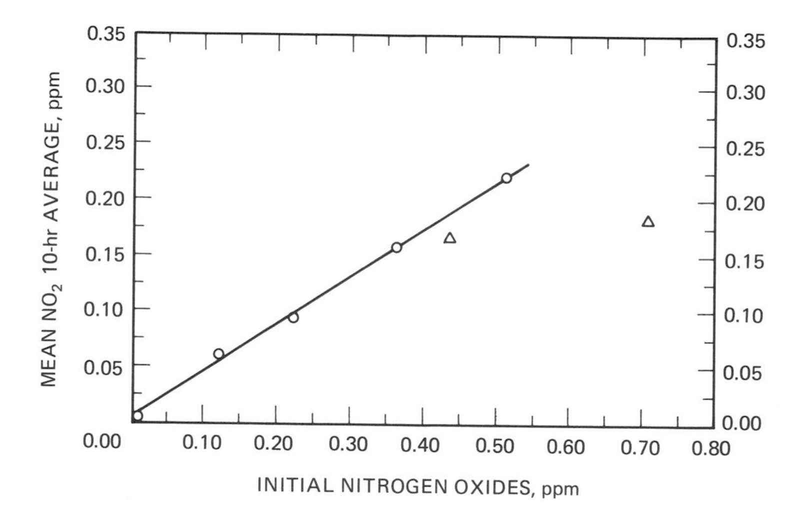

Figure 3.27 Mean 10-hr average NO_2 concentration in a laboratory photochemical system as a function of the mean initial NO_x concentration. Points designated by $\triangle$ had only 4 cases each. The other points had 8 to 51 cases each. Estimates of experimental error are not available. (Source: reference [51])

in the initial HC concentration, the temperature, and the intensity of solar radiation. Figure 3.27 presents the average results of many experiments and, therefore, has no scatter. A statistical analysis of the data presented in the two figures showed that other things remaining unchanged, reductions in the initial NO_x concentration of 25, 50, and 75% would cause reductions in the maximum NO_2 concentration of 23, 47, and 72%, respectively. The corresponding reductions in the average NO_2 concentration would be 23–25%, 46–49%, and 68–75%, depending on the initial NO_x concentration [51].

Although HC is an essential participant in the photochemical reactions that produce NO_2, and only small quantities of NO_2 could be formed without HC, moderate changes in HC emissions appear to have little effect on NO_2 concentrations. For example, in the laboratory system described in the previous paragraph, it was found that, other things remaining unchanged, reductions of 25, 50, and 75% in the initial HC concentration would reduce peak NO_2 concentrations by 4, 9, and 18%, respectively. A 50% reduction in the initial HC concentration was found to have an effect on the average NO_2 concentration that varied from an 11% decrease to a 4% increase, depending on the initial HC and NO_x concentrations [51].

The available evidence suggests that the relation between NO_2 concentrations and precursor emissions in real urban atmospheres is similar to the relation in the laboratory system. This evidence consists of statistical analyses of the dependence of NO_2 concentrations on HC and NO_x emissions in the Los Angeles area, Denver, Chicago, and Houston [49, 52]. These analyses suggest that annual average and peak 1-hr average NO_2 concentrations are both roughly proportional to NO_x emissions, other things remaining unchanged. The analyses also suggest that moderate changes in HC emissions would have little effect on NO_2 concentrations. For example, it was estimated in one study that a 50% reduction in HC emissions would have an effect on annual average NO_2 concentrations that varies from an 11% decrease to a 5% increase, depending on the city [49]. It was also estimated that a 50% reduction in HC emissions would reduce the annual maximum 1-hr average NO_2 concentration by 0–25%, also depending on the city [49]. In another study, it was estimated that a 50% reduction in HC emissions in the Los Angeles area would reduce annual maximum 1-hr average NO_2 concentrations by 5–10% [52].

In areas where significant photochemical activity occurs, NO_2 concentrations tend to exhibit a double-peaked pattern of diurnal variation. This is illustrated in figure 3.28, which shows diurnal variations in seasonal

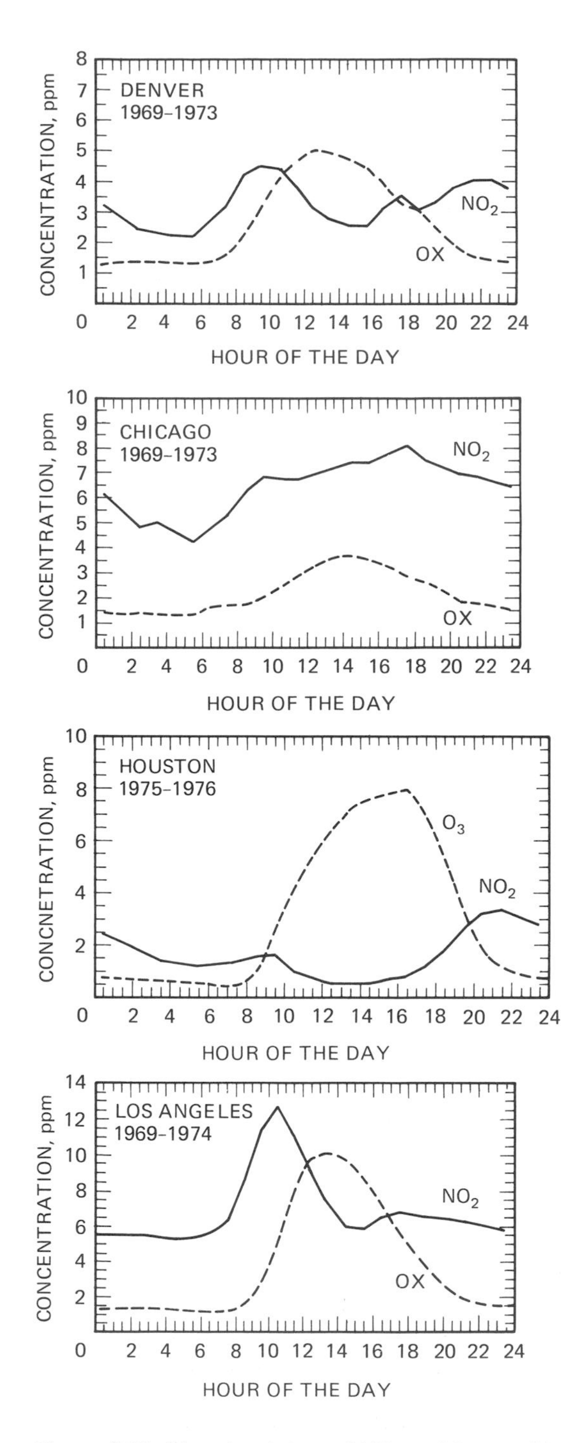

Figure 3.28 Diurnal variations of NO_2 and O_3 or oxidant concentrations in four cities during the third quarter of the year. (Source: reference [49])

mean 1-hr average NO_2 and O_3 or oxidant concentrations in four cities. The morning peak is caused by photochemical oxidation of the NO emitted during the morning traffic peak. Although the cause of the evening NO_2 peak is less well established (particularly if it is a late-evening peak), two likely contributing factors are reaction (3.4) between late-afternoon or evening NO emissions and O_3, and the reductions in wind speeds that frequently occur in the evening. The relative heights of the morning and evening peaks are influenced by meteorological factors, such as wind speed and solar intensity, and by the degree of O_3 generation that occurs during the day. On days when significant photochemical activity does not occur (as indicated by a lack of substantial O_3 or oxidant generation), NO_2 concentrations do not necessarily follow a double-peaked pattern of diurnal variation. This is illustrated in figure 3.29, which shows diurnal variations in seasonal mean 1-hr average NO_2 and oxidant concentrations in Chicago during a season when there is not significant oxidant formation. The diurnal pattern of NO_2 is relatively flat, compared to the patterns that occur when higher concentrations of O_3 or oxidant are formed.

Although NO concentrations tend to vary seasonally in ways that are similar in different cities, NO_2 concentrations do not exhibit such consistency. This is illustrated in figures 3.30 and 3.31, which show seasonal variations in NO and NO_2 concentrations, respec ively, in four

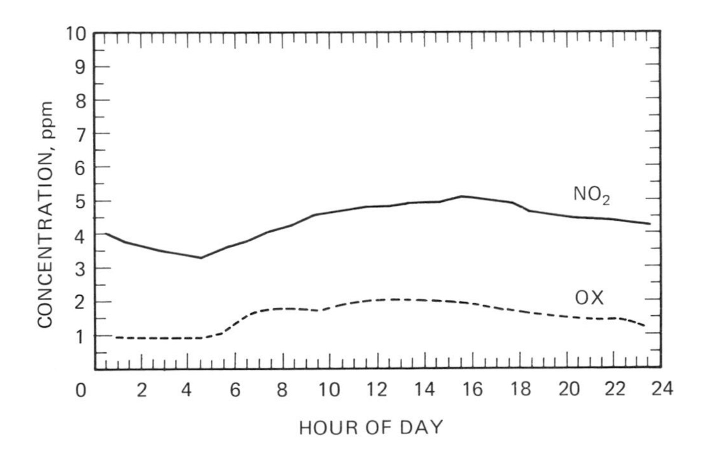

Figure 3.29 Diurnal variation of NO_2 and oxidant concentrations in Chicago, Illinois, during a period of little photochemical activity. Based on measurements during the fourth quarters of the years 1969–1973. (Source: reference [49])

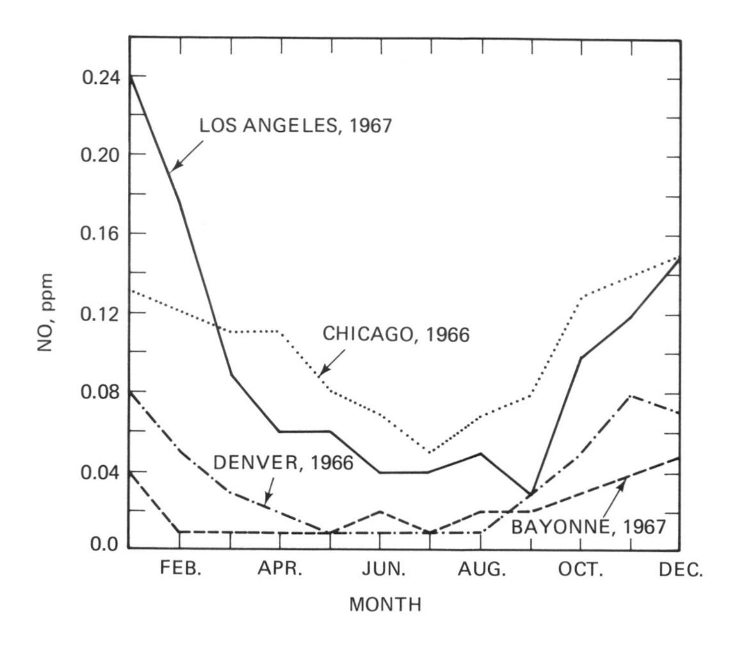

Figure 3.30 Monthly mean NO concentrations at four urban sites. (Source: reference [53])

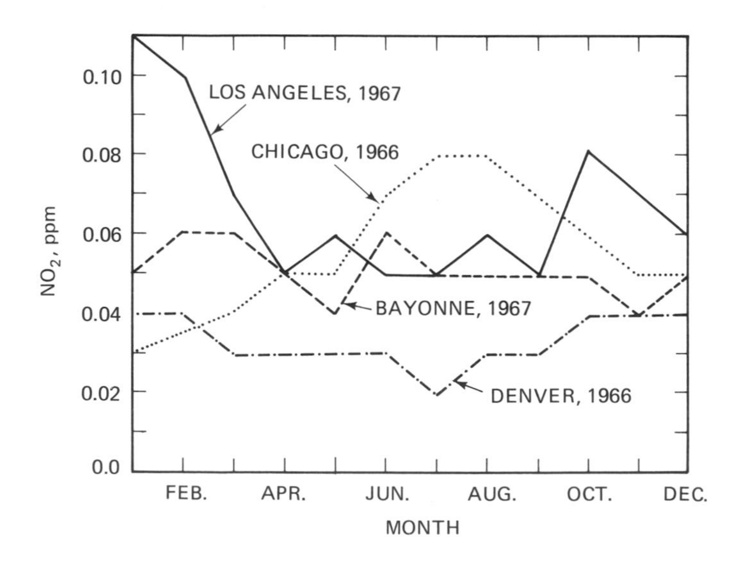

Figure 3.31 Monthly mean NO_2 concentrations at four urban sites. (Source: reference [53])

cities. NO concentrations tend to be higher in winter than at other times of year in all four cities. However, NO_2 concentrations are higher in winter than in summer in two cities, higher in summer than in winter in one city, and relatively constant throughout the year in one city. These differing seasonal patterns in NO_2 concentrations are the result of two factors whose effects tend to counteract one another. Whereas the increases in NO concentrations that occur in winter tend to cause NO_2 concentrations to increase, other things remaining unchanged, the decrease in photochemical activity that occurs in winter tends to reduce NO_2 concentrations. Therefore, seasonal patterns in NO_2 concentrations differ among cities, depending on the relative strengths of these two factors.

NO_2, like O_3, can become entrained in urban photochemical smog plumes. However, NO_2 concentrations tend to decrease relatively quickly downwind of the urban areas and to approach natural background levels within a few tens of kilometers of these areas [54, 55]. This decrease presumably is due to photochemical conversion of NO_2 to nitrates, such as nitric acid and PAN. As a result, NO_2 concentrations tend to be at or near natural background levels in rural areas, even when O_3 concentrations are relatively high [45, 46].

Because of the atmospheric mixing that takes place during the photochemical formation of NO_2 and the spatial smoothing that occurs when concentrations are averaged over many different sets of meteorological conditions, long-term average (e.g., annual average) NO_2 concentrations are likely to vary gradually between locations. Therefore, monitoring networks to measure long-term average NO_2 concentrations need not be dense but should be widely spread over metropolitan areas. However, because of reaction (3.4) between NO and O_3, short-term average (e.g., 1-hr) NO_2 concentrations may be particularly high near large NO sources. Hence, networks to monitor short-term NO_2 concentrations should include measurement sites near these sources. Analogously, reducing long-term average NO_2 concentrations is likely to require reducing NO_x emissions over entire urban areas, whereas reducing short-term NO_2 concentrations may require mainly reducing NO_x emissions from major sources, such as electrical power plants and highways.

3.5 *Summary*

The concentrations of CO, O_3, and NO_2 at a particular location are sensitive in varying degrees to emissions rates of the primary pollutants (CO, HC, and NO_x), the proximity of the location to sources of primary pollutant emissions, and meteorological conditions. The sensitivity of concentrations to meteorological conditions is qualitatively similar in most respects for all three pollutants CO, O_3, and NO_2, although O_3 and NO_2 concentrations are sensitive to more meteorological variables than are CO concentrations. However, CO, O_3, and NO_2 concentrations have greatly different sensitivities to primary pollutant emissions rates and to the proximity of primary pollutant emissions sources. CO concentrations tend to be linearly related to the rate of CO emissions from nearby sources and to be relatively insensitive to emissions from more distant sources. O_3 concentrations are related to HC and NO_x emissions in a highly complex, nonlinear way. Moreover, high O_3 concentrations tend to be the result of HC and NO_x emissions that take place over widespread areas and usually cannot be attributed to specific emissions sources. O_3 can be transported over long distances by winds, with the result that high O_3 concentrations may occur tens or hundreds of miles away from major HC and NO_x sources. NO_2 concentrations tend to be proportional to NO_x emissions but to be insensitive to moderate changes in HC emissions, even though HC is an important participant in the chemical reactions that form NO_2. High long-term average NO_2 concentrations tend to be the result of NO_x emissions over large areas, but high short-term average NO_2 concentrations at a particular location often may be due to a large NO source near that location.

The different spatial properties of CO, O_3, and NO_2 concentrations imply that different approaches to emissions control are needed to reduce excessive concentrations of these pollutants. High CO concentrations at a particular location often can be reduced by CO control measures that affect only emissions sources in the immediate vicinity of that location. Analogous NO_x emissions control measures may be effective in reducing high short-term average NO_2 concentrations at a particular location. However, O_3 and long-term average NO_2 concentrations are unlikely to be affected significantly by emissions control measures whose effects are highly localized. Reducing O_3 and long-term average NO_2 concentrations requires achieving HC and NO_x emissions reductions that are significant on the scale of an entire metropolitan area or, possibly, in the case of O_3, on the scale of a multistate region.

References

1

Brief, R. S., Jones, A. R., and Yoder, J. D., "Lead, Carbon Monoxide and Traffic, a Correlation Study," *Journal of the Air Pollution Control Association*, Vol. 10, pp. 384–388, October 1960.

2
Johnson, K. L., Dworetsky, L. H., and Heller, A. N., "Carbon Monoxide and Air Pollution from Automobile Emissions in New York City," *Science,* Vol. 160, pp. 67–68, 5 April 1968.

3
Colucci, J. M., and Begeman, C. R., "Carbon Monoxide in Detroit, New York and Los Angeles," *Environmental Science and Technology*, Vol. 3, pp. 41–46, January 1969.

4
Ledolter, J., Tiao, G. C., Hudak, G. B., Hsieh, J., and Graves, S. B., *Statistical Analysis of Multiple Time Series Associated with Air Quality Data: New Jersey CO Data*, Technical Report No. 529, Department of Statistics, University of Wisconsin, Madison, WI, June 1978.

5
Tiao, G. C., Box, G. E. P., and Hamming, W. J., "A Statistical Analysis of the Los Angeles Ambient Carbon Monoxide Data 1955–1972," *Journal of the Air Pollution Control Association*, Vol. 25, pp. 1129–1136, November 1975.

6
US Environmental Protection Agency, *Air Quality Criteria for Carbon Monoxide*, Report No. EPA-600/8-79-022, Washington, DC, October 1979.

7
Tiao, G. C., and Hillmer, S. C., "Statistical Models for Ambient Concentrations of Carbon Monoxide, Lead, and Sulfate Based on the LACS Data," *Environmental Science and Technology*, Vol. 12, pp. 820–828, July 1978.

8
Tiao, G. C., and Hillmer, S. C., "Statistical Analysis of the Los Angeles Catalyst Study Data—Rationale and Findings," in *The Los Angeles Catalyst Study Symposium*, Report No. EPA-600/4-77-036, US Environmental Protection Agency, Research Triangle Park, NC, June 1977, pp. 415–460. NTIS Publication No. PB 278305.

9
Godin, G., Wright, G., and Shephard, R. J., "Urban Exposure to Carbon Monoxide," *Archives of Environmental Health*, Vol. 25, pp. 305–313, November 1972.

10
Ranzieri, A. J., *Impact of Transportation Systems on the Air Environment*, State of California, Department of Transportation, Transportation Laboratory, Sacramento, May 1975.

11
Ott, W., and Eliassen, R., "A Survey Technique for Determining the Representativeness of Urban Air Monitoring Stations with Respect to Carbon Monoxide," *Journal of the Air Pollution Control Association*, Vol. 23, pp. 685–690, August 1973.

12
Ramsey, J. M., "Concentrations of Carbon Monoxide at Traffic Intersections in Dayton, Ohio," *Archives of Environmental Health*, Vol. 13, pp. 44–46, July 1966.

13
US Environmental Protection Agency, *Carbon Monoxide Study—Seattle, Washington*, Report No. EPA-910/9-78-054b, Seattle, December 1978. NTIS Publication No. PB 290506.

14
Shelar, E., Ludwig, F. L., and Shigeishi, H., *Analysis of Pollutant and Meteorological Data Collected in the Vicinity of Carbon Monoxide Hot Spots*, report prepared for the US Environmental Protection Agency, Ann Arbor, MI, by SRI International under Contract No. 68-03-2545, May 1979.

15
Johnson, W. B.; Ludwig, F. L., Dabberdt, W. F., and Allen, R. J., "An Urban Diffusion Simulation Model for Carbon Monoxide," *Journal of the Air Pollution Control Association*, Vol. 23, pp. 490–498, June 1973.

16
Dabberdt, W. F., Ludwig, F. L., and Johnson, W. B., Jr., "Validation and Applications of an Urban Diffusion Model for Vehicular Air Pollutants," *Atmospheric Environment*, Vol. 7, pp. 603–618, 1973.

17
Wright, G. R., Jewczyk, S., Onrot, J., Tomlinson, P., and Shephard, R. J., "Carbon Monoxide Concentrations in the Urban Atmosphere," *Archives of Environmental Health*, Vol. 30, pp. 123–129, March 1975.

18
Cortese, A. D., and Spengler, J. D., "Ability of Fixed Monitoring Stations to Represent Personal Carbon Monoxide Exposure," *Journal of the Air Pollution Control Association*, Vol. 26, pp. 1144–1150, December 1976.

19
Delaware Valley Regional Planning Commission, *Transportation Element for Southeastern Pennsylvania of the 1979 State Implementation Plan*, Philadelphia, April 1979.

20
US Environmental Protection Agency, *National Air Quality Monitoring and Emissions Trends Report, 1977*, Report No. EPA-450/2-78-052, Research Triangle Park, NC, December 1978.

21
Central Transportation Planning Staff, *Transportation Element of the State Implementation Plan for the Boston Region*, Boston, December 1978.

22
Southwestern Pennsylvania Regional Planning Commission, *Transportation Component of the State Air Quality Implementation Plan for the Southwestern Pennsylvania Region*, Pittsburgh, January 1979.

23
Falls, A. H., and Seinfeld, J. H., "Continued Development of a Kinetic Mechanism for Photochemical Smog," *Environmental Science and Technology*, Vol. 12, pp. 1398–1406, December 1978.

24
US Environmental Protection Agency, *Uses, Limitations and Technical Basis for Quantifying Relationships between Photochemical Oxidants and Precursors*, Report No. EPA-450/2-77-021a, Research Triangle Park, NC, November 1977. NTIS Publication No. PB 278142.

25
Lonneman, W. A., Bufalini, J. J., and Seila, R. L., "PAN and Oxidant Measurement in Ambient Atmospheres," *Environmental Science and Technology*, Vol. 10, pp. 347–380, April 1976.

26
Cleveland, W. S., Graedel, T. E., and Kleiner, B., "Urban Formaldehyde: Observed Correlation with Source Emissions and Photochemistry," *Atmospheric Environment*, Vol. 11, pp. 357–360, 1977.

27
Bufalini, J. J., Gay, B. W., and Brubaker, K. L., "Hydrogen Peroxide Formation from Formaldehyde Photooxidation and Its Presence in Urban Atmospheres," *Environmental Science and Technology*, Vol. 6, pp. 816–821, September 1972.

28
Kok, G. L., Darnall, K. R., Winer, A. M., and Pitts, J. N., Jr., "Ambient Air Measurements of Hydrogen Peroxide in the California South Coast Air Basin," *Environmental Science and Technology*, Vol. 12, pp. 1077–1080, September 1978.

29
Miller, D. F., and Spicer, C. W., "Measurement of Nitric Acid in Smog," *Journal of the Air Pollution Control Association*, Vol. 25, pp. 940–942, September 1975.

30
Pollack, R. I., Tesche, T. W., Reynolds, S. D., Hillyer, M. J., Jerskey, T. N., and Meldgin, M. J., *Highway Air Quality Impact Appraisals*, Report No. FHWA-RD-78-99, Federal Highway Administration, Washington, DC, June 1978. NTIS Publication No. PB 293798.

31
Evans, G. F., and Rodes, C. E., "Summary of LACS Continuous Data," in *The Los Angeles Catalyst Study Symposium*, Report No. EPA-600/4-77-036, US Environmental Protection Agency, Research Triangle Park, NC, June 1977, pp. 301–414. NTIS Publication No. PB 278305.

32
US Department of Health, Education, and Welfare, *Air Quality Criteria for Photochemical Oxidants*, Report No. AP-63, Washington, DC, 1970. US Government Printing Office Publication No. FS 2.300:AP-63.

33
Jeffries, H. E., Kamens, R., Fox, D. L., and Dimitriades, B., "Outdoor Smog Chamber Studies: Effect of Diurnal Light, Dilution, and Continuous Emission on Oxidant Precursor Relationships," in B. Dimitriades, ed., *International Conference on Photochemical Oxidant Pollution and Its Control: Proceedings*, Vol. 2, Report No. EPA-

600/3-77-001b, US Environmental Protection Agency, Research Triangle Park, NC, pp. 891–902, January 1977. NTIS Publication No. PB 264233.

34

Tiao, G. C., Box, G. E. P., and Hamming, W. J., "Analysis of Los Angeles Photochemical Smog Data: A Statistical Overview," *Journal of the Air Pollution Control Association*, Vol. 25, pp. 260–268, March 1975.

35

Carter, W. P. L., Winer, A. M., Darnall, K. M., and Pitts, J. N., Jr., "Smog Chamber Studies of Temperature Effects in Photochemical Smog," *Environmental Science and Technology*, Vol. 13, pp. 1094–1100, September 1979.

36

Ripperton, L. A., Worth, J. J. B., Vukovich, F. M., and Decker, C. E., "Research Triangle Institute Studies of High Ozone Concentrations in Nonurban Areas," in B. Dimitriades, ed., *International Conference on Photochemical Oxidant Pollution and Its Control: Proceedings*, Vol. 1, Report No. EPA-600/3-77-001a, US Environmental Protection Agency, Research Triangle Park, NC, January 1977, pp. 413–438. NTIS Publication No. PB 264232.

37

Kauper, E. K., and Niemann, B. L., "Transport of Ozone by Upper-Level Land Breeze—An Example of a City's Polluted Wake Upwind from its Center," in B. Dimitriades, ed., *International Conference on Photochemical Oxidant Pollution and Its Control: Proceedings*, Vol. 1, Report No. EPA-600/3-77-001a, US Environmental Protection Agency, Research Triangle Park, NC, January 1977, pp. 227–236. NTIS Publication No. PB 264232.

38

Cleveland, W. S., Kleiner, B., McRae, J. E., and Warner, J. L., "Photochemical Air Pollution: Transport from the New York City Area into Connecticut and Massachusetts," *Science*, Vol. 191, pp. 179–181, 16 January 1976.

39

Cleveland, W. S., and Graedel, T. E., "Photochemical Air Pollution in the Northeast United States," *Science*, Vol. 204, pp. 1273–1278, 22 June 1979.

40

White, W. H., Blumenthal, D. L., Anderson, J. A., Husar, R. B., and Wilson, B. E., "Ozone Formation in the St. Louis Plume," in B. Dimitriades, ed., *International Conference on Photochemical Oxidant Pollution and Its Control: Proceedings*, Vol. 1, Report No. EPA-600/3-77-001a, US Environmental Protection Agency, Research Triangle Park, NC, January 1977, pp. 237–247. NTIS Publication No. PB 264232.

41

Blumenthal, D. L., White, W. H., and Smith, T. B., "Anatomy of a Los Angeles Smog Episode: Pollutant Transport in the Daytime Sea Breeze Regime," *Atmospheric Environment*, Vol. 12, pp. 893–907, 1978.

42

Blumenthal, D. L., White, W. H., Peace, R. L., and Smith, T. B., *Determination of the Feasibility of the Long-Range Transport of Ozone or Ozone Precursors*, Report

No. EPA-450/3-74-061, US Environmental Protection Agency, Research Triangle Park, NC, November 1974. NTIS Publication No. PB 241159.

43

Wolff, G. T., Lioy, P. J., Wright, G. D., Meyers, R. E. and Cederwall, R. T., "An Investigation of Long-Range Transport of Ozone Across the Midwestern and Eastern United States," *Atmospheric Environment*, Vol. 11, pp. 797–802, 1977.

44

Vukovich, F. M., Bach, W. D., Crissman, B. W., and King, W. J., "On the Relationship Between High Ozone in the Rural Surface Layer and High Pressure Systems," *Atmospheric Environment*, Vol. 11, pp. 967–983, 1977.

45

Research Triangle Institute, *Investigation of Ozone and Ozone Precursor Concentrations at Nonurban Locations in the Eastern United States*, Report No. EPA-450/3-74-034, US Environmental Protection Agency, Research Triangle Park, NC, May 1974. NTIS Publication No. PB 236931.

46

Research Triangle Institute, *Investigation of Rural Oxidant Levels as Related to Urban Hydrocarbon Control Strategies*, Report No. EPA-450/3-75-036, US Environmental Protection Agency, Research Triangle Park, NC, March 1975. NTIS Publication No. PB 242299.

47

Wolff, G. T., Lioy, P. J., Meyers, R. E., Cederwall, R. T., Wight, G. D., Pasceri, R. E., and Taylor, R. S., "Anatomy of Two Ozone Transport Episodes in the Washington, D. C., to Boston, Mass., Corridor," *Environmental Science and Technology*, Vol. 11, pp. 506–510, May 1977.

48

Cleveland, W. S., and Kleiner, B., "Transport of Photochemical Air Pollution from Camden-Philadelphia Urban Complex, " *Environmental Science and Technology*, Vol. 9, pp. 869–872, September 1975.

49

Trijonis, J., *Empirical Relationships between Atmospheric Nitrogen Dioxide and Its Precursors*, Report No. EPA 600/3-78-018, US Environmental Protection Agency, Research Triangle Park, NC, February 1978. NTIS Publication No. 278547.

50

Jeffries, H. E., Fox, D. L., and Kamens, R., "Photochemical Conversion of NO to NO_2 by Hydrocarbons in an Outdoor Smog Chamber," *Journal of the Air Pollution Control Association*, Vol. 26, pp. 480–484, May 1976.

51

Jeffries, H., Fox, D., and Kamens, R., *Outdoor Smog Chamber Studies: Effect of Hydrocarbon Reduction on Nitrogen Dixoide*, Report No. EPA 650/3-75-011, US Environmental Protection Agency, Washington, DC, June 1975. NTIS Publication No. PB 245829.

52

Chang, T. Y., Norbeck, J. M., and Weinstock, B., "NO_2 Air Quality–Precursor

Relationship: An Ambient Air Quality Evaluation in the Los Angeles Basin," *Journal of the Air Pollution Control Association,* Vol. 30, pp. 157–162, February 1980.

53

Wayne, L. G., Bryan, R. J., Weisburd, M., and Danchick, R., *Comprehensive Technical Report on All Atmospheric Contaminants Associated with Photochemical Air Pollution,* Report No. EPA-650/4-75-002, US Environmental Protection Agency, Washington, DC, June 1970. NTIS Publication No. PB 239510.

54

Hester, H. E., Evans, R. B., Johnson, F. G., and Martinez, E. L., "Airborne Measurements of Primary and Secondary Pollutant Concentrations in the St. Louis Urban Plume," in B. Dimitriades, ed., *International Conference on Photochemical Oxidant Pollution and Its Control: Proceedings,* Vol. 1, Report No. EPA-600/3-77-001a, US Environmental Protection Agency, Research Triangle Park, NC, January 1977, pp. 259–274. NTIS Publication No. PB 264232.

55

Sickles, J. E., II, Ripperton, L. A., and Eaton, W. C., "Oxidant and Precursor Transport Simulation Studies in the Research Triangle Institute Smog Chambers," in B. Dimitriades, ed., *International Conference on Photochemical Oxidant Pollution and Its Control: Proceedings,* Vol. 1, Report No. EPA-600/3-77-001a, US Environmental Protection Agency, Research Triangle Park, NC, January 1977, pp. 319–328. NTIS Publication No. PB 264232.

4 Harmful Effects of Transportation-Related Air Pollutants

4.1 Introduction

Carbon monoxide, photochemical oxidants, including ozone, and nitrogen dioxide all are harmful to human health at concentrations that are within or near to the ranges of concentrations found in polluted air. In addition, oxidants and nitrogen dioxide are harmful to plants and materials, have distinct odors, and contribute to the formation of aerosols that can significantly reduce visibility.[1]

Two basic approaches to describing the harmful effects of air pollution are discussed in this chapter. The first approach consists of indentifying specific effects and the pollutant concentrations at which they occur. The second approach consists of assigning monetary costs to these effects. Specific harmful effects of transportation-related air pollutants and methods for identifying these effects are discussed in sections 4.2–4.7. Monetary costs of transportation-related air pollution are discussed in section 4.8.

4.2 Human Health Effects: Methods and Limitations

The principal objective of programs to control transportation-related air pollution in the United States is prevention of harm to human health. There are three basic methods for acquiring information on the harmful health effects of air pollution: laboratory studies with animals, human clinical studies, and human epidemiological studies. Laboratory studies with animals and human clinical studies consist of exposing animals of humans to one or more air pollutants under controlled conditions and observing the effects of these exposures on various indicators of the physical or psychological conditions of the exposed individuals. For example, the lung capacities of humans at rest might be measured before and after exposure to a known concentration of O_3 for 1 hr, or laboratory animals might be exposed to O_3 and then exposed to airborne bacteria in order to determine the effects of O_3 exposure on susceptibility to infectious respiratory diseases. In human epidemiological studies, changes in the health of individuals functioning in their normal environments are observed, and efforts are made to relate these changes to air pollution and other factors using statistical methods.

Animal studies have the advantage that animals can be exposed to higher pollutant concentrations and placed at greater risk of incurring serious or irreversible harm than humans can be. Moreover, animals can be subjected to invasive measurement techniques whose use with humans is not permissible. Consequently, it often is easier to observe health effects and to identify the mechanisms of their occurrence in animals than in humans. However, animals and humans differ anatomically and physiologically, so the fact that a particular effect occurs in animals at a particular pollutant concentration does not necessarily imply that the same effect occurs in humans at the same concentration. To establish that an effect occurs in humans and the concentration at which it occurs, it is necessary to observe the effect in humans. Human clinical studies provide one means for doing this.

Because human clinical studies take place under controlled conditions, they offer relatively precise means of investigating the health effects of air pollution. However, clinical studies also have certain important limitations. For example, the controlled laboratory conditions of these studies may eliminate important physiological stresses that people normally experience, thereby altering pollutant dose-response relations. Moreover, it is not possible to expose people to pollutants over long periods of time under laboratory conditions, so only acute effects of pollutant exposures can be investigated in clinical studies. Finally, it is not possible to expose people deliberately to pollutants at concentrations or under conditions that are likely to result in severe or irreversible harmful effects. Hence, clinical studies can investigate directly only relatively minor effects of pollutant exposures. The occurrence of more severe effects must either be inferred from the results of these studies or be investigated by other means.

In epidemiological studies, the individuals being studied perform their normal activities and experience all normal environmental stresses. They also may experience long-term exposures, depending on their activity patterns and where they live, and they may experience any severe or irreversible effects that are produced by exposure to air pollution. Thus, epidemiological studies include the factors and effects that are excluded from clincial studies. However, epidemiological studies lack the control of clinical studies. Because the individuals who are the subjects of epidemiological studies are not constantly observed in a laboratory, it can be difficult to determine what health effects they have experienced. Because these individuals move about in the course of their normal activities, accurate information concerning the pollutant concentrations to which they are exposed can be difficult or impossible to obtain. In addition, it can be difficult to distinguish the effects of exposure to an air pollutant from those of other factors that affect health, including exposure to other air pollutants. Thus, epidemiological studies are less precise than clinical studies. However, epidemiological studies provide the only means of investigating certain types of potential effects of air pollution, such as effects on mortality, and they provide the only means of investigating the effects of air pollution under normal living conditions.

Because of the limitations of the available methods for studying the effects of air pollution on human health, it is very difficult to obtain precise information regarding these effects. Several other factors also contribute to this difficulty. Susceptibilities to air pollution health effects vary greatly across individuals. Moreover, the effects of air pollution on clinical indicators of human health tend to be small compared to normal ranges of variation in these indicators. Consequently, air pollution health effects can be difficult to detect, even under controlled laboratory conditions. Some health effects may occur as the result of concurrent exposures to several air pollutants or exposure to air pollution while other physiological stress factors, such as high temperature or high humidity, are present. This further complicates the problem of measuring the health effects of air pollution and the concentrations at which these effects occur. Records of long-term exposures to air pollution and to other stress factors that cause similar health effects rarely are available. Therefore, information on human health effects of long-term exposures also is rarely available.

It also can be very difficult to assess the policy significance of health effects of air pollution. For example, it will be seen in section 4.3 that when angina pectoris patients exercise on bicycles or treadmills, they experience chest pains sooner after being exposed to CO than they do after breathing pure air. This effect, which has been observed only in controlled clinical studies, occurs because CO exposure reduces the supply of oxygen to the heart. It is possible that CO-induced oxygen deprivation also causes or contributes to cumulative, irreversible heart damage, myocardial infarction, or sudden death in individuals with arteriosclerotic heart disease. However, it is not known whether CO exposure does, in fact, cause these effects. Nor is it known whether exposure to atmospheric CO affects the abilities of angina patients to function in their normal environments. Thus, it is not clear whether the relation between CO exposure and anginal pain should be interpreted as a laboratory artifact or as evidence that CO exposure constitutes a serious danger to public health. Similarly, as will be seen in section 4.4, O_3

exposure causes changes in certain indicators of human pulmonary function. However, the biological mechanisms through which O_3 produces these changes are not known, and it is not clear whether the changes are relatively benign effects or are symptoms of effects that are potentially more serious.

4.3 *Health Effects of CO*

The adverse health effects of CO are caused by its ability to reduce the quantity of oxygen that is delivered by the blood to the tissues and, possibly, to inhibit the utilization of oxygen within the tissues. CO combines with the hemoglobin of the blood to form carboxyhemoglobin (COHb). In so doing, CO occupies the same sites on the hemoglobin molecule as oxygen normally does, thereby reducing the ability of the blood to carry oxygen. CO also inhibits oxygen that is bound to hemoglobin from being released to the tissues, which further reduces the blood's ability to supply the tissues with oxygen. In addition, there is a possibility that CO combines with certain compounds in the tissues and, thereby, reduces the ability of the tissues to use the oxygen that is delivered to them.

The relation between the concentration of COHb in the blood and CO exposure is known and is given by the following equation (the Coburn equation [1])

$$V_b d[\mathrm{COHb}]/dt = \dot{V}_{CO} - B[\mathrm{COHb}]P_{OX}/M[\mathrm{O_2Hb}] + BP_{CO}, \tag{4.1}$$

where

$$B = [1/D_L + (P_B - P_W)/\dot{V}_A]^{-1} \tag{4.2}$$

and

[COHb] = carboxyhemoglobin concentration, expressed as ml of CO/ml of blood,

[O_2Hb] = concentration of oxyhemoglobin (hemoglobin combined with oxygen) in the lung capillaries, expressed as ml of oxygen/ml of blood; this can be estimated by assuming that all of the hemoglobin in the lung capillaries is in the form of either O_2Hb or COHb,

V_b = volume of blood in the body, expressed in ml,

$\dot{V}_{CO}$ = rate of natural CO production by the body in ml/min,

D_L = diffusivity of the lung for CO in ml/min/mm Hg; this is an indicator of the rate at which CO crosses the membranes of the lungs,

P_B = barometric pressure in mm Hg,

P_W = vapor pressure of water at body temperature in mm Hg,
$\dot{V}_A$ = alveolar ventilation rate in ml/min; this measures the rate at which air is drawn into the lungs during respiration,
P_{OX} = partial pressure of oxygen in the lung capillaries in mm Hg,
P_{CO} = partial pressure of CO in the inhaled air in mm Hg.[2]

M is a chemical equilibrium constant (Haldane's constant) whose value is roughly 240 [2]. When COHb and O_2Hb are in equilibrium with CO and O_2, the following equation (the Haldane equation) is true:

$$[COHb]/[O_2Hb] = MP_{CO}/P_{OX}. \quad (4.3)$$

This equation defines M.

Given the values of the physiological and environmental parameters, equation (4.1) can be integrated to obtain the blood COHb concentration that is caused by a given CO exposure. Table 4.1 shows examples of blood COHb concentrations that are produced by various exposures to constant CO concentrations. In accordance with normal practice, the COHb concentrations are expressed as percentages of the saturation concentration.[3]

Table 4.1 Percentage of COHb in the blood of normal individuals expressed as a function of CO exposure[a]

	Exposure time (hr)					
	1		4		8	
CO (ppm)	Resting	Moderate exercise	Resting	Moderate exercise	Resting	Moderate exercise
5.0	0.6	0.6	0.8	0.8	0.9	0.9
9.0	0.7	0.8	1.2	1.3	1.4	1.5
15.0	1.0	1.1	1.8	2.0	2.2	2.4
20.0	1.1	1.4	2.3	2.6	2.9	3.1
25.0	1.3	1.6	2.8	3.2	3.6	3.8
35.0	1.6	2.1	3.8	4.5	4.9	5.3
50.0	2.2	2.9	5.2	6.3	7.0	7.6

a. Source: reference [3]. The COHb concentrations are derived from equation (4.1). The following conditions are assumed: $V_b = 5{,}500$ ml, $V_{CO} = 0.007$ ml/min, $D_L = 30$ ml/min/mm Hg, $P_B = 760$ mm Hg, $P_W = 47$ mm Hg, $V_A = 10$ l/min at rest and 20 l/min at exercise, $P_{XO} = 102.07$ mm Hg, $M = 218$. The hemoglobin content of the blood is 150 g/l, and the initial COHb concentration, prior to exposure, is 0.5%. The CO concentration does not vary during the exposure period.

Because the health effects of CO exposure are related more directly to the COHb concentration in the blood than to the CO concentration in the inhaled air, most health effects investigations are designed to determine the COHb concentrations at which health effects occur. Atmospheric CO concentrations that would produce these COHb concentrations then can be inferred by integrating equation (4.1).

The most important effects of CO exposure concern the cardiovascular and central nervous systems. The cardiovascular effects are more clearly established than the central nervous system effects are and occur at lower COHb concentrations. It has been shown in several experiments that exposure to CO accelerates the onset of chest pain when individuals with angina pectoris exercise. CO exposure also accelerates the onset of intermittent claudication (leg pain and weakness) when individuals with iliofemoral occlusive arterial disease exercise.[4] In the experiments in which these effects were identified, subjects with previously documented angina or iliofemoral occlusive arterial disease exercised on a treadmill or bicycle ergometer after breathing normal air and after breathing air in which the concentration of CO was elevated. The subjects' blood COHb concentrations and the lengths of time they could exercise before the onset of pain were measured. Table 4.2 summarizes the results of several of these experiments. The tabulated results indicate that COHb concentrations in the range 2.5–3.0%, and possibly as low as 1.3%, cause decreases in exercise performance. In another experiment that is not included in table 4.2, angina patients showed decreased exercise performance at a mean COHb concentration of 2% [8]. The electrocardiograms of the exercising subjects were monitored in three of the four experiments that are summarized in table 4.2. CO-induced changes in the electrocardiograms were observed in all three experiments [4–6].

In separate investigations, it has been shown that blood COHb concentrations in the range 2.5–4% cause reductions in the lengths of time that normal individuals and individuals with chronic obstructive pulmonary disease (e.g., asthma, chronic bronchitis, emphysema) can exercise before experiencing exhaustion or severe breathlessness [9–11].

The main significance of the observed cardiovascular effects of CO exposure is their implication that exposure to atmospheric CO may cause or contribute to severe and irreversible cardiovascular damage or sudden death in individuals with arteriosclerotic diseases. However, epidemiological evidence that clearly supports this hypothesis has not yet been presented. Statistical associations between high atmospheric CO concentrations and fatalities from myocardial infarctions have been found

Table 4.2 CO-induced decrements in exercise performance by individuals with angina pectoris and iliofemoral occlusive arterial disease

Disease	No. of patients	COHb concentration (%)		Reduction in time to onset of pain (sec)[a]		Reference
		Mean	Range	Mean	Range	
Angina	10	5.08	3.8–8.0	75.1	42–134	[4]
Angina	10	2.9	1.3–3.8	44.8	−38–110[b]	[5]
Angina	10	2.68	2.5–3.0	36.7	14–86	[6]
Iliofemoral occlusive arterial disease	10	2.77	2.4–3.1	30.3	17–75	[7]

a. Reduction is the difference between the times to onset of pain after breathing normal air and after breathing air with elevated CO.
b. Negative change signifies an increase in the time to onset of pain.

in epidemiological studies conducted in Los Angeles [12–14]. However, the CO concentrations used in these studies were measured at fixed site monitors and, therefore, may not have been indicative of the concentrations to which individuals were exposed (see section 3.2). In addition, neither blood COHb concentrations nor a large variety of potentially confounding factors were measured, so it cannot be established that the CO-associated fatalities were, in fact, caused by CO [15].[5] None of the studies found an association between CO concentrations and hospital admissions for myocardial infarctions. An epidemiological study conducted in Denver found a statistical association between high atmospheric CO concentrations and the number of patients who arrived at the emergency room of a Denver hospital with nontraumatic cardiorespiratory complaints (nontraumatic chest pain, shortness of breath, and wheezing) [16]. However, the symptoms described in this study are not necessarily indicators of cardiovascular disease. Moreover, the study used CO concentrations measured at a single fixed site and did not include measurements of blood COHb or of potentially confounding variables. In an epidemiological study conducted in Baltimore, no associations between atmospheric CO concentrations and either the incidence of myocardial infarctions or the occurrence of sudden deaths from arteriosclerotic heart disease were found [17]. Moreover, there were no differences between the COHb levels in nonsmoking sudden death patients and in nonsmoking living controls.[6] In summary, although there is clear clinical evidence that exposure to atmospheric concentrations of CO produce adverse cardiovascular effects in susceptible individuals, epidemiological evidence for the occurrence of such effects outside of laboratory environments remains weak.

The clinical evidence concerning central nervous system effects of CO exposure is considerably less clear than that concerning cardiovascular effects. The potential central nervous system effects include changes in vigilance (the ability to detect small and unexpected changes in one's environment), sensory function, and psychomotor function. Several clinical studies have found decreases in vigilance, sensory function, and psychomotor function at blood COHb concentrations in the range 3–8% [18–26]. Specific effects include decreases in subjects' abilities to detect changes in the intensities of light and sound pulses, distinguish faint flashes of light against a dim background, recognize letters that are presented in rapid sequence, learn meaningless syllables, and react quickly to visual stimuli. However, other clinical studies have failed to find vigilance, sensory, or psychomotor effects at COHb concentrations in the range 9–20% [27–33].

There are at least two possible causes of these conflicting results. First, the results of vigilance, sensory, and psychomotor tests can vary greatly, depending on the specific tests that are performed [20–22, 24, 26], and there is considerable lack of uniformity in the test procedures that have been used by different investigators. The results of the tests also depend on such variables as temperature, humidity, and the state of excitement of the subjects, which usually are uncontrolled. These uncontrolled variables may explain much of the difficulty that has been encountered in obtaining reproducible results with the same test procedures [31]. Although the available evidence suggests that low COHb concentrations probably do affect the functioning of the central nervous system under appropriate conditions, experiments that are more uniformly designed and carefully controlled than previous ones have been will be needed to establish clearly the nature of these effects and the conditions under which they occur.

The main significance of the observed central nervous system effects of CO exposure is that these effects have occurred at or near COHb concentrations that can be experienced by automobile drivers in heavy traffic. Hence, there is a possibility that exposure to atmospheric CO impairs driving ability and, thereby, contributes to the occurrence of traffic accidents. The clinical studies suggesting that COHb concentrations in the vicinity of 5% cause reductions in speeds of reaction to visual stimuli [20, 24] are particularly noteworthy in this respect, although contradictory clinical results also have been obtained [22, 27, 33]. In addition to these studies, several other clinical studies that were specifically designed to test the effects of CO exposure on driving skills have been performed. These studies have included tests of driving-related sensory and psychomotor functions (e.g., reaction time, depth perception, peripheral vision, night vision, and glare recovery) and tests of subjects' abilities to drive real or simulated vehicles. The latter tests involved such factors as braking and steering abilities, abilities to maintain constant speeds and following distances, and ability to control a vehicle in an emergency situation. The results of the various studies are inconsistent. Some have found decreases in driving skills or driving-related functions at COHb concentrations in the range 5.6–8% [34, 35]. Others have found no important effects at COHb concentrations of 8% or more [32, 36, 37]. An effort to find statistical associations between CO concentrations and traffic accident frequencies in Los Angeles produced negative results [38]. However, the CO concentrations were measured at fixed sites, and several variables that may significantly affect accident frequency, including traffic speed and density, were not measured. Hence, this effort does not provide a strong test of the hypothesis that CO exposure contributes to traffic accidents. In summary, the extent to which CO exposure may impair driving ability remains unclear.

Cigarette smoking is a major source of CO exposure unrelated to air pollution. Cigarette smokers can have COHb concentrations as high as 15%, depending on how heavily they smoke and how long it has been since the last cigarette was smoked. Smokers' COHb levels tend to be higher than equilibrium levels at normal atmospheric CO concentrations, so smokers usually exhale CO into the air, rather than absorbing it. Cardiovascular effects of CO exposure that have been observed in clinical studies that do not involve smoking also have been observed in studies of the effects of smoking [39, 40]. It is not known whether the central nervous system effects of CO exposure are different in smokers and nonsmokers.

Recently, concern has developed over the possibility that exposure of pregnant women to CO may cause fetal damage. The theoretical basis for this concern is that CO may reduce the supply of oxygen to fetal tissues and, thereby, impair fetal development. Empirically, it has been found that when COHb concentrations in the bloodstreams of pregnant laboratory animals are maintained at high levels over long periods of time (e.g., COHb concentrations of 10% or more for periods of 30 days) the offspring have lower birth weights and higher neonatal mortality rates than do the offspring of control animals. However, the high COHb concentrations and long exposures used in the animal studies, in addition to the usual problems of extrapolating the results of animal studies to humans, make it impossible to assess the significance of these studies for human exposure to atmospheric CO. Human epidemiological studies have indicated that children born to cigarette-smoking mothers have lower birth weights and higher neonatal mortality rates than do children born to nonsmoking mothers. Maternal smoking also may impair the subsequent development of the child. However, it is not clear to what extent these adverse effects are due to CO rather than the other toxic components of cigarette smoke. Moreover, smokers' blood COHb levels tend to be considerably higher than the levels likely to result from intermittent exposure to atmospheric CO. Thus, the effects on fetal development of maternal exposure to atmospheric CO are largely matters of speculation.

4.4 Health Effects of O_3 and Other Photochemical Oxidants

No simple mechanism of toxicity, analogous to oxygen deprivation in the case of CO, or physiological exposure indicators, analogous to the COHb concentration in the blood, are known for O_3. However, it has been established through a large number of human clinical studies that O_3 is a strong, acute, pulmonary irritant that can cause measurable physiological changes and considerable discomfort in exposed individuals.

In a typical clinical study, subjects are exposed at intervals of one or more days to filtered, O_3-free air and to one or more known concentrations of O_3 mixed with otherwise clean air. The exposure periods last from 1 to 6 hr, depending on the study, and during these periods the subjects may be required to exercise. The subjects are asked to report any symptoms of illness (e.g., coughing) that they experience during the exposure periods. After each exposure, the subjects are given pulmonary function tests and, in some cases, tests of blood chemistry.[7] The test results obtained following O_3 exposure are then compared with the results obtained following exposure to clean air.

Clinical studies have found that exposure to O_3 concentrations ranging from 294 $\mu g/m^3$ (0.15 ppm) to 1,470 $\mu g/m^3$ (0.75 ppm), depending on the study, can cause measurable decreases in pulmonary function [41–51]. One clinical study reported that exposure to 235 $\mu g/m^3$ (0.12 ppm) of O_3 for 2 hr caused pulmonary function changes [52]. However, this result is based on certain unusual measurement techniques, and its validity is uncertain. At least two clinical studies have found no decreases in pulmonary function following exposures to 196–392 $\mu g/m^3$ (0.10–0.20 ppm) of O_3 [53, 54]. Changes in blood chemistry have been measured with O_3 concentrations of 392 $\mu g/m^3$ (0.20 ppm) or more but not with 294 $\mu g/m^3$ (0.15 ppm) [42, 46, 56]. The effects of O_3 exposure are aggravated significantly by exercise and at an O_3 concentration of 294 $\mu g/m^3$ (0.15 ppm) have been observed only during exercise [41, 43–45]. Moreover, O_3 exposure can impair the abilities of subjects to undertake exercise by producing measurable reductions in their maximum outputs of physical power [41, 57].

Clinical studies also have found that O_3 exposure causes distinct discomfort and symptoms of illness in sensitive individuals. These symptoms include chest pain, coughing, wheezing, breathlessness, headache, and pharyngitis (inflammation of the upper throat area). The symptoms tend to occur simultaneously with measurable changes in pulmonary function and can be sufficiently severe to prevent subjects from carrying out the physiological tests and exercise procedures associated with the clinical studies.

There are substantial differences in individual responses to O_3 exposure. At the same O_3 concentration, some individuals may show no effects, whereas others may experience clinical symptoms of illness and measurable decreases in pulmonary function.

The effects of brief, isolated exposures to O_3 appear to be reversible. Most clinical and physiological symptoms caused by such exposures disappear over periods of roughly 24 hr after exposure ends, although some changes in blood chemistry may still be noticeable after 2 weeks, and the duration of the recovery period seems to increase as the duration of continuous exposure increases [42, 48, 49, 56]. However, there is evidence that repeated brief exposures can cause the development of a tolerance or adaptation to O_3, so that the response to subsequent exposures is diminished. A clinical study that compared the reactions of Canadians and southern Californians to exposures to 725 $\mu g/m^3$ (0.37 ppm) of O_3 for 2 hr found that the Canadians, who rarely encountered high O_3 concentrations in their normal lives, were more reactive than the southern Californians, who frequently encountered high concentrations [42]. Other clinical studies have shown that individuals can become tolerant to O_3 after being exposed to it for 2 hr/day over periods of 1–3 days [58, 59]. Whether tolerance develops appears to depend on the concentration to which an individual is exposed; the tolerance does not develop if the concentration is too low. The development of tolerance presumably also depends on the frequency and duration of O_3 exposure, but clinical studies to test this hypothesis have not yet been performed.

The biological mechanisms through which O_3 exposure causes changes in pulmonary function and blood chemistry are not known, so the significance of these changes for health cannot be assessed unambiguously. However, the physiological changes that accompany O_3 exposure are qualitatively similar to the changes that are produced by certain pulmonary diseases [46]. This observation and the clinical symptoms of illness that usually accompany O_3-induced changes in pulmonary function and blood chemistry suggest that the physiological changes may represent important health effects [46]. The long-term health significance of O_3 tolerance and the relation, if any, of O_3 tolerance to the development of respiratory disease are not known.

It is possible that individuals who have severe or chronic respiratory disease and, therefore, reduced respiratory reserves are more sensitive to O_3 exposure than other individuals are, although the limited clinical evidence that is available tends not to confirm this hypothesis. In a study that was carried out with mild to moderately severe asthmatics, the subjects showed no significant symptoms or changes in pulmonary function after being exposed to 392–490 $\mu g/m^3$ (0.20–0.25 ppm) of O_3

for 2 hr [42]. The subjects did experience changes in blood chemistry. In another study, minor changes in pulmonary function were found in 6 out of 17 asthmatics who were exposed to 490 $\mu g/m^3$ (0.25 ppm) of O_3 for 2 hr, and 4 of the 6 reactive subjects experienced clinical symptoms of O_3 exposure [60]. By themselves, these results suggest that the O_3 concentrations at which asthmatics begin to be affected by O_3 exposure may not be substantially lower than the concentrations at which normally healthy individuals begin to be affected. However, the results may understate asthmatics' repsonses to O_3 because they were obtained in laboratory environments in which common irritants, such as dust, pollen, and other pollutants, were not present. The effects on asthmatics of exposure to O_3 when other irritants are present are not known, but it is possible that O_3 exposure increases asthmatics' sensitivity to such irritants.[8] Thus, the effects of O_3 on individuals with asthma or other chronic respiratory diseases remain uncertain.

The results of epidemiological studies of O_3 exposure are qualitatively similar to the results of the clinical studies, but are more difficult to interpret. Several epidemiological studies have found statistically significant associations between atmospheric O_3 or oxidant concentrations and indicators of pulmonary function or the occurrence of clinical symptoms of O_3 exposure [61–66]. Increases in atmospheric oxidant concentrations also have been found to be associated with reductions in the athletic performances of high school cross-country runners [67]. The ranges of O_3 or oxidant concentrations within which these effects have been observed vary somewhat from study to study, but are roughly 20–588 $\mu g/m^3$ (0.01–0.30 ppm), 1-hr average.

There also have been epidemiological studies that did not find associations between O_3 exposure and pulmonary function indicators or clinical symptoms [68–70]. However, given the large variability in individual sensitivities to O_3 exposure and the inability to control all relevant factors in epidemiological studies, it is to be expected that some studies will fail to find associations between O_3 exposure and indicators of respiratory health, even if such associations exist. Thus, the negative findings of some studies cannot reasonably be considered to contradict the clinical results.

Dose-response relations between atmospheric oxidant concentrations and the incidence of eye discomfort, headache, cough, and chest discomfort were estimated in an epidemiological study of a group of student nurses in Los Angeles [71].[9] Threshold oxidant concentrations for the occurrence of these symptoms also were estimated in this study. Data on symptom rates were obtained from daily symptom diaries that were kept by the participants in the study. Measurements of oxidant concentrations were made at a monitoring station that was located 1–2 mi from the hospitals where the students were in training. The oxidant data used in the study consisted of daily maximum 1-hr average concentrations and varied between roughly 40 $\mu g/m^3$ (0.02 ppm) and 980 $\mu g/m^3$ (0.50 ppm). The dose-response relations and thresholds were estimated by fitting piece-wise-linear functions with two segments to the symptom and oxidant data [72]. The estimated dose-response relations are shown in figure 4.1; the estimated thresholds are shown in table 4.3. Because the cough and chest discomfort thresholds do not have finite upper 95% confidence limits, the results of the student nurse study, by themselves, provide only weak evidence for the existence of a relation between atmospheric oxidant concentrations and the incidence of these symptoms. However, the threshold estimates are consistent with clinical results indicating that symptoms of O_3 exposure begin to be noticeable at O_3 concentrations between 294 $\mu g/m^3$ (0.15 ppm) and 586 $\mu g/m^3$ (0.30 ppm), depending on the sensitivity and level of exercise of the individual [41, 45, 46].

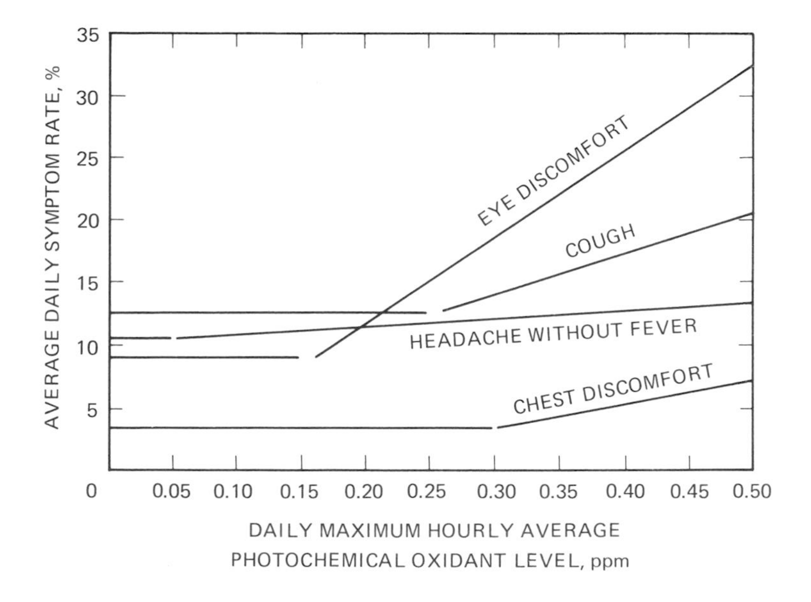

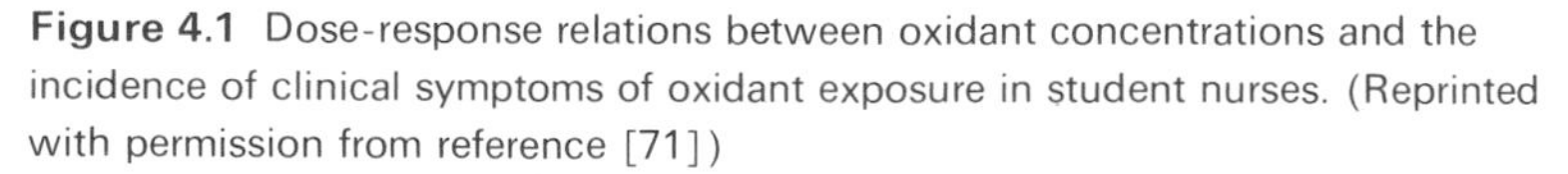
Figure 4.1 Dose-response relations between oxidant concentrations and the incidence of clinical symptoms of oxidant exposure in student nurses. (Reprinted with permission from reference [71])

Table 4.3 Threshold oxidant concentrations for the occurrence of clinical symptoms in student nurses[a]

Symptom	Threshold oxidant concentration	95% confidence limits	
		Lower	Upper
Headache without fever	100	0	390
Eye discomfort	310	280	340
Cough	510	370	∞
Chest discomfort	580	350	∞

a. Reprinted in part with permission from reference [71]. All concentrations are daily maximum 1-hr averages in $\mu g/m^3$.

Epidemiological studies of the relation between atmospheric O_3 concentrations and asthma attacks have yielded contradictory indications of the effects of atmospheric O_3 on asthmatics. One study of asthmatics in Los Angeles found that the incidence of asthma attacks was statistically significantly higher on days when the maximum 1-hr average oxidant concentration was above 490 $\mu g/m^3$ (0.25 ppm) than on other days [73]. A second study conducted in Los Angeles also found an association between increases in oxidant concentrations and increases in asthma attacks [74]. However, a third study conducted in Los Angeles found no association between asthma symptoms and O_3 concentrations [75]. These three studies are not directly comparable, owing to differences in the air quality variables and indicators of asthma symptoms that were used. However, the studies do illustrate the uncertainty that exists concerning the effects of O_3 on asthmatics.

Although the results of the epidemiological studies of the health effects of O_3 exposure are qualitatively consistent with the results of the clinical studies, several factors make it difficult to interpret the epidemiological results in precise terms. Many epidemiological studies have used bivariate statistical methods that are unable to distinguish between effects of O_3 and those of easily identifiable confounding factors, such as variations in weather conditions and in the concentrations of other pollutants. It is unclear whether the effects observed in such studies should be associated with O_3 exposure or with the other factors. In addition, many of the epidemiological results pertain to oxidant, rather than O_3, and concentrations of the non-O_3 constituents of oxidant were not measured in any of the epidemiological studies. There is no way to estimate the extent to which non-O_3 oxidants or other unmeasured substances, such as aldehydes, may have contributed to the health effects observed in these studies.[10] Finally, the epidemiological studies frequently have relied on O_3 or oxidant measurements made at monitoring sites that were somewhat distant from the locations of the individuals under study. Although the monitored O_3 or oxidant concentrations almost certainly were positively correlated with the concentrations actually experienced by these individuals, the use of remote site concentration measurements is, nonetheless, an additional source of potential error in the epidemiological results. Because of these difficulties of interpretation, the epidemiological results by themselves do not provide a firm basis for reaching conclusions concerning quantitative relations between O_3 exposure and health effects, and they provide only tentative support for the conclusion that the relations between exposures and effects observed in the previously discussed clinical studies are transferable quantitatively to nonlaboratory environments.[11]

Experiments performed with laboratory mice have shown that exposure to 157–216 $\mu g/m^3$ (0.08–0.11 ppm) of O_3 for 3 hr increases the susceptibility of these animals to bacterial lung infections [76–80]. The experiments consisted of exposing groups of mice to a bacterial aerosol after first exposing them either to clean air or to air that contained O_3. The subsequent mortality rates of the mice exposed to 157–216 $\mu g/m^3$ of O_3 for 3 hr typically exceeded the mortality rates of the mice exposed to clean air by 20–30% [76, 78, 80]. The mechanism by which O_3 causes increased susceptibility to lung infections appears to be inactivation of alveolar macrophages (cells that ingest and destroy bacteria) [77, 81–83]. Although it is not possible to extrapolate the animal results directly to humans, owing to anatomical and physiological differences between mice and humans, it is likely that O_3 at sufficient concentrations also impairs the functioning of human alveolar macrophages and, thereby, increases the susceptibility of humans to respiratory infections.[12] However, clinical infectivity studies of humans to determine O_3 concentrations at which this effect occurs cannot be performed for legal and ethical reasons, and convincing epidemiological evidence for the occurrence of O_3-related increases in the incidence of respiratory diseases has not been presented.[13] Thus, it is not known whether, or by how much, exposure to O_3 at atmospheric concentrations increases the susceptibility of humans to respiratory infections.

There have been few studies of the health effects of exposure to the non-O_3 components of oxidant. The results of one set of clinical studies indicated that healthy subjects' abilities to exercise and their physiological responses to exercise were unaffected by exposures of less than 45 min to

PAN at concentrations that greatly exceeded atmospheric concentrations [11, 85]. In addition, PAN and aldehydes (which are not oxidants) are known to be eye irritants. In other respects, the health effects of exposure to non-O_3 oxidants are unknown.

4.5 *Health Effects of NO_2*

NO_2, like O_3, is an acute pulmonary irritant. However, most human clinical studies in which subjects have been exposed on alternate days to pure air and to NO_2 for periods of 3 hr or less have found minimal symptoms of illness and alterations of pulmonary function at NO_2 concentrations below 2,800 $\mu g/m^3$ (1.5 ppm) [86–90].[14] These findings apply to individuals with asthma and chronic bronchitis as well as to normally healthy individuals, although the frequency with which minor symptoms of exposure (e.g., chest tightness, slight headache, thin nasal discharge) occur appears to be greater in asthmatics and chronic bronchitics than in healthy individuals [89]. The clinical symptoms of NO_2 exposure are reported to be milder than the symptoms typically found in clinical studies of O_3 exposure [63].

A clinical study whose design was substantially different from that of the others found measurable pulmonary effects in mild asthmatics who had been exposed to 210 $\mu g/m^3$ (0.11 ppm) of NO_2 for 1 hr [91]. In this study, subjects were exposed to carbachol, a substance that constricts the airways and thereby simulates the effects of natural respiratory irritants, after being exposed to either NO_2 or pure air. NO_2 exposure reduced the carbachol dosage needed to produce a 100% increase in airway resistance in 13 of the 20 asthmatics studied. This result suggests that exposure to NO_2 at concentrations frequently found in polluted air may increase some asthmatics' sensitivity to natural irritants significantly and that the NO_2 in polluted air may contribute to increases in the frequency or severity of asthma attacks. There have been no reports of efforts to test this hypothesis epidemiologically.

Animal infectivity studies similar to those carried out with O_3 have shown that NO_2 exposure increases the susceptibility of laboratory animals to bacterial and viral respiratory infections [78, 92–99]. These studies have used NO_2 concentrations or exposure durations that are greater than those likely to result from exposure to atmospheric NO_2. Increased susceptibility to infections has been found to occur both as a consequence of brief exposures to relatively high concentrations [e.g., 3,800 $\mu g/m^3$ (2 ppm) for 3 hr] and of prolonged exposures to lower concentrations [e.g., 940 $\mu g/m^3$ (0.5 ppm) for more than 1 week]. The mechanism of acute NO_2 toxicity appears to be impairment of the activity of pulmonary macrophages. In prolonged exposures other mechanisms, involving anatomical and biochemical changes in lung tissue, also may be important [77, 92, 95, 96].

Epidemiological studies of the relation between NO_2 exposure and the incidence of respiratory symptoms of disease in humans have yielded conflicting results. In a group of studies performed in Chattanooga, Tennessee, in 1968 and 1969, it was possible to define geographical areas of high, intermediate, and low NO_2 concentrations, according to the locations of the areas relative to a large trinitrotoluene (TNT) plant that was a major source of NO_2 emissions. The annual average NO_2 concentrations measured at fixed sites in these areas during 1968 were 190 $\mu g/m^3$ (0.10 ppm), 110 $\mu g/m^3$ (0.06 ppm), and 60 $\mu g/m^3$ (0.03 ppm) in the high, intermediate, and low areas, respectively [100].[15] Second-grade school children in the high-NO_2 area had slightly reduced forced expiratory volumes (an indicator of pulmonary function) compared to children in the intermediate and low-NO_2 areas, and acute upper respiratory illness rates in children and adults were higher in the high-NO_2 area than in the other two areas [101, 102]. In addition, school children who had resided in the high- or intermediate-NO_2 areas for at least 2 yr and preschool children who had resided in one of these areas for 3 yr had higher rates of bronchitis than did school and preschool children who had resided in the low-NO_2 area for comparable periods of time [103]. Differences among the bronchitis rates of children residing in the high and intermediate areas were inconsistent and not statistically significant [103].

The relation between respiratory illness and NO_2 concentrations in Chattanooga was studied again in 1972 and 1973 [100]. Although the annual average NO_2 concentrations measured in the high- and intermediate-NO_2 areas had decreased to less than 100 $\mu g/m^3$ (0.05 ppm), owing to installation of air pollution control devices and reduced production at the TNT plant, NO_2 concentrations still tended to be higher in the high-NO_2 area than in the other two areas. The incidence of upper respiratory diseases in children and adults and the incidence of lower respiratory diseases in children also tended to be higher in the high-NO_2 area than in the other areas. It has not been reported whether the illness rates in the high- and intermediate-NO_2 areas decreased from 1968–1969 to 1972–1973 along with the mean NO_2 concentrations in these areas.

Another group of epidemiological studies has dealt with the relation between indoor NO_2 concentrations and the incidence of respiratory disease. NO_2 concentrations tend to be higher in homes where gas ranges are used for cooking than in homes where electric ranges are used, owing to the oxidation of atmospheric nitrogen that occurs in gas flames. For example, 7-day average NO_2 concentrations of 8–317 $\mu g/m^3$ (0.004–0.169 ppm) have been reported in nonkitchen rooms of homes with gas ranges, compared to 6–70 $\mu g/m^3$ (0.003–0.037 ppm) in nonkitchen rooms of homes with electric ranges [105, 106]. Several epidemiological studies have sought to determine whether respiratory symptom or disease rates are higher and pulmonary function indicators are lower in persons living in homes with gas ranges than in persons living in homes with electric ranges. The results of these studies are mixed. Four studies found statistically significant associations between respiratory symptom or disease rates in children and either the type of range used for cooking or indoor NO_2 concentrations [107–110], but two other studies did not find such associations [111–113]. One study found that pulmonary function was statistically significantly poorer in children from gas-cooking households than in children from electric-cooking ones [110], but another study found that pulmonary function in girls was slightly better in gas-cooking households [109], and a third study found that a group of children and adults from gas-cooking households had slightly better pulmonary function than did a similar group from electric-cooking households [111]. In addition, one study found no association between range type and pulmonary function in children and adults [113], and one found no association between indoor NO_2 concentrations and pulmonary function in children [109]. Only two studies have attempted to find an association between respiratory disease rates in adults and range type [111–113]. Neither study found such an association.

Several of the epidemiological studies that were noted in connection with O_3 also attempted to find associations between atmospheric NO_2 concentrations and indicators of respiratory health. One set of studies found an association between NO_2 and pulmonary function in children [62, 63], and another study found an association between NO_2 and the incidence of mild respiratory disease in college students [61]. However, other studies failed to find associations between NO_2 concentrations and respiratory symptoms in junior high school students [64] and in adults with chronic respiratory diseases [68, 69].

The results of the various epidemiological studies of NO_2 exposures are summarized in table 4.4. There are no consistent patterns in these results, and it is not clear how they should be interpreted. In contrast to the situation with O_3, there have been no reports of laboratory or clinical studies of the effects of exposure to low concentrations of NO_2 that could help to clarify the epidemiological results. The inability of the

Table 4.4 Summary of the results of epidemiological studies of the relation between NO_2 exposure and respiratory disease[a]

Reference	Increased incidence of respiratory symptoms or illness in				Changes in pulmonary function
	Adults	All children	Children 3 yr old or younger	Children at least 5 yr old	
[61]	+ College students				
[62, 63]					Increases in NO_2 concentrations associated with reductions in some pulmonary function indicators in children
[64]				– Junior high school students	
[68]					No association between NO_2 and pulmonary function in adults with chronic respiratory diseases
[69]	– Adults with chronic respiratory diseases				No association between NO_2 and pulmonary function in adults with chronic respiratory diseases
[101–103][b]	+	+	+	+	Pulmonary function poorer in children in high-NO_2 areas than in children in low-NO_2 areas

[107]				+ Only for girls in urban areas	
[108]				+ Only in urban areas	
[109]				+	Pulmonary function not associated with indoor NO_2 but slightly better in girls from gas-cooking homes than in girls from electric-cooking homes
[110]			+	–	Pulmonary function poorer among children from gas-cooking homes than among children from electric-cooking homes
[111, 112]	–	–			Pulmonary function slightly better in a group of children and adults from gas-cooking homes than in a similar group from electric-cooking homes
[113]	–	–			Pulmonary function in children and adults not associated with range type

a. + denotes a statistically significant association between the indicated effect and range type, indoor NO_2 concentration or outdoor NO_2 concentration, depending on the study. – denotes the absence of a significant association. Blank signifies that the effect was not studied. Significance levels vary, depending on the study and the number of comparisons that were made.
b. The Chattanooga studies are treated as a single group for the purposes of this table.

epidemiological studies to obtain consistent associations between NO_2 concentration or range type and respiratory illness or pulmonary function may simply reflect the difficulty of detecting small health effects when the available data are imperfect and all potentially relevant factors cannot be controlled.[16] Alternatively, it may indicate that the associations found in some studies were caused by unmeasured and uncontrolled factors whose correlations with NO_2 concentrations and range type are different in different populations. Although all of the epidemiological studies attempted in various ways to control or compensate for variations in potentially confounding factors, in some studies it is impossible to distinguish with confidence between the effects of NO_2 and those of other pollutants, and there can be no assurance that any of the studies achieved adequate control over or compensation for confounding factors. Systematically varying uncontrolled factors could have created spurious associations or masked real associations between range type or NO_2 concentration and indicators of respiratory illness in any of the studies.

In summary, the results of one clinical study suggest that exposure to NO_2 at atmospheric concentrations may contribute to increased frequencies of asthma attacks. The results of human epidemiological studies and the results of animal infectivity studies performed at relatively high NO_2 concentrations suggest that exposure to NO_2 at atmospheric concentrations may increase the incidence of respiratory symptoms or diseases in humans. However, the available information does not enable a conclusion to be reached as to whether exposure to atmospheric NO_2 does, in fact, cause these effects.

4.6 Effects of Oxidants and NO_2 on Plants, Materials, Odor, and Visibility

O_3, PAN, and NO_2 are harmful to plants at atmospheric or near-atmospheric concentrations. The effects of these pollutants on plants are reviewed comprehensively in references [114–116]. CO at atmospheric concentrations has no known effects on plants. Plant damage caused by O_3, PAN, and NO_2 includes visible leaf injury (e.g., bleaching or other forms of discoloration), reduced growth, reduced flowering of ornamental and fruit-bearing plants, and reduced crop yields from agricultural plants. Sensitive plants, such as carnations and certain varieties of soybeans and sweet corn, are harmed by exposure to 98–116 $\mu g/m^3$ (0.05–0.10 ppm) of O_3 or 175 $\mu g/m^3$ (0.035 ppm) of PAN. At these concentrations, harmful effects are most noticeable when the exposures occur for 6–8 hr/day over periods of several weeks or months, although some plants show effects after a single 2-hr exposure [117]. Reductions in flower or crop yields following prolonged exposure to 98–196 $\mu g/m^3$ (0.05–0.10 ppm) of O_3 can be as much as 50% in some plants [114, 118]. Exposure to 128 $\mu g/m^3$ (0.068 ppm) of NO_2 for a period of 8 weeks has been found to reduce the growth of a species of meadow grass [119]. Long-term average NO_2 concentrations in the vicinity of 128 $\mu g/m^3$ occur in some cities. Damage to other sensitive plants (e.g., tomatoes, pinto beans, navel oranges) has been found to occur as a consequence of exposure to 270–620 $\mu g/m^3$ (0.14–0.33 ppm) of NO_2 for periods ranging from 10 days to 8 months, depending on the plant [120, 121]. These concentrations are somewhat higher than atmospheric NO_2 concentrations usually are for the given averaging times.

Damage to materials caused by O_3 includes cracking of rubber, fading of dyes in fabrics, and accelerated erosion of paints. The rate of formation of cracks in stressed rubber was used as an indicator of atmospheric O_3 concentrations before more precise instrumentation was developed. NO_2 causes dye fading and discoloration of white fabrics. The rate at which damage to materials from O_3 or NO_2 occurs is an increasing function of the average pollutant concentration during the exposure period. Any concentration above zero will cause damage, but the damage becomes noticeable more rapidly at high concentrations than at low ones.

The odors of O_3 and NO_2 are perceptible at concentrations above roughly 200 $\mu g/m^3$ (0.10 ppm of O_3 and 0.11 ppm of NO_2), depending on the sensitivity of the individual. O_3 and NO_2 are important contributors to the unpleasant odor that photochemically polluted air often has.

NO_2 is a brown gas, and concentrations of it in the vicinity of 400 $\mu g/m^3$ (0.21 ppm) or more cause the horizon in urban areas to appear brownish. In addition, the atmospheric photochemical processes that produce O_3 and NO_2 also produce fine aerosols. The aerosols scatter light and, thereby, can cause substantial reductions in visibility. The combination of NO_2 and photochemical aerosols is at least partially responsible for the thick, brownish haze that usually accompanies photochemical smog.

CO has no odor and does not affect materials or visibility.

4.7 *Relation of Harmful Effects to the US Air Quality Standards*

The air quality standards that were in effect in the United States during 1980 are shown in table 1.1. Legally, these standards were required to be set at concentrations below which no harmful effects of air pollution occur. However, strict observance of this requirement almost certainly would have made it necessary to set some of these standards at concentrations of zero, thereby depriving the standards of practical value. For example, O_3 and NO_2 damage materials at all concentrations above zero.

At the CO concentrations specified by the 8-hr and proposed 1-hr CO standards, exposure to atmospheric CO would be unlikely to cause most peoples' blood COHb concentrations to exceed 2% [122]. This is somewhat higher than the COHb concentration that would be inferred from table 4.1 because it incorporates effects of exposure to time-varying CO concentrations and of variations among individuals in the values of the relevant physiological parameters. CO-induced reductions in the ability of angina patients to exercise have been observed at a blood COHb concentration of 2% [8]. Noncardiovascular harmful effects of CO exposure have not been observed at COHb concentrations below 3%. It is possible, though not established, that there is no threshold for some of the cardiovascular effects of CO exposure, and, therefore, that no CO standard above zero could provide complete protection against these effects.

The O_3 standard corresponds to a concentration below those at which most human clinical and epidemiological studies have found harmful effects. The results of one clinical study [52] and of certain epidemiological studies [67, 71] suggest the possibility that O_3 may cause changes in pulmonary function, reduce athletic performance, or induce headaches at concentrations of 235 $\mu g/m^3$ or less, but the occurrence of these effects at such low concentrations cannot be considered to be established. Similarly, the results of animal infectivity studies in which exposure to O_3 concentrations below 235 $\mu g/m^3$ has been found to increase the susceptibilities of animals to respiratory infections cannot be extrapolated quantitatively to humans. Although O_3 can affect plants at concentrations below 235 $\mu g/m^3$, low concentrations of O_3 cause serious reductions in flowering or crop yields only when exposures occur for several hours per day over periods of several weeks or months. Given the episodic nature of periods of high O_3 concentrations and the fact that the air quality standard is aimed at concentration peaks, achievement of the 1-hr average O_3 standard is likely to prevent serious O_3-induced plant damage.

At the time that the NO_2 air quality standard was established, the results of the Chattanooga studies [101–103] provided the only available information on the human health effects of exposure to low concentrations of NO_2. The standard corresponds to an average concentration that is slightly below the concentration that was measured in the intermediate-NO_2 area during the studies. Most plants are not harmed by NO_2 at the concentrations that normally occur in the atmosphere.

4.8 *Economic Evaluation of the Harmful Effects of Air Pollution*

There are essentially two ways of assigning monetary values to the harm caused by air pollution or, equivalently, to the benefits of air pollution control. One way is to enumerate and assign a monetary value to each of the adverse physical effects that occurs at the levels of air pollution that are of interest. For example, if the pollutants and concentrations that are being evaluated cause some people to feel uncomfortable or ill, then these feelings of discomfort or illness are evaluated in monetary terms; if crop damage occurs, then the resulting monetary losses are estimated. This will be called the "enumeration-and-evaluation method" in the subsequent discussion. The other way of assigning monetary values to the harm caused by air pollution is to observe the effects of air pollution on the market price of (or, possibly, the demand for) a good or service whose value is likely to be affected by air quality. The observed relation between price (or demand) and air quality is then used to infer the amount of money that people would be willing to pay to avoid the harmful effects of air pollution. This will be called the "market approach." Most applications of the market approach are based on comparisons of housing prices in areas with clean air and polluted air. The basic concept underlying this form of the market approach is that if people value clean air, then they are likely to be willing to pay more for housing in areas where the air is clean than for equivalent housing in polluted areas and, moreover, that the differences between the resulting prices of equivalent housing in clean and polluted areas can be used to infer the value that people place on clean air.[17]

Ideally, the enumeration-and-evaluation method estimates the amount of money that society should be willing to pay for clean air if people had perfect information concerning the adverse effects of air pollution, whereas the market approach estimates the amount of money that society is willing to pay, given peoples' presumably imperfect information. However, in practice both methods fall considerably short of these ideals.

The practical difficulties associated with the enumeration-and-evaluation method are particularly severe, owing mainly to the unavailability of most of the information on harmful effects that is needed to implement the method. As discussed in sections 4.3–4.5, the existences of many potentially important health effects of transportation-related air pollution, such as O_3- or NO_2-induced increases in the incidence of respiratory diseases, are highly uncertain, and many effects whose existences are well established, such as O_3-induced changes in pulmonary function and blood chemistry, are of uncertain significance. Therefore, it is virtually impossible to enumerate the health effects of transporation-related air pollution in a way that permits meaningful monetary values to be assigned to them. Moreover, information on individuals' exposures to the various air pollutants (as opposed to measurements of pollutant concentrations at fixed sites) is largely unavailable, as are population dose-response relations that would enable the incidence of air pollution health effects to be computed as a function of the exposures that individuals receive. Consequently, the prevalence in the population of these health effects cannot be estimated reliably. Finally, it is not necessarily clear what monetary values should be assigned to air pollution health effects, even when the nature of the effects seems to be reasonably well understood. For example, to assign a monetary value to O_3-induced headaches and other forms of discomfort, it is necessary to know how much people would be willing to pay to avoid these symptoms and to what extent the symptoms cause losses of productivity in affected workers. Such information is not now available. The evaluation of air pollution-induced damages to vegetation and materials and air pollution-induced reductions in visibility presents problems that are similar to, although perhaps not as severe as, the problems of evaluating the health effects of air pollution.

Possibly as a consequence of these difficulties, there have been few applications of the enumeration-and-evaluation method to transportation-related air pollution.[18] A committee of the National Academy of Sciences (NAS) used the method as part of an attempt to estimate the benefits and costs of controlling automobile emissions throughout the United States [124]. The NAS analysis included only two health effects because there were only two for which dose-response relations could be obtained. These effects were the oxidant exposure symptoms observed in the Los Angeles student nurse study [71] and increases in mortality due to NO_2 exposure. The oxidant dose-response relations obtained in the student nurse study were assumed to be applicable to the national population. Dose-response relations for NO_2-induced mortality were obtained from an unpublished statistical analysis of the relation between air pollution and mortality. The NAS analysis did not include any adverse effects of CO exposure.

The assumptions that atmospheric NO_2 contributes to increased mortality and that the student nurse results can be applied nationally are both risky. No reliable evidence for the existence of a relation between atmospheric NO_2 and mortality has been published. The student nurse group, which consisted of healthy women of college age, was not representative of the national population, and the chemical compositions of the oxidant mixtures to which the student nurses were exposed were not necessarily the same as the compositions of the mixtures to which individuals in the national population are exposed. Moreover, the student nurse study used the oxidant concentrations measured at a remote site as a surrogate for the concentrations to which individual students were exposed, and the relation between this surrogate and individual exposures that is implicit in the student nurse results is not necessarily transferable to the rest of the nation.

The NAS estimated that a nationwide 90% reduction in automobile emissions from their 1973 levels would prevent 200 million person days/yr of oxidant exposure symptoms and between 800 and 4,400 NO_2-induced deaths/yr. It arbitrarily estimated the value of oxidant exposure symptoms to be between $1 and $10/person day of discomfort and the value of a human life to be $200,000 (all in 1973 dollars). Hence, the estimated health benefits of a 90% reduction in automobile emissions were $0.36–3 billion/yr. In addition, the NAS estimated that a 90% reduction in automobile emissions would prevent between $0.6 and $1.8 billion/yr in damages to vegetation and materials. Therefore, the NAS enumeration-and-evaluation estimate of the total benefits of a nationwide 90% reduction in automobile emissions was $0.96–4.8 billion/yr in 1973 dollars, or roughly $1.6–8.2 billion/yr in 1980 dollars.

The enumeration-and-evaluation method also has been used to estimate the economic value of reducing the daily maximum 1-hr average concentrations of oxidants, NO_2, and particulates in the Los Angeles area by roughly 45, 30, and 20%, respectively, from their 1975 levels [126].[19] The evaluation portion of the analysis was carried out by means of a survey of a sample of Los Angeles area residents. The surveyed individuals were shown photographs that illustrated changes in visibility that would accompany the improvements in air quality, and they were given descriptions of the reductions in adverse health effects that the researchers believed could be expected.[20] They were then asked a series of questions concerning how much they would be willing to pay for the air quality improvements that had been described to them. The results of this procedure indicated that Los Angeles area residents would be willing to pay an average of roughly $350/household/yr, or a total of $0.65 billion/yr for the Los Angeles area as a whole (in 1978 dollars), for these air quality improvements. Inflated to 1980 dollars, these estimates correspond roughly to $410/household/yr and $0.75 billion/yr for the total. The estimates do not include the values of reductions in damage to plants and materials or of any benefits or losses to businesses that would accompany the air quality improvements, as these were not described in the survey instrument.

The NAS and Los Angeles enumeration-and-evaluation estimates of the value of improved air quality are not comparable because they apply to different geographical areas and different air quality improvements. Although the health effects enumerations and other aspects of the procedures used to develop both estimates can be criticized easily (see the preceding discussion of the NAS estimates, for example), it is not possible to make a quantitative assessment of the accuracy of either estimate; the information on the adverse effects of air pollution that would be needed to make such an assessment does not yet exist.

Compared to the enumeration-and-evaluation method, the market approach has the considerable advantage of not requiring specific harmful effects of air pollution to be identified or evaluated. This is because the market approach is based on observations of peoples' economic responses to variations in air quality, given their existing information and beliefs about air pollution, rather than on evaluation of physical effects or consideration of how people might behave if their information were perfect. However, the market approach also has certain disadvantages. One of these is that when, as is nearly always the case, the approach is based on comparisons of housing prices in areas with clean and polluted air, it tends to underestimate peoples' willingness to pay for clean air. This is because housing prices reflect peoples' willingness to pay for clean air in the vicinities of their homes and not necessarily their willingness to pay for clean air at work, shopping, and recreational locations. Nor do housing prices reflect producers' willingness to pay for such things as the reductions in damages to crops and materials that cleaner air would bring about. Another possible disadvantage of the market approach is that it does not provide information on what people think the benefits of clean air will be. Therefore, it is difficult to estimate the extent to which estimates of willingness to pay based on the market approach reflect the true social cost of air pollution (i.e., the willingness to pay of a perfectly informed population).

In addition to these conceptual difficulties, there are certain practical problems associated with the market approach. A market approach analysis that is based on housing prices proceeds in essentially two phases. First the relation between air quality and housing prices is inferred statistically from data on housing prices and characteristics, air quality, and other relevant variables. Then the relation between housing prices and willingness to pay for clean air is inferred, also using statistical methods.[21]

The validity of these inferences depends on having a correct understanding of the determinants of housing prices and accurate data on the behavior of the housing market. For example, suppose that people who live in polluted areas tend to buy central air conditioners for their homes in order to avoid the effects of air pollution and that this increases the values of houses in polluted areas relative to the values of houses in clean-air areas, where people tend not to buy central air conditioners. If this behavior is not taken into account when the relation between housing prices and air quality is estimated, the analysis will tend to underestimate peoples' willingness to pay for improved air quality. Similarly, if, as usually happens, air quality in the vicinities of individual houses cannot be measured, but must be estimated using models or interpolated from measurements made at possibly distant monitoring sites, then the resulting errors in the air quality estimates will tend to induce errors in the estimates of willingness to pay. Finally, it often is difficult to identify the separate influences of different pollutants on housing prices. For example, neighborhoods that have high O_3 concentrations often tend also to have high NO_2 concentrations, whereas neighborhoods that have low O_3 concentrations tend to have low NO_2 concentrations. This covariation of O_3 and NO_2 concentrations can make it difficult or impossible to distinguish the influence of O_3 on housing prices from that of NO_2. If the distinction cannot be made, then the market approach based on housing prices will be able to estimate willingness to pay only for air quality improvements that preserve the existing relations between O_3 concentrations and NO_2 concentrations.

Table 4.5 summarizes the results of several market approach studies of peoples' willingness to pay for the control of transportation-related air pollution. The tabulated estimates all were obtained from housing price analyses and, presumably, reflect mainly peoples' willingness to pay for air pollution control at home. The estimates obtained from different cities and different studies (with the possible exception of the two Boston estimates) pertain to different pre- and postcontrol levels of air quality and, therefore, are not directly comparable, except, perhaps, to obtain very crude indications of the effects of these levels on willingness to pay. It is likely that the postcontrol pollutant concentrations associated with the Los Angeles estimates exceed the precontrol concentrations associated with the Boston and Washington estimates. The ranges given for the Boston and Los Angeles estimates illustrate the effects of uncertainties regarding the functional form of the relation between housing prices and air quality. Different plausible functional forms yielded estimates of willingness to pay that spanned the tabulated ranges. Although these ranges are quite broad

Table 4.5 Market approach estimates of willingness to pay for air quality improvements as inferred from housing prices

		Percentage reductions in concentration of				Willingness to pay ($/household/yr) in		
City	Emissions control measures	O_3 or oxidant	NO_2	HC	Particulates	Study year dollars (date)	1980 dollars	Reference
Boston	Emissions from new automobiles reduced 90% from model year 1970–71 levels	Not stated	10[a]	60[a]	Not stated	11–239 (1970)	22–470	[124]
Boston	Same as above	Not stated	Not stated	Not stated	Not stated	47–118 (1970)	93–230	[127]
Los Angeles	Same as above	Not stated	30[a]	50[a]	Not stated	57–409 (1970)	110–810	[124]
Los Angeles	Not stated	45[b]	30[b]	Not stated	20[b]	318–763 (1977)	400–970	[126]
Washington, DC	Automobile and truck emissions reduced 90% from 1968 levels	45[c]	Not stated	Not stated	0	27 (1972)	49	[128]

a. The reductions are relative to 1970 concentrations.
b. The reduction in oxidant concentrations is not stated in reference [126], but has been estimated from data presented there. The reductions are relative to 1975 concentrations.
c. The reduction is relative to 1968 concentrations. This study used housing data from the 1970–1972 period and pollution data from 1968.

(varying from roughly a factor of 2.4 to over a factor of 20, depending on the study), they may understate the true uncertainties of the estimates. This is because the ranges do not include the effects of using modeled or interpolated values of pollutant concentrations (rather than measured values) in the analyses of housing prices or of most of the studies' inability to separate the influences on housing prices of transportation and nontransportation pollutants. Identifying the separate influences of transportation pollutants and particulates has proven to be especially difficult in housing price studies.

4.9 Conclusions

It is clear that transportation-related air pollutants at atmospheric concentrations can have harmful effects on human health, plants, materials, and visibility. However, the uncertainties that currently exist regarding the nature and severity of these effects are substantial. Because mitigation or avoidance of harmful effects is the ultimate justification of air pollution control programs, these uncertainties are major causes of controversies over how much air pollution control is necessary or desirable. It is frequently suggested that the implementation of costly or disruptive pollution control measures should be delayed until improved information on harmful effects is available. Although it is beyond the scope of this book to discuss the question of how much certainty concerning harmful effects is needed to justify air pollution control programs, it is worth noting that the current uncertainties regarding the effects of transportation-related air pollutants are unlikely to be reduced significantly in the foreseeable future. Hence, for practical purposes it is not possible to delay the implementation of measures to control these pollutants for a predictable time period so that better information on the harmful effects of the pollutants can be developed. Delaying the implementation of control measures for these pollutants until substantially improved information on harmful effects is available is equivalent to delaying the implementation of the measures indefinitely.

References

1

Coburn, R. F., Forster, R. E., and Kane, P. P., "Considerations of the Physiological Variables That Determine the Blood Carboxyhemoglobin Concentration in Man," *Journal of Clinical Investigation*, Vol. 44, pp. 1899–1910, 1965.

2

Roughton, F. J. W., "The Equilibrium of Carbon Monoxide with Human Hemoglobin in Whole Blood," *Annals of the New York Academy of Sciences*, Vol. 174, pp. 177–188, 1970.

3

US Environmental Protection Agency, *Preliminary Assessment of Adverse Health Effects from Carbon Monoxide and Implications for Possible Modifications of the Standard*, staff paper, Research Triangle Park, NC, 1 June 1979.

4

Aronow, W. S., Harris, C. N., Isbell, M. W., Rokaw, S. N., and Imparato, B., "Effect of Freeway Travel on Angina Pectoris," *Annals of Internal Medicine*, Vol. 77, pp. 669–676, 1972.

5

Anderson, E. W., Andelman, R. J., Strauch, J. M., Fortuin, N. J., and Knelson, J. H., "Effect of Low-Level Carbon Monoxide Exposure on Onset and Duration of Angina Pectoris," *Annals of Internal Medicine*, Vol. 79, pp. 46–50, 1973.

6

Aronow, W. S., and Isbell, M. W., "Carbon Monoxide Effect on Exercise-Induced Angina Pectoris," *Annals of Internal Medicine*, Vol. 79, pp. 392–395, 1973.

7

Aronow, W. S., Stemmer, E. A., and Isbell, M. W., "Effect of Carbon Monoxide Exposure on Intermittent Claudication," *Circulation*, Vol. XLIX, pp. 415–417, March 1974.

8

Aronow, W. S., "Aggravation of Angina Pectoris by 2 Percent Carboxyhemoglobin," *Chest*, Vol. 78, p. 507, September 1980.

9

Aronow, W. S., and Cassidy, J., "Effect of Carbon Monoxide on Maximal Treadmill Exercise: A Study of Normal Persons," *Annals of Internal Medicine*, Vol. 83, pp. 496–499, 1975.

10

Aronow, W. S., Ferlinz, J., and Glauser, F., "Effect of Carbon Monoxide on Exercise Performance in Chronic Obstructive Pulmonary Disease," *The American Journal of Medicine*, Vol. 63, pp. 904–908, December 1977.

11

Drinkwater, B. L., Raven, P. B., Horvath, S. M., Gliner, J. A., Ruhling, R. O., Bolduan, N. W., and Taguchi, S., "Air Pollution, Exercise, and Heat Stress," *Archives of Environmental Health*, Vol. 28, pp. 177–181, April 1974.

12

Goldsmith, J. R., and Landaw, S. A., "Carbon Monoxide and Human Health," *Science*, Vol. 162, pp. 1352–1359, 20 December 1968.

13

Cohen, S. I., Deane, M., and Goldsmith, J. R., "Carbon Monoxide and Survival from Myocardial Infarction," *Archives of Environmental Health*, Vol. 19, pp. 510–517, October 1969.

14
Hexter, A. C., and Goldsmith, J. R., "Carbon Monoxide: Association of Community Air Pollution with Mortality," *Science*, Vol. 172, pp. 265–267, 16 April 1971.

15
National Research Council, *Carbon Monoxide*, Report No. EPA 600/1-77-034, US Environmental Protection Agency, Washington, DC, September 1977. NTIS Publication No. PB 274965.

16
Kurt, T. L., Mogielnicki, R. P., Chandler, J. E., and Hirst, K., "Ambient Carbon Monoxide Levels and Acute Cardiorespiratory Complaints: An Exploratory Study," *American Journal of Public Health*, Vol. 69, pp. 360–363, April 1979.

17
Kuller, L. H., Radford, E. P., Swift, D., Perper, J., and Fisher, R., "The Relationship between Ambient Carbon Monoxide Levels, Postmortem Carboxyhemoglobin, Sudden Death and Myocardial Infarction," *Archives of Environmental Health*, Vol. 30, pp. 477–482, October 1975.

18
Halperin, M. H., McFarland, R. A., Niven, J. I., and Roughton, F. J. W., "The Time Course of the Effects of Carbon Monoxide on Visual Thresholds," *Journal of Physiology*, Vol. 146, pp. 583–593, 1959.

19
Horvath, S. M., Dahms, T. E., and O'Hanlon, J. F., "Carbon Monoxide and Human Vigilance: A Deleterious Effect of Present Urban Concentrations," *Archives of Environmental Health*, Vol. 23, pp. 343–347, November 1971.

20
Ramsey, J. M., "Carbon Monoxide, Tissue Hypoxia, and Sensory Psychomotor Response in Hypoxaemic Subjects," *Clinical Science*, Vol. 42, pp. 619–625, 1972.

21
Bender, W., Gothert, M., and Malorny, G., "Effect of Low Carbon Monoxide Concentrations on Psychological Functions," *Staub-Reinhaltung der Luft*, Vol. 32, pp. 54–60, April 1972.

22
Fodor, G. G., and Winneke, G., "Effect of Low CO Concentrations on Resistance to Monotony and on Psychomotor Capacity," *Staub-Reinhaltung der Luft*, Vol. 32, pp. 46–54, April 1972.

23
Groll-Knapp, E., Wagner, H., Hauck, H., and Haider, M., "Effects of Low Carbon Monoxide Concentrations on Vigilance and Computer-Analyzed Brain Potentials," *Staub-Reinhaltung der Luft*, Vol. 32, pp. 64–68, April 1972.

24
Ramsey, J. F., "Effects of Single Exposures of Carbon Monoxide on Sensory and Psychomotor Response," *American Industrial Hygiene Association Journal*, Vol. 34, pp. 212–216, May 1973.

25
Salvatore, S., "Performance Decrement Caused by Mild Carbon Monoxide Levels on Two Visual Functions," *Journal of Safety Research*, Vol. 6, pp. 131–134, September 1974.

26
Wright, G. R., and Shephard, R. J., "Carbon Monoxide Exposure and Auditory Duration Discrimination," *Archives of Environmental Health*, Vol. 33, pp. 226–235, September/October 1978.

27
Stewart, R. D., Petersen, J. E., Baretta, E. D., Bachand, R. T., Hosko, M. J., and Herrmann, A. A., "Experimental Human Exposure to Carbon Monoxide," *Archives of Environmental Health*, Vol. 21, pp. 154–164, August 1970.

28
O'Donnell, R. D., Chikos, P., and Theodore, J., "Effect of Carbon Monoxide Exposure on Human Sleep and Psychomotor Performance," *Journal of Applied Physiology*, Vol. 31, pp. 513–518, October 1971.

29
Stewart, R. D., Newton, P. E., Hosko, M. J., and Peterson, J. E., "Effect of Carbon Monoxide on Time Perception," *Archives of Environmental Health*, Vol. 27, pp. 155–160, September 1973.

30
Haider, M., Groll-Knapp, E., Holler, H., Neuberger, M., and Stidlo, H., "Effects of Moderate CO Dose on the Central Nervous System—Electrophysiological and Behavior Data and Clinical Relevance," in E. J. Finkel and W. C. Duel, eds., *Clinical Implications of Air Pollution Research*, Publishing Sciences Group, Inc., Acton, MA, 1976.

31
Benignus, V. A., Otto, D. A., Prah, J. D., and Benignus, G., "Lack of Effects of Carbon Monoxide on Human Vigilance," *Perceptual and Motor Skills*, Vol. 45, pp. 1007–1014, 1977.

32
McFarland, R. A., "Low Level Exposure to Carbon Monoxide and Driving Performance," *Archives of Environmental Health*, Vol. 27, pp. 355–359, December 1973.

33
Luria, S. M., and McKay, C. L., "Effects of Low Levels of Carbon Monoxide on Visions of Smokers and Nonsmokers," *Archives of Environmental Health*, Vol. 34, pp. 38–43, January/February 1979.

34
Wright, G., Randell, P., and Shephard, R. J., "Carbon Monoxide and Driving Skills," *Archives of Environmental Health*, Vol. 27, pp. 349–354, December 1973.

35
Rummo, N., and Sarlanis, K., "The Effect of Carbon Monoxide on Several Measures of Vigilance in a Simulated Driving Task," *Journal of Safety Research*, Vol. 6, pp. 126–130, September 1974.

36
Weir, F. H., and Rockwell, T. H., *An Investigation of the Effects of Carbon Monoxide on Humans in the Driving Task*, Report No. EPA 650/1-73-003, US Environmental Protection Agency, January 1973. NTIS Publication No. PB 224646.

37
Wright, G. R., and Shephard, R. J., "Brake Reaction Time—Effects of Age, Sex, and Carbon Monoxide," *Archives of Environmental Health*, Vol. 33, pp. 141–150, May/June 1978.

38
Ury, H. K., Perkins, N. M., and Goldsmith, J. R., "Motor Vehicle Accidents and Vehicular Pollution in Los Angeles," *Archives of Environmental Health*, Vol. 25, pp. 314–322, November 1972.

39
Aronow, W. S., and Rokaw, S. N., "Carboxyhemoglobin Caused by Smoking Nonnicotine Cigarettes: Effects in Angina Pectoris," *Circulation*, Vol. 44, pp. 782–788, November 1971.

40
Aronow, W. S., Cassidy, J., Vangrow, J. S., March, H., Kern, J. C., Goldsmith, J. R., Khemka, M., Pagano, J., and Vawter, M., "Effect of Cigarette Smoking and Breathing Carbon Monoxide on Cardiovascular Hemodynamics in Anginal Patients," *Circulation*, Vol. 50, pp. 340–347, August 1974.

41
DeLucia, A. J., and Adams, W. C., "Effects of O_3 Inhalation During Exercise on Pulmonary Function and Blood Biochemistry," *Journal of Applied Physiology; Respiratory, Environmental and Exercise Physiology*, Vol. 43, pp. 75–81, 1977.

42
Hackney, J. D., Linn, W. S., Karuza, S. K., Buckley, R. D., Law, D. C., Bates, D. V., Hazucha, M., Pengelly, L. D., and Silverman, F., "Effects of Ozone Exposure in Canadians and Southern Californians," *Archives of Environmental Health*, Vol. 32, pp. 110–116, May/June 1977.

43
Silverman, F., Folinsbee, L. J., Barnard, J., and Shephard, R. J., "Pulmonary Function Changes in Ozone—Interaction of Concentration and Ventilation," *Journal of Applied Physiology*, Vol. 41, pp. 859–864, 1976.

44
Folinsbee, L. J., Silverman, F., and Shephard, R. J., "Exercise Responses Following Ozone Exposure," *Journal of Applied Physiology*, Vol. 38, pp. 996–1001, June 1975.

45
Hackney, J. D., Linn, W. S., Mohler, J. G., Pedersen, E. E., Breisacher, P., and Russo, A., "Experimental Studies on Human Health Effects of Air Pollutants, II," *Archives of Environmental Health*, Vol. 30, pp. 379–384, August 1975.

46
Hackney, J. D., Linn, W. S., Law, D. C., Karuza, S. K., Greenberg, S., Buckley, R. D.,

and Pedersen, E. E., "Experimental Studies on Human Health Effects of Air Pollutants, III," *Archives of Environmental Health*, Vol. 30, pp. 385–390, August 1975.

47

Hazucha, M., Silverman, F., Parent, C., Field, S., and Bates, D. V., "Pulmonary Function in Man after Short-Term Exposure to Ozone," *Archives of Environmental Health*, Vol. 27, pp. 183–188, September 1973.

48

Bates, D. V., Bell, G. M., Burnham, C. D., Hazucha, M., Mantha, J., Pengelly, L. D., and Silverman, F., "Short-Term Effects of Ozone on the Lung," *Journal of Applied Physiology*, Vol. 32, pp. 176–181, February 1972.

49

Kerr, H. D., Kulle, T. J., McIlhany, M. L., and Swidersky, P., "Effects of Ozone on Pulmonary Function in Normal Subjects," *American Review of Respiratory Disease*, Vol. 111, pp. 763–773, 1975.

50

Ketcham, B., Lassiter, S., Haak, E., and Knelson, J. H., "Effects of Ozone Plus Moderate Exercise on Pulmonary Function in Healthy Young Men," in B. Dimitriades, ed., *International Conference on Photochemical Oxidant Pollution and Its Control: Proceedings*, Vol. I, Report No. EPA 600/3-77-001a, US Environmental Protection Agency, Research Triangle Park, NC, January 1977, pp. 495–504. NTIS Publication No. PB 264232.

51

Hazucha, M., Parent, C., and Bates, D. V., "Development of Ozone Tolerance in Man," in B. Dimitriades, ed., *International Conference on Photochemical Oxidant Pollution and Its Control: Proceedings*, Vol. I, Report No. EPA 600/3-77-001a, US Environmental Protection Agency, Research Triangle Park, NC, January 1977, pp. 527–541. NTIS Publication No. PB 264232.

52

Nieding, G. v., and Wagner, H. M., "Experimental Studies on the Short-Term Effects of Air Pollutants on Pulmonary Function in Man: Two-Hour Exposure to NO_2, O_3 and SO_2 Alone and in Combination," *Proceedings of the Fourth International Clean Air Congress*, The Japanese Union of Air Pollution Prevention Associations, Tokyo, 1977, pp. 5–8.

53

Higgins, E. A., Lategola, M. T., McKenzie, J. M., Melton, C. E., and Vaughan, J. A., *Effects of Ozone on Exercising and Sedentary Adult Men and Women Representative of the Flight Attendant Population*, Report No. FAA-AM-79-20, Federal Aviation Administration, Washington, DC, October 1979. NTIS Publication No. AD-A080 045/8.

54

Horvath, S. M., Gliner, J. A., Folinsbee, L. G., and Bedi, J. F., *Physiological Effects of Exposure to Different Concentrations of Ozone*, report to the California Air Resources Board and the National Institutes of Health on work performed during the time period 1 June 1975–31 May 1978, Institute of Environmental Stress, University of California, Santa Barbara, CA.

55
Linn, W. S., Buckley, R. D., Spier, C. E., Blessey, R. L., Jones, M. P., Fischer, D. A., and Hackney, J. D., "Health Effects of Ozone Exposure in Asthmatics," *American Review of Respiratory Disease*, Vol. 117, pp. 835–843, 1978.

56
Buckley, R. D., Hackney, J. D., Clark, K., and Posin, C., "Ozone and Human Blood," *Archives of Environmental Health*, Vol. 30, pp. 40–43, January 1975.

57
Folinsbee, L. J., Silverman, F., and Shephard, R. J., "Decrease of Maximum Work Performance Following Ozone Exposure," *Journal of Applied Physiology: Respiratory, Environmental and Exercise Physiology*, Vol. 42, pp. 531–536, 1977.

58
Hackney, J. D., Linn, W. S., Mohler, J. G., and Collier, C. R., "Adaptation to Short-Term Respiratory Effects of Ozone in Men Exposed Repeatedly," *Journal of Applied Physiology: Respiratory, Environmental and Exercise Physiology*, Vol. 43, pp. 82–85, 1977.

59
Hazucha, M., Parent, C., and Bates, D. V., "Development of Ozone Tolerance in Man," in B. Dimitriades, ed., *International Conference on Photochemical Oxidant and Its Control: Proceedings*, Vol. I, Report No. EPA-600/3-77-001a, US Environmental Protection Agency, Research Triangle Park, NC, January 1977, pp. 527–541. NTIS Publication No. PB 264232.

60
Silverman, F., "Asthma and Respiratory Irritants (Ozone)," *Environmental Health Perspectives*, Vol. 29, pp. 131–136, 1979.

61
Durham, W. H., "Air Pollution and Student Health," *Archives of Environmental Health*, Vol. 28, pp. 241–254, May 1974.

62
Kagawa, J., and Toyama, T., "Photochemical Air Pollution: Its Effects on Respiratory Function of Elementary School Children," *Archives of Environmental Health*, Vol. 30, pp. 117–122, March 1975.

63
Kagawa, J., Toyama, T., and Nakaza, M., "Pulmonary Function Test in Children Exposed to Air Pollution," in A. J. Finkel and W. J. Duel, eds., *Clinical Implications of Air Pollution Research*, Publishing Sciences Group, Inc., Acton, MA, 1976.

64
Mizoguchi, I., Makino, K., Kudou, S., and Mikami, R., "On the Relationship of Subjective Symptoms to Photochemical Oxidant," in B. Dimitriades, ed., *International Conference on Photochemical Oxidant Pollution and Its Control: Proceedings*, Vol. I, Report No. EPA 600/3-77-001a, US Environmental Protection Agency,

Research Triangle Park, NC, January 1977, pp. 477–494. NTIS Publication No. PB 264232.

65

Zagraniski, R. T., Leaderer, B. P., and Stolwijk, J. A. J., "Ambient Sulfates, Photochemical Oxidants, and Acute Adverse Health Effects: An Epidemiological Study," *Environmental Research*, Vol. 19, pp. 306–320, 1979.

66

Ury, H. K., and Hexter, A. C., "Relating Photochemical Air Pollution to Human Physiological Reactions under Controlled Conditions: Statistical Procedures," *Archives of Environmental Health*, Vol. 18, pp. 473–480, April 1969.

67

Wayne, W. S., Wehrle, P. F., and Carroll, R. E., "Oxidant Air Pollution and Athletic Performance," *Journal of the American Medical Assocation*, Vol. 199, pp. 151–154, March 1967.

68

Rokaw, S. N., and Massey, F., "Air Pollution and Chronic Respiratory Disease," *American Review of Respiratory Disease*, Vol. 86, pp. 703–704, November 1962.

69

Schoettlin, C. E., "The Health Effect of Air Pollution on Elderly Males," *American Review of Respiratory Disease*, Vol. 86, pp. 878–897, December 1962.

70

McMillan, R. S., Wiseman, D. H., Hanes, B., and Wehrle, P. F., "Effects of Oxidant Air Pollution on Peak Expiratory Flow Rates in Los Angeles School Children," *Archives of Environmental Health*, Vol. 18, pp. 941–949, June 1969.

71

Hammer, D. I., Hasselblad, V., Portnoy, B., and Wehrle, P. F., "Los Angeles Student Nurse Study," *Archives of Environmental Health*, Vol. 28, pp. 255–260, May 1974.

72

Quandt, R. E., "The Estimation of the Parameters of a Linear Regression System Obeying Two Separate Regimes," *Journal of the American Statistical Association*, Vol. 53, pp. 873–880, 1958.

73

Schoettlin, C. E., and Landau, E., "Air Pollution and Asthmatic Attacks in the Los Angeles Area," *Public Health Reports*, Vol. 76, pp. 545–548, June 1961.

74

Whittemore, A. S., and Korn, E. L., "Asthma and Air Pollution in the Los Angeles Area," *American Journal of Public Health*, Vol. 20, pp. 687–696, July 1980.

75

Kurata, J. H., Glovsky, M. M., Newcomb, R. L., and Easton, J. G., "A Multifactorial Study of Patients with Asthma, Part 2: Air Pollution, Animal Dander and Asthma Symptoms," *Annals of Allergy*, Vol. 37, pp. 398–409, December 1976.

76

Coffin, D. L., and Blommer, E. J., "Alteration of the Pathogenic Role of Streptococci

Group C in Mice Conferred by Previous Exposure to Ozone," in I. H. Silver, ed., *Aerobiology*, Academic Press, New York, 1970, pp. 54–61.

77

Gardner, D. E., and Graham, J. A., "Increased Pulmonary Disease Mediated through Altered Bacterial Defenses," in C. L. Sanders, R. P. Schneider, G. E. Dagle, and H. A. Ragan, eds., *Pulmonary Macrophage and Epithelial Cells*, Proceedings of the 16th Annual Hanford Biology Symposium, Richland, WA, September 1976, pp. 1–21. NTIS Publication No. CONF-760927.

78

Ehrlich, R., Findlay, J. C., Fenters, J. D., and Gardner, D. E., "Health Effects of Short-Term Inhalation of Nitrogen Dioxide and Ozone Mixtures," *Environmental Research*, Vol. 14, pp. 223–231, 1977.

79

Ehrlich, R., Findlay, J. C., Fenters, J. D., and Gardner, D. E., "Health Effects of Short-Term Exposures to NO_2-O_3 Mixtures," in B. Dimitriades, ed., *International Conference on Photochemical Oxidant Air Pollution and Its Control*, Vol. I, Report No. EPA 600/3-77-001a, US Environmental Protection Agency, Research Triangle Park, NC, January 1977, pp. 565–575. NTIS Publication No. PB 264232.

80

Miller, F. J., Illing, J. W., and Gardner, D. E., "Effect of Urban Ozone Levels on Laboratory-Induced Respiratory Infections," *Toxicology Letters*, Vol. 2, pp. 163–169, 1978.

81

Coffin, D. L., Holzman, R. S., and Wolock, F. J., "Influence of Ozone on Pulmonary Cells," *Archives Environmental Health*, Vol. 16, pp. 633–636, May 1968.

82

Weissbecker, L., Carpenter, R. D., Luchsinger, P. C., and Osdene, T. S., "In Vitro Alveolar Macrophage Viability," *Archives of Environmental Health*, Vol. 18, pp. 756–759, May 1969.

83

Coffin, D. L., and Gardner, D. E., "Interaction of Biological Agents and Chemical Air Pollutants," *Annals of Occupational Hygiene*, Vol. 15, pp. 219–234, 1972.

84

Peterson, M. L., Harder, S., Rummo, N., and House, D., "Effect of Ozone on Leukocyte Function in Exposed Human Subjects," *Environmental Research*, Vol. 15, pp. 485–493, 1978.

85

Raven, P. B., Drinkwater, B. L., Ruhling, R. O., Bolduan, N., Taguchi, S., Gliner, J., and Horvath, S. M., "Effect of Carbon Monoxide and Peroxyacetyl Nitrate on Man's Maximal Aerobic Capacity," *Journal of Applied Physiology*, Vol. 36, pp. 288–293, March 1974.

86

Hackney, J. D., Thiede, F. C., Linn, W. S., Pedersen, E. E., Spier, C. E., Law, D. C., and Fischer, D. A., "Experimental Studies on Human Health Effects of Air Pollutants,

IV: Short-Term Physiological and Clinical Effects of Nitrogen Dioxide Exposure," *Archives of Environmental Health*, Vol. 33, pp. 176–181, July/August 1978.

87

Posin, C., Clark, K., Jones, M. P., Patterson, J. V., Buckley, R. D., and Hackney, J. D., "Nitrogen Dioxide Inhalation and Human Blood Biochemistry," *Archives of Environmental Health*, Vol. 33, pp. 318–324, November/December 1978.

88

Folinsbee, L. J., Horvath, S. M., Bedi, J. F., and Delehunt, J. C., "Effect of 0.62 ppm NO_2 on Cardiopulmonary Function in Young Male Nonsmokers," *Environmental Research*, Vol. 15, pp. 199–205, 1978.

89

Kerr, H. D., Kulle, T. J., McIlhany, M. L., and Swidersky, P., "Effects of Nitrogen Dioxide on Pulmonary Function in Human Subjects: An Environmental Chamber Study," *Environmental Research*, Vol. 19, pp. 392–404, 1979.

90

Nieding, G. von, and Wagner, H. M., "Effects of NO_2 on Chronic Bronchitics," *Environmental Health Perspectives*, Vol. 29, pp. 137–142, 1979.

91

Orehek, J., Massari, J. P., Gayrard, P., Grimand, C., and Charpin, J., "Effect of Short-Term, Low-Level Nitrogen Dioxide Exposure on Bronchial Sensitivity of Asthmatic Patients," *The Journal of Clinical Investigation*, Vol. 57, pp. 301–307, February 1976.

92

Ehrlich, R., and Henry, M. C., "Chronic Toxicity of Nitrogen Dioxide," *Archives of Environmental Health*, Vol. 17, pp. 860–865, December 1968.

93

Henry, M. C., Ehrlich, R., and Blair, W. H., "Effect of Nitrogen Dioxide on Resistance of Squirrel Monkeys to Klebsiella Pneumoniae Infection," *Archives of Environmental Health*, Vol. 18, pp. 580–587, April 1969.

94

Henry, M. C., Findlay, J., Spangler, J., and Ehrlich, R., "Chronic Toxicity of NO_2 in Squirrel Monkeys," *Archives of Environmental Health*, Vol. 20, pp. 566–570, May 1970.

95

Coffin, D. L., Gardner, D. E., and Blommer, E. J., "Time-Dose Response for Nitrogen Dioxide Exposure in an Infectivity Model System," *Environmental Health Perspectives*, Vol. 13, pp. 11–15, 1976.

96

Gardner, D. E., Miller, F. J., Blommer, E. J., and Coffin, D. L., "Influence of Exposure Mode on the Toxicity of NO_2," *Environmental Health Perspectives*, Vol. 30, pp. 23–29, 1979.

97

Gardner, D. E., Coffin, D. L., Pinigin, M. A., and Sidorenko, G. I., "Role of Time as a Factor in the Toxicity of Chemical Compounds in Intermittent and Continuous

Exposures, Part I. Effects of Continuous Exposure," *Journal of Toxicology and Environmental Health*, Vol. 3, pp. 811–820, 1977.

98

Coffin, D. L., Gardner, D. E., Sidorenko, G. I., and Pinigin, M. A., "Role of Time as a Factor in the Toxicity of Chemical Compounds in Intermittent and Continuous Exposures, Part II. Effects of Intermittent Exposure," *Journal of Toxicology and Environmental Health*, Vol. 3, pp. 821–828, 1977.

99

Goldstein, E., Eagle, M. C., and Hoeprich, P. D., "Effect of Nitrogen Dioxide on Pulmonary Bacterial Defense Mechanisms," *Archives of Environmental Health*, Vol. 26, pp. 202–204, April 1973.

100

Shy, C. M., and Love, G. J., "Recent Evidence on the Human Health Effects of Nitrogen Dioxide," paper prepared for presentation at the Symposium on Nitrogen Oxides, American Chemical Society Annual Meeting, Honolulu, Hawaii, 4–5 April 1979.

101

Shy, C. M., Creason, J. P., Pearlman, M. E., McClain, K. E., and Benson, F. B., "The Chattanooga School Children Study: Effects of Community Exposure to Nitrogen Dioxide, I: Methods, Description of Pollutant Exposure, and Results of Ventilatory Function Testing," *Journal of the Air Pollution Control Association*, Vol. 20, pp. 539–545, August 1970.

102

Shy, C. M., Creason, J. P., Pearlman, M. E., McClain, K. E., and Benson, F. B., "The Chattanooga School Children Study, II: Incidence of Acute Respiratory Illness," *Journal of the Air Pollution Control Association*, Vol. 20, pp. 582–588, September 1970.

103

Pearlman, M. E., Finklea, J. F., Creason, J. P., Shy, C. M., Young, M. M., and Horton, R. J. M., "Nitrogen Dioxide and Lower Respiratory Illness," *Pediatrics*, Vol. 47, pp. 391–398, February 1971.

104

Hauser, T. R., and Shy, C. M., "Position Paper: NO_x Measurement," *Environmental Science and Technology*, Vol. 6, pp. 890–894, October 1972.

105

Palmes, E. D., Tomczyk, C., and DiMattio, J., "Average NO_2 Concentrations in Dwellings with Gas or Electric Stoves," *Atmospheric Environment*, Vol. 11, pp. 869–872, 1977.

106

Goldstein, B. D., Melia, R. J. W., Chinn, S., Florey, C. du V., Clark, D., and John, H. H., "The Relation between Respiratory Illness in Primary School-Children and the Use of Gas for Cooking, II: Factors Affecting Nitrogen Dioxide Levels in the Home," *International Journal of Epidemiology*, Vol. 8, pp. 339–345, 1979.

107
Melia, R. J. W., Florey, C. du V., Altman, D. G., and Swan, A. V., "Association between Gas Cooking and Respiratory Disease in Children," *British Medical Journal*, Vol. 2, pp. 149–152, 1977.

108
Melia, R. J. W., Florey, C. du V., and Chinn, S., "The Relation between Respiratory Illness in Primary Schoolchildren and the Use of Gas for Cooking, I: Results from a National Study," *International Journal of Epidemiology*, Vol. 8, pp. 333–338, 1979.

109
Florey, C. du V., Melia, R. J. W., Chinn, S., Goldstein, B. D., Brooks, A. G. F., John, H. H., Craighead, I. B., and Webster, X., "The Relation between Respiratory Illness in Primary Schoolchildren and the Use of Gas for Cooking, III: Nitrogen Dioxide, Respiratory Illness and Lung Function," *International Journal of Epidemiology*, Vol. 8, pp. 347–353, 1979.

110
Speizer, F. E., Ferris, B., Jr., Bishop, Y. M. M., and Spengler, J., "Respiratory Disease Rates and Pulmonary Function in Children Associated with NO_2 Exposure," *American Review of Respiratory Disease*, Vol. 121, pp. 3–10, 1980.

111
Keller, M. D., Lanese, R. R., Mitchell, R. I., and Cote, R. W., "Respiratory Illness in Households Using Gas and Electricity for Cooking, I: Survey of Incidence," *Environmental Research*, Vol. 19, pp. 495–503, 1979.

112
Keller, M. D., Lanese, R. R., Mitchell, R. I., and Cote, R. W., "Respiratory Illness in Households Using Gas and Electricity for Cooking, II: Symptoms and Objective Findings," *Environmental Research*, Vol. 19, pp. 504–515, 1979.

113
Hosein, H. R., and Bouhuys, A., "Possible Environmental Hazards of Gas Cooking," *British Medical Journal*, Vol. 1, p. 125, January 1979.

114
Heck, W. W., and Brandt, C. S., "Effects on Vegetation: Native, Crops, Forests," in A. C. Stern, ed., *Air Pollution*, 3rd ed. Vol. II, Academic Press, New York, 1977.

115
US Environmental Protection Agency, *Air Quality Criteria for Ozone and Other Photochemical Oxidants*, Report No. EPA-600/8-78-004, Research Triangle Park, NC, April 1978. NTIS Publication No. PB80-124753.

116
US Environmental Protection Agency, *Air Quality Criteria for Nitrogen Oxides*, Report No. AP-84, Washington, DC, January 1971. US Government Printing Office Publication No. FS2,300: AP-84.

117
Adedipe, N. O., Barrett, R. E., and Ormrod, D. P., "Phytotoxicity and Growth Responses of Ornamental Bedding Plants to Ozone and Sulfur Dioxide," *Journal of the American Society for Horticultural Science*, Vol. 97, pp. 341–345, 1972.

118
Neely, G. E., Tingey, D. T., and Wilhour, R. G., "Effects of Ozone and Sulfur Dioxide Singly and in Combination on Yield, Quality and N-Fixation of Alfalfa," in B. Dimitriades, ed., *International Conference on Photochemical Oxidant Pollution and Its Control: Proceedings*, Vol. II, Report No. EPA 600/3-77-001b, US Environmental Protection Agency, Research Triangle Park, NC, January 1977, pp. 663–673. NTIS Publication No. PB 264233.

119
Ashenden, T. W., "The Effects of Long-Term Exposures to SO_2 and NO_2 Pollution on the Growth of *Dactylis Glomerata* L. and *Poa Pratensis* L.," *Environmental Pollution*, Vol. 18, pp. 249–258, 1979.

120
Taylor, O. C., and Eaton, F. M., "Suppression of Plant Growth by Nitrogen Dioxide," *Plant Physiology*, Vol. 41, pp. 132–135, 1966.

121
Thompson, C. R., Hensel, E. G., Katz, G., and Taylor, O. C., "Effects of Continuous Exposure of Navel Oranges to Nitrogen Dioxide," *Atmospheric Environment*, Vol. 4, pp. 349–355, 1970.

122
US Environmental Protection Agency, "Proposed Revisions to the National Ambient Air Quality Standard for Carbon Monoxide," *Federal Register*, 45 FR 55066, 15 August 1980.

123
Rubinfeld, D. L., "Market Approaches to the Measurement of the Benefits of Air Pollution Abatement," in A. F. Friedlaender, ed., *Approaches to Controlling Air Pollution*, The MIT Press, Cambridge, MA, 1978.

124
National Academy of Sciences, "The Costs and Benefits of Automobile Emission Control, " in *Air Quality and Automobile Emission Control*, Vol. 4, report by the Coordinating Committee on Air Quality Studies to the Committee on Public Works of the United States Senate, Serial No. 93–24, US Government Printing Office, September 1974.

125
Waddell, T. E., *The Economic Damages of Air Pollution*, Report No. EPA 600/5-74-012, US Environmental Protection Agency, Washington, DC, May 1974. NTIS Publication No. PB 235701.

126
Brookshire, D. S., d'Arge, R. C., Schulze, W. D., and Thayer, M. A., *Methods Development for Assessing Tradeoffs in Environmental Management*, Report No. EPA 600/6-79-001b, US Environmental Protection Agency, Washington, DC, February 1979. NTIS Publication No. PB 293616.

127
Harrison, D., Jr., and Rubinfeld, D. L., "Hedonic Housing Prices and the Demand

for Clean Air," *Journal of Environmental Economics and Management*, Vol. 5, pp. 81–102, 1978.

128

Nelson, J. P., *The Effects of Mobile-Source Air and Noise Pollution on Residential Property Values*, report prepared for the US Department of Transportation, Washington, DC, April 1975. NTIS Publication No. PB 241570.

5 *Motor Vehicle Emissions and Emissions Control Technology*

5.1 *Introduction*

This chapter is concerned with the relation between motor vehicle emissions and the design and operation of motor vehicles. The chapter discusses the causes of motor vehicle emissions, motor vehicle emissions control technology and its effects, the effects of vehicle maintenance on emissions, and the dependence of emissions rates on driving conditions. In addition, several means for encouraging vehicle owners to maintain emissions-related components of their vehicles properly are described. The discussion is oriented mainly toward conventional, gasoline-powered vehicles—particularly automobiles—as these are the predominant sources of motor vehicle emissions in most places, and controlling emissions from gasoline-powered vehicles is the principal means by which aggregate emissions from all motor vehicles can be reduced. The control of emissions from diesel vehicles is discussed briefly in section 5.4. However, detailed discussion of the design and operation of alternatives to the conventional, gasoline-powered engine and of variants of this engine (such as stratified charge engines) is beyond the scope of this book.

5.2 *The Formation of Emissions*[1]

Emissions from vehicles with conventional, gasoline-powered, internal combustion engines arise from three sources: the crankcase, the fuel system, and the exhaust. The crankcase and the fuel system are sources of hydrocarbons, whereas the exhaust contains hydrocarbons, carbon monoxide, and nitrogen oxides (mainly nitric oxide, NO).

To understand the formation and control of emissions, it is necessary to have a rudimentary understanding of the operation of internal combustion engines. In conventional, gasoline-powered engines, pistons move in and out of cylinders in repeating cycles. The motion of the pistons is transmitted through connecting rods to a crankshaft and, ultimately, to the wheels. In the most common form of gasoline-powered motor vehicle engine, including the automobile engine, each piston operates in a four-stroke cycle consisting of an intake stroke, a compression stroke, a power stroke, and an exhaust stroke. This cycle is illustrated in figure 5.1. In the

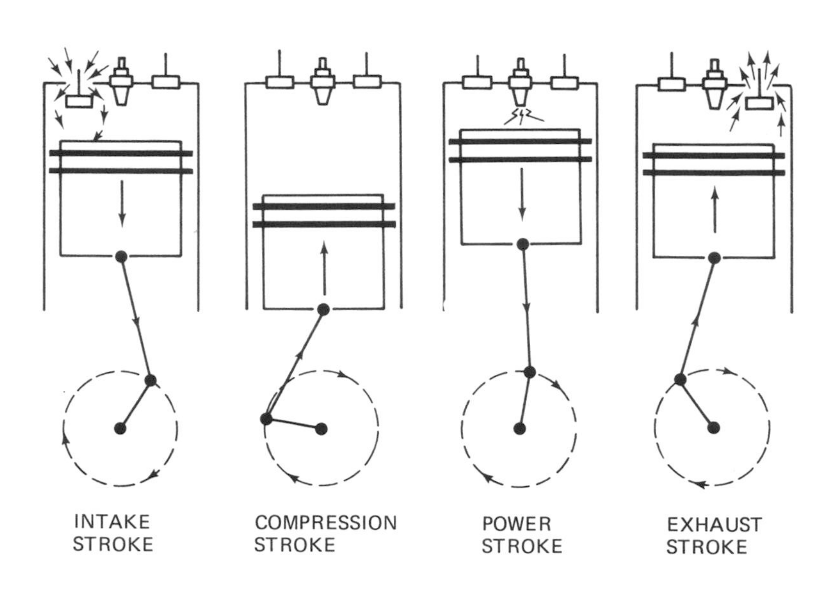

Figure 5.1 Four-stroke cycle of an internal combustion engine.

intake stroke, the piston moves downward in the cylinder. This causes a mixture of air and fuel to be drawn into the cylinder through the open intake valve. The air and fuel are mixed together in the carburetor, which controls the air-fuel ratio (the ratio of the mass of air in the mixture to the mass of fuel). At the bottom of the intake stroke, the intake valve closes, and the piston begins to move upward in the cylinder, compressing the mixture of air and fuel. This is the compression stroke. An electrical current causes the spark plug to create a spark in the cylinder near the top of the compression stroke. The spark ignites the compressed air-fuel mixture, and the expansion of the burning gases forces the piston to move downward, thereby delivering power to the crankshaft. This is the power stroke. At the end of the power stroke, the exhaust valve opens, and the piston begins to rise in the cylinder, forcing the combustion products out through the exhaust valve. Following this exhaust stroke, the four-stroke cycle repeats itself.

During the compression and power strokes, some of the gases in the cylinders escape past the pistons and into the crankcase. This escape of gases is the source of crankcase, or blowby, emissions. The crankcase is the space underneath the pistons that contains the connecting rods and the crankshaft, among other engine parts. The escaping gases consist mostly of unburned air-fuel mixture and, since gasoline is largely a mixture of hydrocarbons, have high HC concentrations. In vehicles without emissions controls, the crankcase is vented to the air, thus giving rise to crankcase HC emissions. These emissions account for roughly 20% of the HC emissions of uncontrolled automobiles.

Emissions from the fuel system are caused by evaporation of gasoline from the fuel tank and the carburetor. These emissions consist entirely of HC. The emissions from the fuel tank are caused by gradual heating of the tank that occurs on warm days and by spillage of vapors from the tank during refueling. The heating on warm days causes the fuel and fuel vapors in the tank to expand, and some of the vapors spill out of the tank and into the air. The resulting emissions, which sometimes are called "diurnal evaporations," are, at least to a first approximation, independent of any use that a vehicle receives during a day. In refueling, the entering fuel displaces the gasoline vapors that are in the tank and forces them into the air. These refueling losses are localized at filling stations and usually are considered to be associated with gasoline marketing operations, rather than with individual vehicles. Refueling emissions will not be discussed further in this book.

Emissions from the carburetor occur mainly while the engine is still hot after having been recently turned off at the end of a trip. The fuel left in the carburetor at this time is hot, and its more volatile constituents evaporate rapidly. The resulting emissions often are called "hot soak" emissions. Although fuel also evaporates from the carburetor when the engine is operating, the resulting vapors are drawn into the cylinders and do not reach the outside air. Accordingly, carburetor emissions tend to be associated primarily with the terminations of trips and to be independent of trip lengths.

Evaporations from the fuel system (excluding refueling losses) account for roughly 20% of HC emissions from automobiles without emissions controls.

Exhaust emissions account for all CO and NO_x emissions and, in automobiles without emissions controls, about 60% of HC emissions. The organic constituents of the exhaust include aldehydes and traces of alcohols and other products of partial oxidation of hydrocarbons, in addition to true hydrocarbons. If the fuel supplied to the cylinders burned completely, then it would be oxidized to carbon dioxide (CO_2) and water, and there would be no exhaust HC or CO emissions. However, as described below, a variety of conditions prevent this from happening in internal combustion engines.

In properly functioning vehicles, exhaust HC emissions are caused mainly by the inability of the flames in the cylinders to propagate in the vicinities of the cylinder walls. This effect, which is known as "wall quenching," is caused partly by chemical reactions that occur in the layers of air-fuel mixture that are adjacent to the walls and partly by cooling of these layers by the walls. As a result of wall quenching, the air-fuel mixture near the walls does not burn, and some of the hydrocarbons contained in this mixture enter the exhaust stream. (The rest of the hydrocarbons stay in the cylinders.) HC also can enter the exhaust stream if the part of the air-fuel mixture that is not near the cylinder walls fails to burn completely. This can occur during transient conditions, such as warm-up, when the fuel entering the cylinders may be inadequately atomized and mixed with air, or during idle or deceleration, when the cylinders may contain excessive quantities of residual exhaust gases from the previous piston cycles. Incomplete combustion also can be caused by engine malfunctions or maladjustments. For example, if an ignition system malfunction prevents one or more spark plugs from producing a spark at the proper time or if a carburetor malfunction causes the air-fuel ratio to be too high or too low, the combustion will be incomplete and HC emissions will be excessive.

CO is formed when carbon-containing substances, such as gasoline, are burned with an inadequate supply of oxygen. In an internal combustion engine, low air-fuel ratios tend to produce high CO emissions, whereas CO emissions are lower at high air-fuel ratios.

NO_x is formed by oxidation of atmospheric nitrogen inside the cylinders. Formation of NO_x is promoted by high combustion temperatures and a good supply of oxygen.

Exhaust emissions rates are sensitive to a variety of engine adjustments and design parameters, including air-fuel ratio, spark timing, valve overlap, cylinder surface-to-volume ratio, and compression ratio, among others. The following is a brief explanation of the influence of each of these factors on emissions. Of course, many other factors also influence emissions. Some of these additional factors, such as driving conditions and the installation of special emissions control devices, are discussed later in this chapter.

Air-Fuel Ratio

The effect of the air-fuel ratio on exhaust emissions is illustrated in figure 5.2. At a low air-fuel ratio, there is insufficient oxygen to enable complete oxidation of the fuel to occur. Consequently, HC and CO emissions are both relatively high, and NO_x emissions are relatively low. The air-fuel

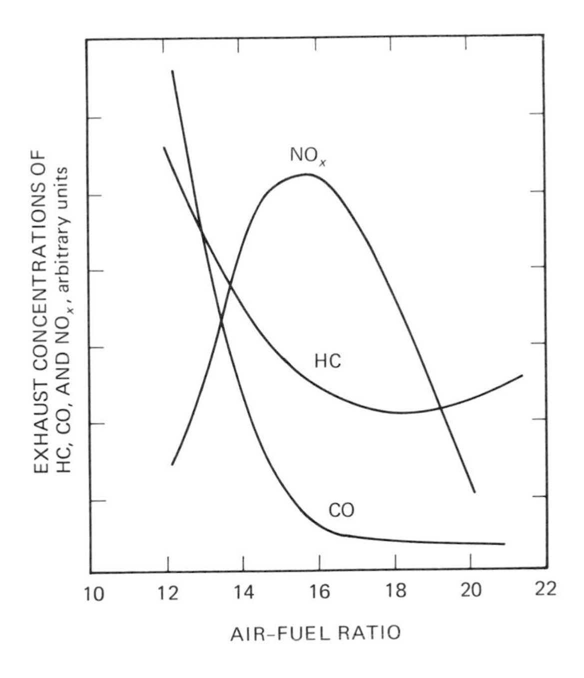

Figure 5.2 Dependence of exhaust concentrations of HC, CO, and NO_x on the air-fuel ratio. (Source: reference [3])

ratio that would provide just enough oxygen to enable complete oxidation to occur under ideal combustion conditions is called the "stoichiometric ratio." This ratio varies somewhat, depending on the fuel composition, but is usually in the vicinity of 15. Since combustion conditions in an internal combustion engine are not ideal, complete oxidation does not occur in an engine when the mixture is stoichiometric. However, as the air-fuel ratio increases from a low value, the increased supply of oxygen causes better oxidation of the fuel to occur, and emissions of HC and CO decrease. The increased supply of oxygen also causes NO_x emissions to increase until the air-fuel ratio reaches a value of roughly 16. Increasing the air-fuel ratio beyond 16 causes NO_x emissions to decrease because the combustion temperature begins to decrease as the mixture becomes leaner. At air-fuel ratios above roughly 18, the mixture becomes too lean to burn well, and HC emissions tend to increase due to misfires.

Spark Timing

To achieve maximum power output and fuel economy, it is desirable for the spark that ignites the air-fuel mixture in a cylinder to be timed to occur before the piston reaches the top of its compression stroke. This enables combustion to develop in the cylinder before the power stroke begins and maximizes the pressure that can be applied to the piston during the power stroke. The optimal degree of spark advance depends on engine operating conditions, such as engine speed and intake manifold pressure, but can be as early as 30° of crankshaft rotation before the top of the compression stroke (30° before top dead center). However, by retarding the spark relative to the timing needed for optimal fuel economy, it is possible to reduce HC and NO_x emissions (although at the cost of reduced fuel economy). This is because retarding the spark tends to reduce the peak combustion temperature and to increase the exhaust temperature. The reduced combustion temperature reduces NO_x formation, whereas the increased exhaust temperature helps to cause HC oxidation in the exhaust stream. In addition, retarding the spark causes the combustion period to extend further into the power stroke than it otherwise would, thereby reducing the cylinder's surface-to-volume ratio during combustion. As will be discussed shortly, reducing the surface-to-volume ratio reduces the wall quenching effect and, therefore, HC emissions. Spark timing has little effect on CO emissions.

Valve Overlap

This refers to a condition in which the intake and exhaust valves of a cylinder are partially open simultaneously. Valve overlap typically occurs shortly before the end of the exhaust stroke and shortly after the beginning of the intake stroke because the valves cannot be opened and closed instantaneously, and engine efficiency would be lost if the exhaust valve began to close sufficiently early in the exhaust stroke and the intake began to open sufficiently late to prevent overlap. As a result of valve overlap, some of the exhaust gases are drawn back (reinducted) into the cylinder during the intake stroke. Depending on the air-fuel ratio, increasing the degree of valve overlap can reduce HC and NO_x emissions. This is because the reinducted fraction of the exhaust usually contains much HC, and the reinduction enables this HC to be burned in the subsequent cycle. Moreover, the inert constituents of the reinducted exhaust gases dilute the air-fuel mixture and reduce its peak combustion temperature, which reduces NO_x emissions. However, if the valve overlap and resulting dilution are too great, then combustion is inhibited, and HC emissions increase. The degree of valve overlap that minimizes HC emissions tends to decrease as the air-fuel ratio increases. Valve overlap has little effect on CO emissions if the air-fuel ratio remains constant as the valve overlap changes.

Surface-to-Volume Ratio

The proportion of air-fuel mixture that is adjacent to cylinder walls decreases as the cylinder surface-to-volume ratio decreases. Therefore, because of the importance of wall quenching in the formation of HC emissions, cylinders that are designed to have relatively low surface-to-volume ratios tend to produce lower HC emissions than cylinders that are designed to have relatively high surface-to-volume ratios.

Compression Ratio

Let V_b denote the volume of the space between the piston and the top of the cylinder at the beginning of the compression stroke, and let V_a denote the volume of the same space at the end of the compression stroke. Then the compression ratio is defined to be V_b/V_a. Reducing the compression ratio tends to reduce the surface-to-volume ratio and the peak combustion temperature, whereas it tends to increase the exhaust temperature. Therefore, reducing the compression ratio tends to reduce HC and NO_x emissions. However, it also tends to reduce the thermal efficiency of combustion, thereby reducing fuel economy.

5.3 Emissions Measurements for Emissions Standards

The establishment of legally enforceable motor vehicle emissions standards has been a major motivating factor in the development of motor vehicle emissions control technology. To be meaningful, such standards must be accompanied by reproducible procedures for measuring motor vehicle emissions. It will be seen in section 5.6 that a vehicle's HC, CO, and NO_x emissions rates are highly sensitive to driving conditions, including the sequences of driving modes (accelerations, decelerations, cruises, and idles) through which the vehicle is driven and, if the engine is started cold, the air temperature. Therefore, to establish motor vehicle emissions standards and to determine whether vehicles are in compliance with these standards, it is necessary to define standard sets of driving conditions.

The driving conditions and emissions measurement procedures that are used in connection with the motor vehicle emissions standards that the US federal government has established are known as federal test procedures (FTPs). The FTP driving conditions for automobiles, light trucks (under 8,500 lb gross weight), and motorcycles whose engine displacements are at least 170 cc have evolved over time in an effort to develop a realistic representation of a typical set of urban driving conditions. In the FTP for these vehicles that was in use in 1980 (known as the 1975 FTP because it was first used to define emissions standards for 1975 model year automobiles), a vehicle's exhaust emissions are measured while it is being driven through a prescribed sequence of driving modes (called a driving cycle) on a chassis dynamometer. A chassis dynamometer is a device with large rollers on which the drive wheels of a vehicle are placed. When the vehicle's engine causes the drive wheels to rotate, the rollers of the dynamometer also rotate. The rollers are connected to flywheels and to power absorption devices that enable the dynamometer to simulate the load that the weight of the vehicle and frictional resistance to the vehicle's movement place on the engine in actual driving. The FTP driving cycle is based on data obtained from driving instrumented vehicles over a route in Los Angeles (the LA-4 route) and simulates driving conditions on this route. These conditions include both freeway and surface street driving, as well as periods of idle. The simulated length of the cycle is 7.5 mi. The cycle lasts for approximately 23 min and has an average speed of roughly 19.6 mi/hr.

Before being driven on the dynamometer in an exhaust emissions test, an automobile, light truck, or motorcycle with a displacement of at least 170 cc stands with its engine off for a period of at least 12 hr (8 hr for certain types of motorcycles) at a temperature between 20° and 30°C (68–86°F). This ensures that the engine is fully cooled before the emissions test begins. The vehicle is then placed on the dynamometer without starting its engine, and the instruments for measuring exhaust emissions are connected. The engine is started (a cold start), and the vehicle is driven through the driving cycle while on the dynamometer. At the end of the driving cycle, the engine is turned off, and the vehicle stands for 10 min. Then the engine is started again (a hot start), and the first 505 sec of the driving cycle are repeated. This completes the exhaust emissions test. Measurements are made of the masses of HC, CO, and NO_x emitted in the exhaust during the first 505 sec of the driving cycle following the cold start (the cold transient period), the remainder of the 23-min cycle before the 10-min engine-off period (the hot stabilized part of the cycle), and the first 505 sec after the hot start (the hot transient period). Total HC, CO, and NO_x emissions for the test are expressed as weighted averages of these measurements. The weights are 0.43, 0.57, and 1.0 for the cold transient, hot transient, and hot stabilized periods, respectively. Thus, for each pollutant average exhaust emissions per mile, E, during the dynamometer test are given by

$$E = (0.43\text{CT} + \text{HS} + 0.57\text{HT})/7.5, \tag{5.1}$$

where CT, HS, and HT are the masses of HC, CO, or NO_x emitted during the cold transient, hot stabilized, and hot transient periods, respectively.

The effect of this weighting procedure is to cause the emissions test to simulate a driving pattern in which 43% of trips begin with cold starts and 57% of trips begin with hot starts.

Although the FTP driving cycle provides a good representation of driving conditions along a particular route in Los Angeles, a single driving cycle clearly cannot represent all urban driving conditions. Emissions rates during driving conditions that differ from those represented in the FTP can be estimated from data obtained from vehicles that are operated on dynamometers under non-FTP conditions. Procedures for carrying out this estimation are described in section 5.7.

The FTP exhaust emissions test for motorcycles with 50–170 cc displacement is like that for cars, light trucks, and larger motorcycles, except that a different driving cycle is used and vehicles must stand with their engines off for only 6 hr prior to testing.[2] The driving cycle for small motorcycles simulates a 6.8-mi trip. The cycle lasts for approximately 23 min and has an average speed of roughly 18 mi/hr. This cycle does not include the high-speed freeway driving that is present in the cycle that is used for automobiles, light trucks, and larger motorcycles.

Because it is difficult to operate heavy-duty vehicles (vehicles whose gross weights exceed 8,500 lb) on chassis dynamometers, exhaust emissions standards for these vehicles apply to the engines, rather than to the entire vehicles. The emissions are measured with the engines mounted on engine dynamometers, rather than installed in vehicles. Engine dynamometers are devices that enable engines to be operated under load. The FTP exhaust emissions test for a heavy-duty vehicle engine consists of measuring exhaust emissions while the engine is operated through a sequence of driving modes on an engine dynamometer. The test conditions do not necessarily represent conditions encountered in real driving, and the emissions rates are expressed in mass per unit of energy output by the engine (specifically, mass per brake horsepower hour), rather than in units of mass per mile traveled.

In addition to exhaust emissions testing, vehicles for which there are evaporative emissions standards are tested for evaporative emissions. Hot soak evaporations are measured as the mass of HC emitted during the hour following the end of the exhaust emissions test. The engine is hot but not running during this test. The diurnal evaporative emissions test consists of measuring the mass of HC emitted during an hour-long period in which the temperature of the fuel in the fuel tank is raised by approximately 13°C (24°F). The engine is cold during this test. The evaporative emissions standards apply to the sum of hot soak and diurnal evaporations.

5.4 *Emissions Standards, Emissions Control Devices, and Their Effects*

New motor vehicles sold in the United States have been subject to increasingly stringent emissions control requirements since the early 1960s. In 1961, as a result of action taken by the state of California, crankcase emissions control devices began to be installed on new automobiles sold in that state. Crankcase emissions controls were installed on new automobiles sold nationwide beginning with the 1963 model year. Exhaust emissions standards for automobiles and for trucks under 6,000 lb gross weight first came into effect in the 1966 model year in California and in the 1968 model year nationwide. Evaporative emissions standards for automobiles and for trucks under 6,000 lb gross weight were first implemented in the 1971 model year. During the late 1960s and the 1970s, emissions standards also were established for heavier trucks and for motorcycles. The gradually increasing stringency of motor vehicle emissions standards is illustrated in table 5.1, which shows the evolution of exhaust emissions standards for automobiles. Because the numerical values of emissions standards depend on their associated test procedures and these procedures have changed considerably since 1966, all of the tabulated standards have been converted to equivalent 1975 FTP values to establish comparability across years. Emissions standards applicable to vehicles other than automobiles and to crankcase and evaporative emissions are published in references [5, 6].

The various standards are enforced by the US Environmental Protection Agency (EPA) through tests of the emissions of preproduction prototype vehicles and engines and through exhaust emissions tests that are performed on samples of production vehicles taken from the ends of assembly lines. The latter procedure is called "assembly line testing" or, in legal terms, "selective enforcement auditing" (SEA). There also are legal and administrative procedures for carrying out manufacturers' recalls if large numbers of vehicles in consumer use are found to have excessive emissions due to manufacturing defects.

Motor vehicle manufacturers have achieved compliance with the new-motor-vehicle emissions standards by modifying the design and adjustment of new vehicles and by installing special emissions control devices on these vehicles. The earliest motor vehicle emissions controls were for crankcase emissions. Crankcase emissions from gasoline-powered engines

Table 5.1 Exhaust emissions standards for automobiles[a]

Model year	Region	HC	CO	NO_x
Pre-1966	All		No standards	
1966–1967	Nationwide		No standards	
	California	5.9	51	No standard
1968–1969	Nationwide	5.9	51	No standard
	California	5.9	51	No standard
1970	Nationwide	3.9	33	No standard
	California	3.9	33	No standard
1971	Nationwide	3.9	33	No standard
	California	3.9	33	6.2
1972	Nationwide	3.0	28	No standard
	California	2.8	28	4.6
1973	Nationwide	3.0	28	3.1
	California	2.8	28	3.1
1974	Nationwide	3.0	28	3.1
	California	2.8	28	2.1
1975–1976	Nationwide	1.5	15	3.1
	California	0.9	9.0	2.0
1977–1979	Nationwide	1.5	15	2.0
	California	0.41	9.0	1.5
1980	Nationwide	0.41	7.0	2.0
	California	0.41	9.0	1.0
1981	Nationwide	0.41	3.4[b]	1.0[c]
	California Option A[d]	0.41	3.4	1.0
	California Option B[d]	0.41	7.0	0.7
1982	Nationwide	0.41	3.4[b]	1.0[c]
	California Option A[d]	0.41	7.0	0.4
	California Option B[d]	0.41	7.0	0.7
1983 and later	Nationwide	0.41	3.4	1.0
	California	0.41	7.0	0.4

a. Standards are specified as of December 1980 and are expressed in units of g/mi. Standards for pre-1975 vehicles have been converted to the 1975 FTP using the conversion factors of reference [4]. Nationwide standards are not applicable in California.
b. For certain classes of vehicles the nationwide CO standard for model years 1981 and 1982 is 7.0 g/mi.
c. During model years 1981 and 1982 the nationwide NO_x standard for vehicles manufactured by American Motors Corporation is 2.0 g/mi. During the 1981 and 1982 model years the nationwide NO_x standard for diesel vehicles is 1.5 g/mi.
d. Manufacturers may choose between the California options for model years 1981 and 1982. The same option (A or B) must be chosen in both years.

can be eliminated by closing the crankcase vent to the air and recirculating the blowby gases to the intake manifold. The system that does this is called "positive crankcase ventilation" (PCV). Emissions control regulations in the United States prohibit crankcase emissions from all new gasoline-powered automobiles, trucks, and motorcycles with at least 50 cc displacement sold in this country. Crankcase emissions from diesel engines are minimal, even without emissions controls, and are not regulated.

Evaporative emissions from gasoline-powered vehicles can be controlled by directing fuel vapors from the carburetor and fuel tank to the crankcase or a special cannister, where the vapors are stored until they can be returned to the engine intake and burned. By the mid-1980s evaporative emissions standards will be in effect for all new gasoline-powered vehicles sold in the United States except motorcycles if current law remains unchanged. Evaporative emissions tests of automobiles in consumer use have indicated that diurnal and hot soak evaporative emissions as measured by the FTP are both approximately 80% less in model year 1978 vehicles than in uncontrolled vehicles [7, 8].[3] Although comparable data are not yet available for model year 1980 and later automobiles, the evaporative emissions standards applicable to these vehicles require their combined diurnal and hot soak evaporations to be reduced by approximately 95% relative to evaporations from uncontrolled vehicles. Diesel fuel is not highly volatile, and evaporative emissions from diesels are low. Diesel evaporations are not regulated.

Exhaust emissions controls for automobiles through the 1974 model year consisted of modifications of engine design and operating parameters. The modifications included increased air-fuel ratios, retarded spark timing, reduced compression ratios, reduced surface-to-volume ratios of cylinders, and injection of air into the exhaust manifold to promote oxidation of HC and CO in the exhaust gases. In addition, since the 1973 model year, exhaust gas recirculation (EGR) has been used to reduce NO_x emissions. EGR consists of recirculating a portion of the exhaust gases to the intake manifold. This dilutes the air-fuel mixture without adding oxygen to it and reduces the peak combustion temperature. The emissions controls that were installed on new model year 1974 automobiles reduced their exhaust emissions rates by roughly 50% (HC and CO) and 20% (NO_x) as measured by the current FTP, relative to the emissions rates of pre-controlled new automobiles (i.e., new automobiles from model years before there were exhaust emissions standards) [5].

The emissions standards that took effect in the 1975 model year required exhaust emissions of HC and CO from new automobiles to be reduced by roughly 50% relative to emissions from new 1974 model year vehicles. To achieve the required emissions reductions, catalytic converters were installed in the exhaust streams of approximately 85% of the model year 1975 automobiles sold in the United States [9]. A catalytic converter consists of a mixture of noble metals, such as platinum and palladium, that is supported by a ceramic base. When heated to a sufficiently high temperature, the metals catalyze the oxidation of HC and CO to CO_2 and water. Lead additives in gasoline destroy the activity of noble metal catalysts, and vehicles equipped with these devices must be operated on lead-free fuel.

By carefully controlling the oxygen content of the exhaust stream, it is possible to cause a catalytic converter to catalyze the reduction of NO_x to nitrogen as well as to oxidize HC and CO. Catalytic converters that operate in this way are called "three-way catalysts" and are being used to meet the exhaust emissions standards applicable to model year 1981 (1980 in California) and later automobiles.

Emissions control technology for motorcycles and gasoline-powered trucks is similar to that for automobiles. Diesel engines operate at very high air-fuel ratios (in the vicinity of 30) and, therefore, tend to have low HC and CO emissions. Most diesels do not need emissions control equipment to achieve compliance with applicable HC and CO exhaust emissions standards. However, diesels will need NO_x controls to comply with the NO_x emissions standards that are scheduled to take effect in the mid-1980s. NO_x control techniques for diesels include retarding the timing of the injection of fuel into the cylinders, thereby reducing the combustion temperature, and EGR. Diesel engines tend to have considerably higher particulate matter emissions than do gasoline-powered engines. The EPA has established particulate matter or smoke emissions standards for new diesel vehicles and engines sold in the United States.

The most stringent of the currently scheduled motor vehicle emissions standards took effect on many classes of new vehicles beginning with the 1981 model year and will take effect on the remaining classes during the mid-1980s. These standards require exhaust HC and CO emissions and evaporative HC emissions from most new vehicles sold in the United States to be reduced by roughly 90% (and in some cases, such as evaporative emissions from automobiles, by more than 90%) as measured by the relevant FTPs, relative to emissions from precontrolled new vehicles.[4] NO_x emissions reductions of roughly 70% are required.

Emissions from vehicles in actual use do not decrease as rapidly or as much as emissions from new vehicles do. There are several reasons for this. First, the emissions performance of new vehicles affects average emissions rates of in-use vehicles mainly through replacement of older vehicles by new ones. Depending on the vehicle type (e.g., automobile, diesel truck), it takes roughly 7–14 yr for 90% of the vehicles in use at a given initial time to be replaced [5]. Consequently, reductions in average emissions rates of in-use vehicles lag behind reductions in emissions rates of new vehicles. Second, motor vehicles' emissions rates tend to increase as the vehicles age. This phenomenon, which is called deterioration, occurs partly because of normal wear and tear on engine parts and emissions control equipment, and partly because of inadequate or improper maintenance. Deterioration causes the average emissions rates of vehicles in use to be higher than those of new vehicles, even after the effects of gradual introduction of new vehicles into the in-use fleet are taken into account. Finally, the vehicle population and vehicle usage are growing. This growth causes reductions in aggregate emissions from all vehicles to be less than reductions in average emissions per vehicle.

The effects of new-vehicle emissions standards, fleet turnover, deterioration, and growth in vehicle use on average emissions per vehicle mile traveled (VMT) and on aggregate emissions from all vehicles are illustrated in figure 5.3. This figure shows estimates of changes in average emissions per VMT (as measured by the FTP) and aggregate emissions from all vehicles between 1970 and 1999. A VMT growth rate of 3%/yr has been used in constructing the estimates of aggregate emissions.[5] Average emissions per VMT reach a minimum in the mid-1990s, approximately 10 yr after the most stringent emissions standards for new vehicles go into effect. Average emissions per VMT are then approximately 80% (HC), 75% (CO), and 60% (NO_x) less than 1970 levels. Aggregate emissions reach a minimum in the early 1990s but start to increase before 1995, due to the effects of growth. The minimum levels of aggregate emissions are approximately 60% (HC), 50% (CO), and 20% (NO_x) below 1970 levels.

Emissions control requirements have added roughly $200 to the price of a new 1980 model year automobile and $550 to the price of a 1981 model year automobile, relative to estimates of the prices of hypothetical equivalent vehicles without emissions controls [9, 10]. In addition, emissions control requirements may increase the lifetime maintenance costs of model year 1980 and 1981 automobiles by $20–100. All costs are in 1980 dollars.

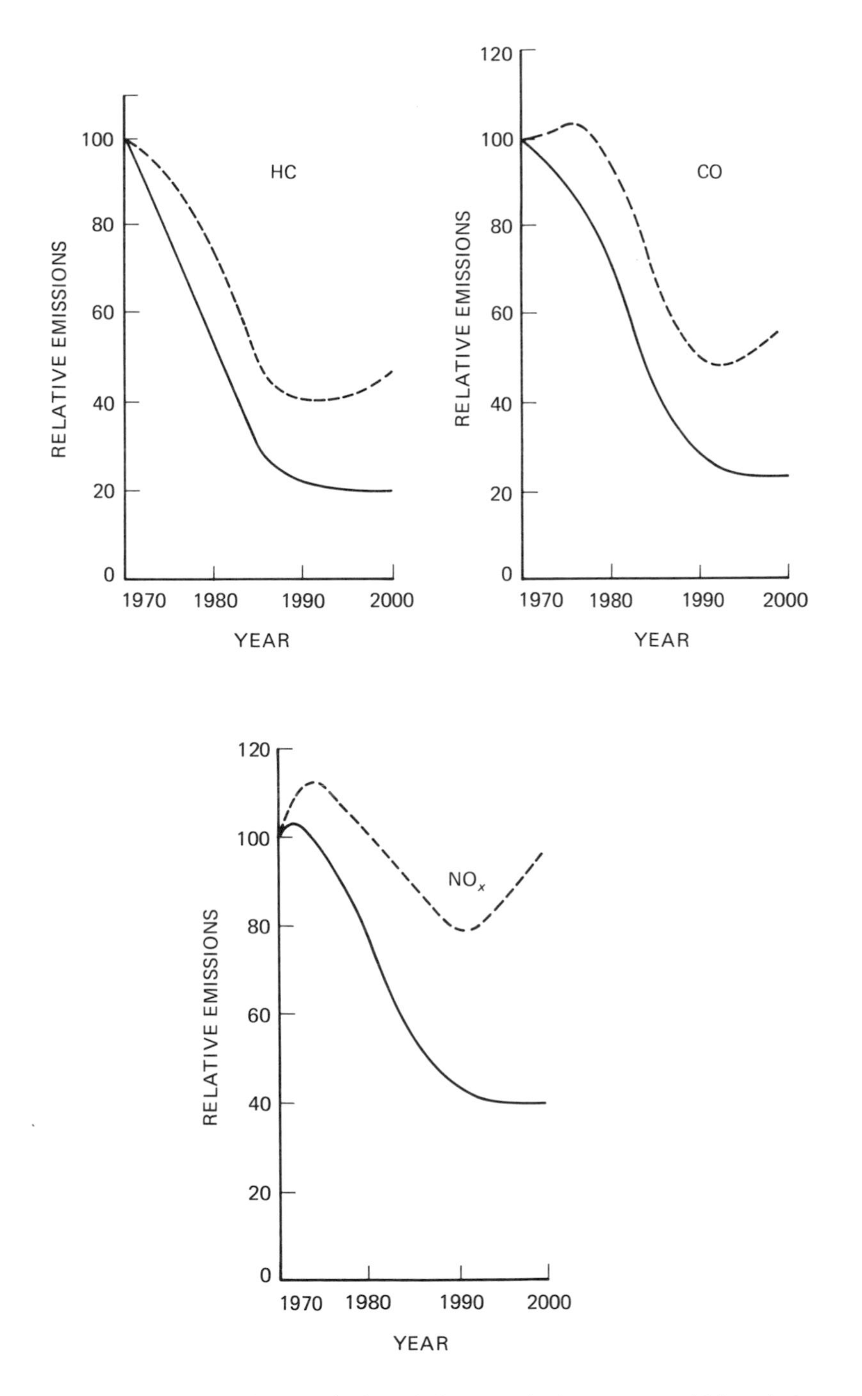

Figure 5.3 Effects of new-vehicle emissions standards on motor vehicle emissions. Key: ———, relative emissions per VMT (1970 = 100); ---, relative aggregate emissions from all vehicles if VMT increase at a rate of 3%/yr (1970 = 100). (Based on emissions estimates from reference [5])

Table 5.2 Effects of gasohol use on emissions and fuel economy[a]

Parameter	Percentage change 100(gasohol − gasoline)/gasoline
CO	−20 to −50
Exhaust HC	−26 to +31
NO_x	−6 to +18
Diurnal HC	+56 to +890
Hot soak HC	+16 to +71
Fuel economy	0 to −7

a. Source: references [15–20]. Emissions were measured by the 1975 FTP and fuel economy by the EPA method.

Emissions controls installed on 1973 and 1974 model year automobiles caused a reduction of approximately 1% in the weight- and mileage-adjusted average fuel economy of those vehicles, as measured by the EPA method, relative to the average fuel economy of precontrolled automobiles [11]. This estimate includes both city and highway driving conditions. The fuel economy loss for city driving conditions alone was approximately 4%, whereas there was a 4% improvement in highway fuel economy [11]. Weight and mileage adjustments are needed to compensate for the effects of vehicles' weights and ages on fuel economy. Several earlier analyses that did not include mileage adjustment estimated the emissions control-related fuel economy loss in 1973 and 1974 model year automobiles to be approximately 10% in city driving, 0% in highway driving, and 5% in combined city and highway driving [12, 13]. These estimates are superseded by the mileage-adjusted estimates. The installation of catalytic converters on 1975 model year and later automobiles made it possible to tune vehicles for improved fuel economy while still complying with emissions control requirements. The weight- and mileage-adjusted average fuel economy of model year 1979 automobiles was 22% higher than that of precontrolled automobiles in both city and highway driving [11].[6]

In recent years, interest has developed in the possibility of using gasohol, a blend of gasoline and any of several alcohols, as a motor vehicle fuel so as to reduce petroleum consumption. Several studies of the effects of gasohol fuels on motor vehicle emissions and fuel economy have been conducted [15–20]. These studies used small groups of vehicles and a variety of different gasohol fuels. Although there is considerable variation among the results obtained from different studies and with different fuels, these results do give some indication of the potential effects of gasohol use on emissions and fuel economy. The results of the various studies are summarized in table 5.2. The tabulated results suggest that gasohol use causes a substantial reduction in CO emissions and a substantial increase in evaporative HC emissions. The effects of gasohol use on exhaust HC emissions and on NO_x emissions are ambiguous, ranging from modest decreases to relatively large increases, depending on the study and the fuel. The changes in fuel economy resulting from gasohol use range from none to small decreases.

5.5 *Automobile Inspection and Maintenance*

Many automobiles in consumer use do not comply with the exhaust emissions standards that they were designed to meet. The failure of automobiles in use to meet the applicable standards is illustrated in tables 5.3 and 5.4, which summarize the results of exhaust emissions tests that were performed in 1978 on a sample of 1972–1978 model year vehicles. Within each model year, the tested vehicles had average exhaust emissions of HC and CO that exceeded the applicable emissions standards. In some cases, average HC or CO emissions were more than a factor of two over the standards. Roughly half of the 1975–1978 model year vehicles that were tested failed to comply with the applicable HC standard, and roughly half failed to comply with the applicable CO standard. Nearly 70% of the 1975–1978 model year vehicles that were tested failed to comply with at least one of the applicable exhaust emissions standards for HC, CO, and NO_x.

There are at least three possible explanations for vehicles' failures to comply with exhaust emissions standards: the vehicles may be improperly designed or assembled and, therefore, incapable of meeting the standards; the emissions-related components of vehicles may not be maintained properly by vehicle owners; and vehicle owners may deliberately disable or otherwise tamper with emissions-related components. There is evidence that improper maintenance and tampering are the predominant causes of poor emissions performance of in-use automobiles. Table 5.5 shows the results of tests of emissions and engine adjustments that were performed on 300 1975 and 1976 model year automobiles as part of a study of the effects of maintenance on emissions and fuel economy [21]. Among the vehicles that were tested, 72% had one or more engine maladjustments, and 67% of the maladjusted vehicles failed to comply with one or more of the applicable exhaust emissions standards. Only 30% of the properly adjusted vehicles failed to comply with the emissions standards. Vehicles

Table 5.3 Average exhaust emissions of a sample of automobiles in 1979[a]

Model year	Number of vehicles tested	Average odometer reading (mi)	HC emissions		CO emissions		NO_x emissions	
			Average	Standard	Average	Standard	Average	Standard
1972	36	75,981	5.81	3.0	65.52	28	3.34	None
1973	40	57,200	6.05	3.0	54.89	28	2.88	3.1
1974	32	50,840	3.92	3.0	52.71	28	3.05	3.1
1975	32	42,325	2.13	1.5	33.05	15	2.47	3.1
1976	75	35,020	2.00	1.5	22.40	15	2.31	3.1
1977	150	24,246	1.79	1.5	24.21	15	1.82	2.0
1978	24	10,847	1.53	1.5	22.54	15	1.48	2.0

a. Source: reference [8]. Emissions and standards are in units of g/mi as measured by the 1975 FTP.

Table 5.4 Numbers and percentages of automobiles failing to meet exhaust emissions standards applicable to their respective model years[a]

Model year	Number of vehicles tested	HC failures		CO failures		NO_x failures		Failure to meet one or more standards	
		Number	Percent	Number	Percent	Number	Percent	Number	Percent
1975	32	22	69	24	75	8	25	28	88
1976	75	37	49	33	44	11	15	43	57
1977	150	61	41	80	53	41	27	106	71
1978	24	9	38	14	58	1	4	16	67
All	281	129	46	151	54	61	22	193	69

a. Source: reference [8]. Based on tests conducted in 1978.

Table 5.5 Relation between engine maladjustments and failures of automobiles to comply with exhaust emissions standards[a]

Engine adjustment	Number of maladjusted vehicles	Number of maladjusted vehicles failing to comply with one or more emissions standards	Number of properly adjusted vehicles	Number of properly adjusted vehicles failing to comply with one or more emissions standards	Percentage of emissions standards failures attributable to maladjusted vehicles
Spark timing	105	72	195	103	41
Idle rpm	106	67	194	108	38
Idle CO[b]	118	106	182	69	61
At least one of the above	217	145	83[c]	30	83

a. Source: reference [21]. Based on tests of 1975 and 1976 model year vehicles.
b. The concentration of CO in the exhaust at idle is an indicator of the idle air-fuel ratio.
c. Total number of properly adjusted vehicles.

with at least one engine maladjustment accounted for 83% of the failures to comply with the exhaust emissions standards.

In a separate survey that was carried out by the EPA in 1978, it was found that emissions control devices had been tampered with in 19% of 1,953 in-use vehicles from model years 1973–1978 that were inspected [22]. The components most frequently tampered with were EGR systems and spark timing.

To evaluate the effects of maintenance on emissions and fuel economy, the 300 vehicles whose emissions and mechanical conditions are summarized in table 5.5 were subjected to the following sequence of maintenance events and emissions tests. After the initial emissions testing that led to table 5.5, all maladjustments except the exhaust CO concentration at idle (an indicator of the idle air-fuel ratio) and idle rpm were corrected, and the emissions testing was repeated. Vehicles that did not comply with the emissions standards on the second emissions test and whose idle CO or idle rpm were incorrect had their idle adjustments corrected and were retested for compliance with the emissions standards. Vehicles that failed to comply with the emissions standards on the third test or that failed on the second test but had correct idle adjustments received major tune-ups, including repair of defective emissions control devices, and then had their emissions measured once again. Fuel economy testing on the urban (LA-4) driving cycle was included with each emissions test. The results of the emissions and fuel economy tests are shown in table 5.6. The tabulated results at each stage of maintenance pertain to all 300 vehicles, regardless of the number of vehicles that actually received maintenance at any given stage. Maintenance reduced the proportion of vehicles failing to comply with at least one exhaust emissions standard from 58 to 19%. In addition, average HC, CO, and NO_x emissions from all vehicles were reduced by 34, 62, and 10%, respectively. Average fuel economy in urban driving improved by 2%.

Because of the sensitivity of a vehicle's exhaust emissions to its mechanical condition, several regulatory methods for encouraging vehicle owners to maintain their vehicles in good condition have been proposed. These methods include periodic mandatory tune-ups, periodic mandatory inspection of engine components followed by readjustments or repairs as found to be necessary, and periodic mandatory emissions testing followed by maintenance of vehicles whose emissions are found to be excessive. Emissions testing followed, as necessary, by maintenance is at least as effective as and less costly than the other methods [23–25]. Accordingly, this method, which is known as "inspection and maintenance" (I/M), is

Table 5.6 Cumulative effects of maintenance on exhaust emissions and fuel economy of automobiles[a]

Maintenance performed	Average exhaust emissions (g/mi)			Number of vehicles failing to comply with one or more exhaust emissions standards	Average urban fuel economy (mi/gal)
	HC	CO	NO_x		
None	1.32	20.27	2.82	175	13.74
Correction of all maladjustments except idle CO and rpm	1.25	18.44	2.65	153	13.75
Correction of idle CO and rpm	0.90	8.13	2.69	81	13.98
Tune-up and repair of emissions control devices	0.87	7.65	2.55	56	13.95

a. Source: reference [21]. Based on tests of 300 model year 1975 and 1976 automobiles. The maintenance sequence is described in the text.

the only one being implemented on a large scale. In the remainder of this section, the main concepts underlying I/M are discussed, and the results of a study of the emissions reduction effectiveness of I/M are presented.

To carry out I/M, it is necessary to have a suitable emissions test procedure. The FTP is too costly and time consuming to be used for routine testing of large numbers of in-use vehicles. Therefore, I/M programs rely on any of several available short exhaust emissions tests. The simplest short test consists of measuring the concentrations of HC, CO, and NO_x in the exhaust while the engine is idling.[7] Other short tests include measurements of exhaust HC, CO, and NO_x concentrations or mass emissions rates while the vehicle is operated through a brief sequence of driving modes (less than 10) on a chassis dynamometer [26]. Depending on their complexity, the short tests last from 30 sec to 2 min [26].

The short tests are not equivalent to the FTP. Some vehicles that comply with exhaust emissions standards as measured by the FTP may fail short tests (erroneous failures), whereas other vehicles that do not meet the FTP standards may pass short tests (erroneous passes). The proportions of erroneous failures and erroneous passes among tested vehicles depend on the short test procedure that is used, the ages and designs of the vehicles that are being tested, and the stringency of the test. A standard measure of the stringency of a test is the proportion of vehicles in the population under consideration that would fail an initial short test (i.e., the first test of an I/M program). This proportion is called the "stringency factor" of the test.

Increasing the value of the stringency factor of a test tends to increase the proportion of erroneous failures and decrease the proportion of erroneous passes. The resulting trade-off between erroneous failures and erroneous passes is illustrated in figure 5.4 for two short tests applied to a fleet of model year 1975 and 1976 vehicles that were less than 1 yr old at the time of the tests. The short tests are the federal short-cycle test and the two-speed idle test. The former test consists of measuring a vehicle's mass emissions rates while the vehicle is operated through nine driving modes on a chassis dynamometer. The latter test consists of measuring the concentrations of pollutants in the exhaust while the engine is idling and while it is operating at 2,250 rpm with the transmission in neutral. In the examples shown in figure 5.4, the federal short-cycle test is more effective than the two-speed idle test in separating high-emitting vehicles from low-emitting ones. With any proportion of erroneous failures, the short-cycle test has a lower proportion of erroneous passes for each pollutant than the two-speed idle test does. However, in each test, decreases in the

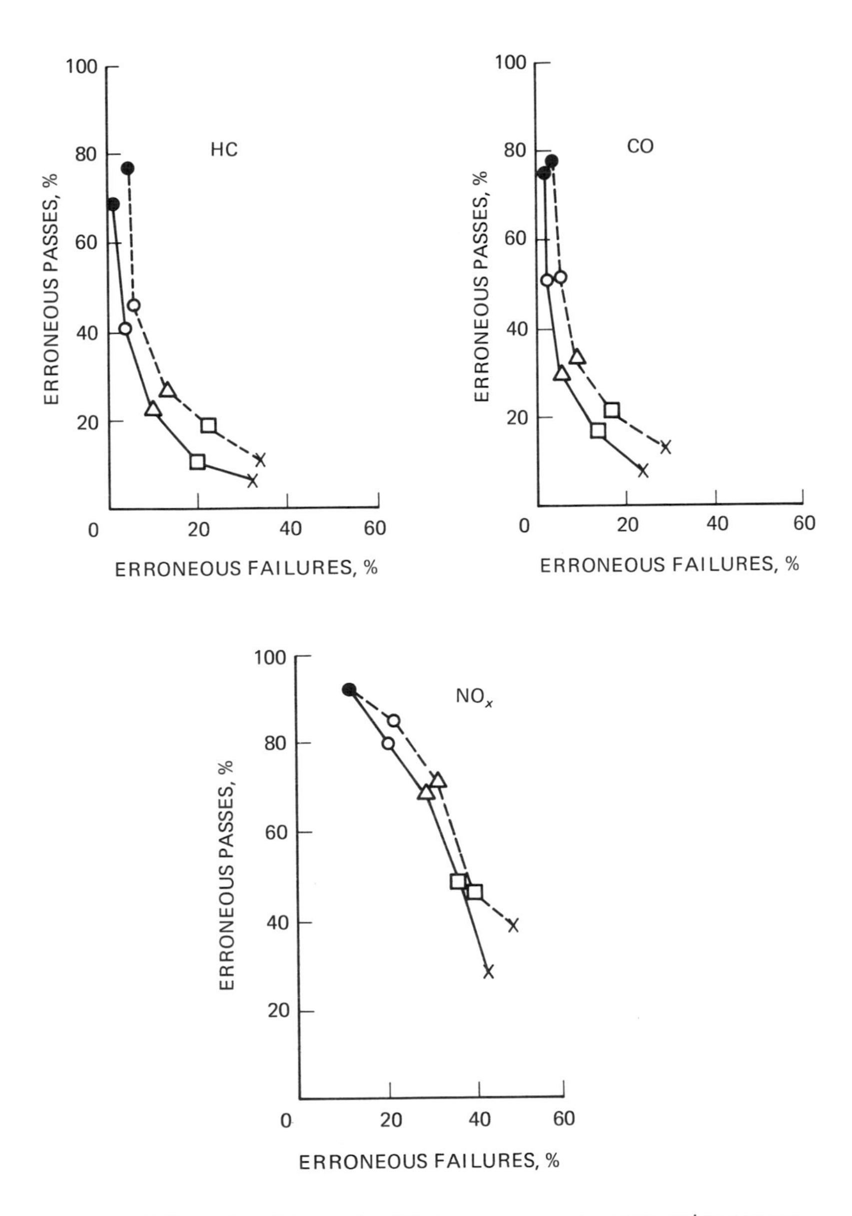

Figure 5.4 Examples of the trade-off between erroneous passes and erroneous failures for two short emissions tests. Based on tests of model year 1975 and 1976 cars that were less than 1 yr old at the time of testing. Key: – – –, two-speed idle test; ———, federal short-cycle test; ●, 10% stringency factor; ○, 20% stringency factor; △, 30% stringency factor; □, 40% stringency factor; ×, 50% stringency factor. (Data from reference [21])

Table 5.7 Average FTP emissions of automobiles that passed and failed two short emissions tests[a]

Test procedure	Stringency factor (%)	Pass			Fail		
		HC	CO	NO_x	HC	CO	NO_x
Federal short cycle	10	1.16	15.01	2.83	2.81	67.37	2.78
	20	1.01	11.64	2.75	2.60	54.71	3.12
	30	0.89	9.01	2.68	2.33	46.49	3.15
	40	0.83	7.67	2.56	2.07	39.14	3.22
	50	0.77	6.96	2.45	1.88	33.56	3.20
Two-speed idle	10	1.21	16.80	2.71	2.19	45.61	2.70
	20	1.06	13.20	2.73	2.39	46.70	2.68
	30	0.88	10.01	2.69	2.31	42.25	2.76
	40	0.82	8.46	2.58	2.03	36.51	2.90
	50	0.79	7.68	2.53	1.82	31.68	Not available

a. Source: reference [21]. Emissions in g/mi. Based on tests of model year 1975 and 1976 vehicles that were less than 1 yr old at the time of testing. The stringency factor equals the percentage of vehicles that failed the test.

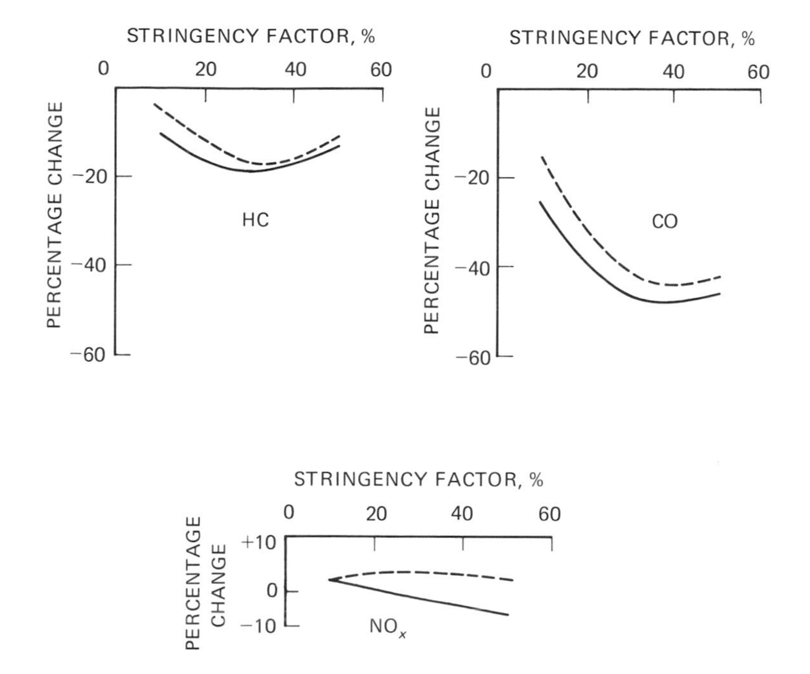

Figure 5.5 Examples of the emissions reduction effectiveness of I/M with two different emissions tests. Based on table 5.7. Percentage changes in emissions refer to the percentage differences between fleet-average FTP emissions measured immediately before and after vehicles that fail inspection are repaired. Key: ———, federal short-cycle test; – – –, two-speed idle test.

proportions of erroneous failures are associated with increases in the proportions of erroneous passes and vice versa.

The ability of short emissions tests to distinguish between vehicles that should and should not receive maintenance has an important influence on the emissions reduction effectiveness of I/M. This is illustrated by the following hypothetical example, which is based on the same vehicles and short tests that were used in constructing figure 5.4. Table 5.7 shows the average FTP emissions of vehicles that passed and failed each of the short tests at various values of the stringency factors. Assume that the tabulated emissions levels apply to the vehicles in an I/M program and that vehicles that fail inspection are repaired so that their emissions coincide with the exhaust emissions standards for the 1975 and 1976 model years. The percentage reductions in fleet average emissions of HC, CO, and NO_x

that result from repairing the failed vehicles can be computed from table 5.7. These reductions are shown as functions of the stringency factors of the short tests in figure 5.5.

Referring to figure 5.5, it can be seen that at most values of the stringency factors, the federal short-cycle test achieves greater reductions in the emissions of all pollutants than the two-speed idle test does. This is a consequence of the greater ability of the former test to distinguish between high-emitting and low-emitting vehicles. With both short tests, the HC and CO emissions reduction effectiveness of the hypothetical I/M program increases with increasing stringency factor until the value of this factor reaches roughly 30–40%. However, further increases in the value of the stringency factor tend to decrease the effectiveness of the program. The reason for this is that increasing the value of the stringency factor causes increasing numbers of vehicles whose emissions of one or more pollutants are in compliance with the emissions standards to fail the inspection test. These vehicles include erroneously failing vehicles and, possibly, vehicles that have moderately excessive emissions of one or two pollutants but whose emissions of the remaining pollutants are within the standards. Since maintenance following failure of inspection is assumed to cause emissions of all pollutants to coincide with the emissions standards, this maintenance increases the emissions of pollutants whose preinspection emissions levels are below the standards. At sufficiently high stringency factor values (30–40% in the case of this example) these increases in emissions from erroneously failing vehicles and vehicles that comply with some of the emissions standards become sufficiently large to cause the emissions reduction effectiveness of the tests to decrease with further increases in the value of the stringency factor. Similar relations between the value of the stringency factor and the emissions reduction effectiveness of I/M have been observed in field tests of I/M effectiveness [23–25].

In general, the stringency factor value at which the emissions reduction effectiveness of an I/M program begins to decrease depends on the emissions levels of vehicles after repair as well as on the proportions of erroneous failures and erroneous passes. Decreasing the postrepair emissions levels increases the stringency factor value that can be reached before the emissions reduction effectiveness of a test begins to decline.

The I/M program illustrated in figure 5.5 has a very small effect on NO_x emissions, regardless of which short test is used, and, depending on the test and the stringency factor, may increase NO_x emissions. This reflects the inability of the two tests being considered to distinguish effectively between vehicles with high and low NO_x emissions.

In addition to depending on the abilities of short emissions tests to correctly identify vehicles whose emissions are excessive and that should receive maintenance, the emissions reduction effectiveness of I/M programs depends on vehicle design, the immediate emissions reductions brought about by maintenance (i.e., the differences between vehicles' emissions just before and just after maintenance), and the rates at which emissions increase after maintenance. Vehicle design affects I/M effectiveness by influencing the frequency, severity, and nature of maladjustments and failures of emissions-related engine components. For example, the proper operation of three-way catalysts requires careful control of the air-fuel ratio, and many model year 1981 automobiles use electronic systems to achieve the necessary control. The electronic systems eliminate maladjustments of many conventional engine parameters as causes of excessive emissions, but introduce malfunctions of the electronic controls as new causes of excessive emissions. This change in vehicles' failure modes may gradually alter the emissions reduction effectiveness of I/M programs as the prevalence of electronically controlled vehicles in the automobile population increases [27, 28].

The sensitivity of I/M effectiveness to immediate emissions reductions and postmaintenance deterioration is illustrated in figure 5.6, which presents two hypothetical examples of the effects of I/M on a vehicle's emissions. In these examples it is assumed that when the vehicle is new, it emits a certain pollutant at the rate of 1 g/mi, and that without I/M the vehicle's emissions of this pollutant increase by 0.2 g/mi/yr as the vehicle ages. The resulting emissions rate as a function of the vehicle's age is indicated by the solid lines in parts a and b of figure 5.6.[8] When I/M is in effect, it is assumed that the vehicle is required to pass an annual emissions test and that in order to do this it receives annual maintenance of emissions-related engine adjustments and components. In figure 5.6a it is assumed that this maintenance causes an immediate 20% reduction in the vehicle's emissions and that during the year following maintenance, emissions increase to the same level that they would reach without the I/M program. Thus, the effects of maintenance are completely lost in 1 yr. Moreover, the level to which the vehicle's emissions can be reduced by maintenance increases as the vehicle ages because in each year emissions prior to maintenance are higher than they were the year before. The resulting changes in emissions as a function of the vehicle's age are indicated by the dashed line in figure 5.6a. Figure 5.6a corresponds to a

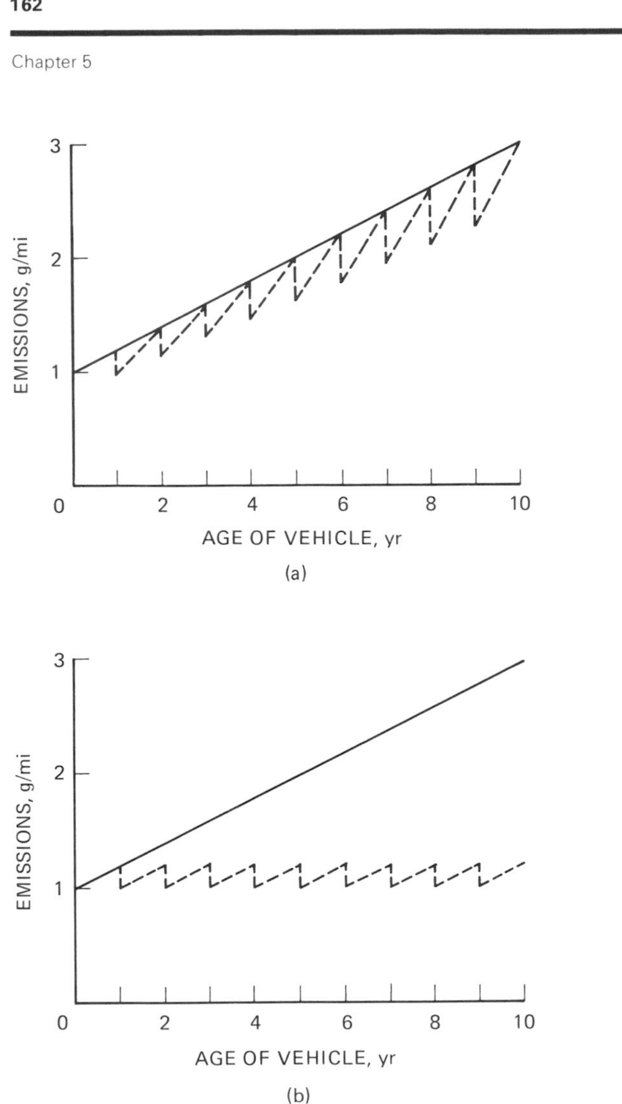

Figure 5.6 Two models of the long-run effects of I/M on emissions. Key: ———, emissions without I/M; – – –, emissions with I/M. (*a*) Deterioration is due to a combination of reversible and irreversible changes in the performances of emissions-related components. (*b*) Deterioration is due only to reversible changes in the performances of emissions-related components.

model of emissions deterioration in which the increases in a vehicle's emissions that occur as it ages are due partly to reversible changes in the performances of emissions-related engine components (e.g., reversible changes in engine adjustments) and partly to irreversible changes (e.g., wearing out of emissions-related components).

In figure 5.6b it is assumed that maintenance reduces the vehicle's emissions to the new-vehicle level of 1 g/mi, regardless of the vehicle's age, and that after maintenance the rate of increase of emissions is the same as it would be without the I/M program, 0.2 g/mi/yr. Thus, 1 yr after each maintenance event, the vehicle's emissions still are less than they would be without I/M. The resulting changes in emissions as a function of the vehicle's age are indicated by the dashed line in figure 5.6b. Figure 5.6b corresponds to a model of emissions deterioration in which the increase in a vehicle's emissions with age is due entirely to reversible changes in the performance of emissions-related engine components.

In the model of figure 5.6a, I/M slows emissions deterioration but does not stop it. The total mass of the vehicle's emissions in each year following the initiation of I/M is approximately 10% less than it would be without I/M. In the model of figure 5.6b, I/M stops deterioration. Consequently, the emissions reduction effectiveness of I/M increases as the vehicle ages. In the first year following the initiation of I/M, the total mass of the vehicle's emissions is approximately 15% less than it would be without I/M, but in the ninth year of I/M, the mass of emissions is more than 60% less than it otherwise would be.

There have been no empirical studies of either the long-run effects of I/M on emissions deterioration or the long-run emissions reduction effectiveness of I/M. The emissions reduction effectiveness of I/M during a 1-yr period has been studied in Portland, Oregon, where a program of annual I/M for HC and CO is in effect [29, 30]. The Portland I/M program uses an idle emissions test and has a stringency factor of approximately 35%. Figure 5.7 shows the average FTP emissions of HC and CO by model year 1975–1977 automobiles in Portland during the year of the study. The emissions are plotted as functions of the average odometer readings of the vehicles. The emissions measurements were made when all vehicles were inspected, immediately after maintenance of vehicles that failed inspection (this measurement was made only on failing vehicles) and at 3-month intervals during the year following inspection. Figure 5.7 also shows the average HC and CO emissions during the study year of a fleet of 1975–1977 model year vehicles in Eugene, Oregon, where there is no

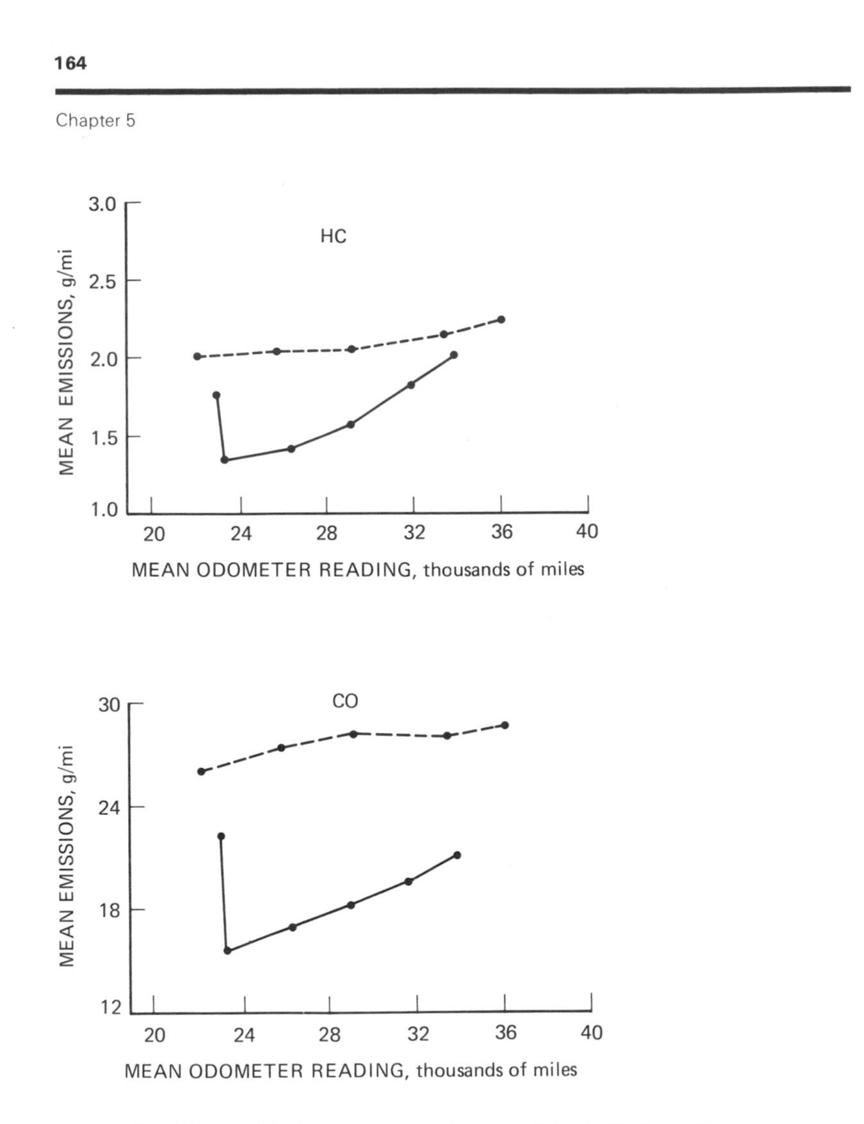

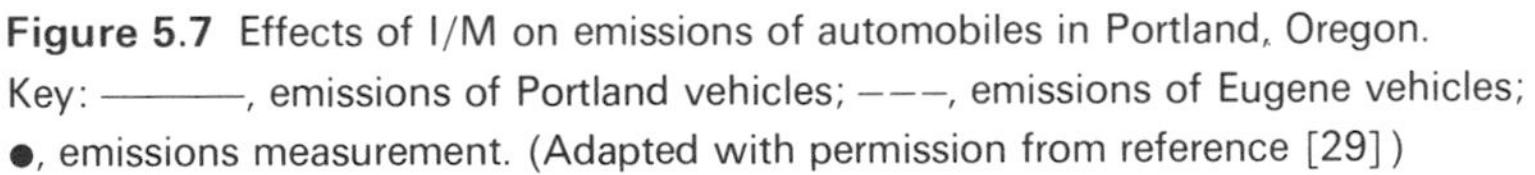

Figure 5.7 Effects of I/M on emissions of automobiles in Portland, Oregon. Key: ———, emissions of Portland vehicles; – – –, emissions of Eugene vehicles; ●, emissions measurement. (Adapted with permission from reference [29])

I/M. I/M caused immediate, postinspection-and-maintenance reductions of approximately 25% (HC) and 35% (CO) in the average emissions of Portland vehicles. One year after inspection and maintenance, average CO emissions were still approximately 5% below the preinspection level, and average HC emissions were approximately 15% above the preinspection level. The differences between the actual average HC and CO emissions rates of Portland vehicles during the study year and the emissions rates that would have occurred if I/M had not been in effect can be estimated by assuming that if there had been no I/M in Portland, then the emissions of Portland vehicles would have been the same as those of Eugene vehicles.[9] The resulting estimates suggest that I/M reduced the average rates of HC and CO emissions from the Portland vehicles during the year of the study by roughly 20% (HC) and 36% (CO) as estimated by FTP emissions measurements.

The cost of inspection in I/M programs is approximately $5–10/vehicle, depending on the test procedure that is used. The cost of repairing vehicles that fail the inspection typically is in the vicinity of $30/vehicle, although a small proportion of the vehicles that fail inspection may require extensive repairs and incur considerably greater costs. These cost estimates are based on reference [26] but have been inflated to 1980 dollars. The most common types of repairs that failing vehicles need are idle adjustments (idle rpm, air-fuel ratio, and spark timing) and tune-ups [26, 29].

5.6 *The Dependence of Motor Vehicle Emissions on Driving Conditions*

Motor vehicles' emissions rates are sensitive to changes in a large number of variables that are related to driving conditions. Variables to which emissions are particularly sensitive include driving mode, engine temperature (whether the engine is cold or warmed up) and, in the case of cold engines, air temperature. Other variables that affect emissions rates but to which there is less sensitivity include air pressure, relative humidity, the operations of air conditioners, variations in the weight of the passengers or cargo carried by the vehicle, and variations in frictional resistance to movement of the vehicle.

The dependence of emissions rates and fuel economy on driving mode is illustrated in table 5.8. HC and CO emissions are relatively high and fuel economy is relatively low in acceleration modes, some deceleration modes, and idles. Relatively low HC and CO emissions and high fuel economy

Table 5.8 Average emissions rates and fuel economy of 27 model year 1975 automobiles in various driving modes[a]

Type of mode	Initial speed (mph)	Final speed (mph)	Duration (sec)	Distance (mi)	Emissions (g/min) HC	Emissions (g/min) CO	Emissions (g/min) NO_x	Emissions (g/mi) HC	Emissions (g/mi) CO	Emissions (g/mi) NO_x	Fuel economy (mi/gal)
Acceleration	0	30	12	0.0602	0.89	17.84	1.84	2.96	59.29	6.11	7.22
	0	15	8	0.0201	0.52	12.77	0.35	3.44	84.70	2.33	6.65
	15	30	11	0.0705	0.55	9.14	1.17	1.43	23.76	3.05	11.17
	30	60	17	0.2183	1.90	51.20	4.93	2.46	66.46	6.40	9.75
Deceleration	30	0	13	0.0592	0.33	7.46	0.36	1.22	27.31	1.31	16.75
	15	0	8	0.0173	0.31	8.30	0.12	2.41	63.97	0.91	10.04
	30	15	9	0.0579	0.25	5.63	0.39	0.64	14.58	1.02	23.93
	60	30	18	0.2382	0.50	4.89	1.84	0.63	6.16	2.32	23.10
Idle	0	0	Any	0	0.36	7.85	0.04		Infinite		90.43[b]
Cruise	15	15	Any	Any	0.27	5.30	0.13	1.07	21.20	0.52	16.65
	30	30	Any	Any	0.26	1.87	0.56	0.52	7.19	1.12	23.03
	60	60	Any	Any	0.25	3.40	3.73	0.25	3.40	3.73	19.28

a. Source: reference [31].
b. Fuel economy at idle expressed in units of min/gal.

occur during steady-state cruises and during certain decelerations. NO_x emissions do not exhibit a simple dependence on driving mode.

As a result of the dependence of emissions rates on driving mode, the rate of emissions of a traffic stream depends on the detailed sequences of accelerations, decelerations, cruises, and idles that each of the vehicles in the stream experiences. Since it is not possible to make direct measurements of emissions from roadway traffic, these emissions usually are estimated by combining information on traffic flow conditions with laboratory data on vehicles' emissions rates in various driving modes. Information on the sequences of driving modes that occur in a stream of traffic can be obtained by driving instrumented vehicles in the stream or from traffic flow simulation models. However, these methods are too costly and time consuming to be used in most applications. Therefore, it is customary in most applications to estimate motor vehicle emissions from the average speed of traffic flow. Standardized speed-emissions relations have been developed for this purpose [5]. The estimated emissions rates at any given average speed consist of appropriately weighted averages of the emissions rates in the driving modes that occur in typical flows of traffic at that average speed. Thus, the resulting speed-emissions relations include the effects of accelerations, decelerations, and idles as well as steady-state operation. The development of speed-emissions relations is discussed in more detail in section 5.7. Figure 5.8 shows examples of speed-emissions relations for HC, CO, and NO_x. The relations shown in the figure applied to model year 1975 automobiles in 1975, but the general shapes of the curves are the same for cars of all model years and ages. HC and CO emissions (expressed in terms of the mass of emissions per unit of distance traveled) decrease with increasing average speed, whereas NO_x emissions tend to increase.

When a vehicle's engine is cold, the fuel tends not to vaporize well, and the effects of wall quenching are accentuated. Thus, much unburned or incompletely burned fuel may pass through the engine and into the exhaust. In addition, catalytic converters do not function when they are cold. Consequently, cold vehicles have higher HC and CO emissions rates and lower fuel economies than warmed-up vehicles do. The effects of cold starts on emissions and fuel economy are illustrated in table 5.9, which shows the average HC, CO, and NO_x emissions rates and fuel economy of a group of 1978 model year vehicles during the cold transient, hot stabilized, and hot transient phases of the FTP. The cold and hot transient phases are identical, except that the former phase has a cold start and the latter phase has a hot start. HC emissions are approximately three times higher, CO emissions are approximately four times

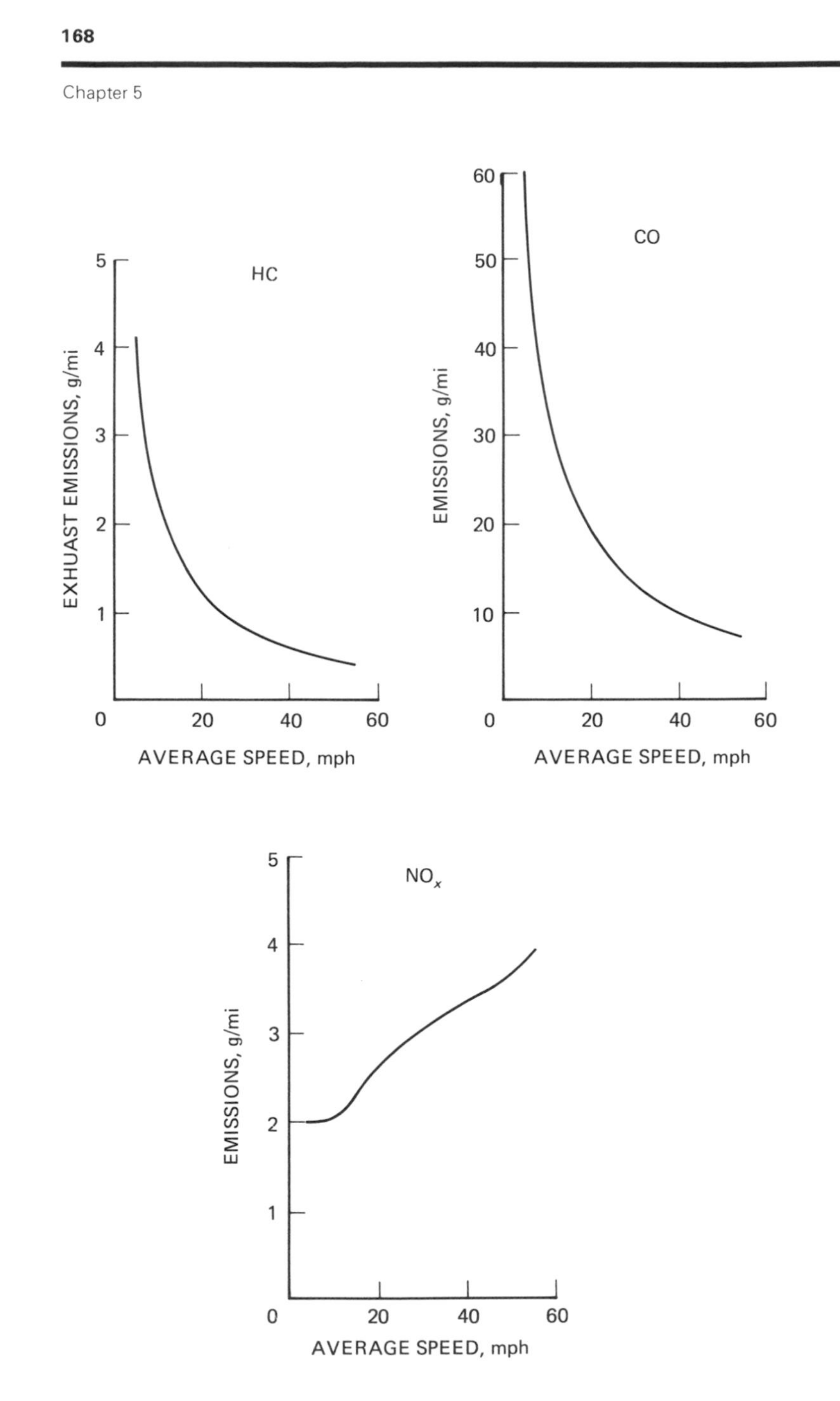

Figure 5.8 Speed-emissions relations for warmed-up 1975 model year automobiles in 1975. (Based on data from reference [5])

Table 5.9 Average emissions rates and fuel economy of 24 model year 1978 automobiles during the three phases of the FTP[a]

	Emissions (g/mi)			
Phase	HC	CO	NO_x	Fuel economy (mi/gal)
Cold transient	2.74	47.18	2.12	14.64
Hot stabilized	0.89	14.12	1.43	16.37
Hot transient	0.93	11.00	2.00	18.02

a. Source: reference [8]. Based on measurements made in 1978.

higher, and fuel economy is approximately 20% lower in the cold transient phase than in the hot transient phase. NO_x emissions in the two phases are not appreciably different. Most of the excess emissions and fuel consumption that are associated with cold starts occur within 3–5 min after starting the engine.

The hot stabilized phase of the FTP has a lower average speed (16 mph) than the other two phases do (25.6 mph each). Therefore, in table 5.9 the hot stabilized phase has somewhat higher CO emissions and lower NO_x emissions and fuel economy than the hot transient phase does. However, CO emissions are not as high and fuel economy is not as low in the hot stabilized phase as they are in the cold transient phase. HC emissions in table 5.9 are approximately equal in the hot stabilized and hot transient phases, despite the difference in the speeds of the two phases. The reasons for the relatively high rate of HC emissions during the hot transient phase are not well understood at present. However, most of the vehicles represented in the table have catalytic converters, and during the engine-off period preceding the hot transient phase, the catalysts may cool sufficiently to reduce the effectiveness with which they oxidize hydrocarbons. In addition, the effectiveness of the catalysts during the hot start period may be reduced as a result of the relatively low air-fuel ratio that occurs during start-up.

The temperature of a vehicle's engine during a cold start is approximately equal to the temperature of the surrounding air, and the air temperature has a significant effect on the magnitudes of cold start emissions of HC and CO. Decreases in the air temperature cause these emissions to increase considerably. This is illustrated in table 5.10, which shows the

Table 5.10 Effects of air temperature on average exhaust emissions rates of two groups of test vehicles[a]

FTP phase	Temperature (°F)	Emissions (g/mi) HC	CO	NO_x
	4 vehicles with catalytic converters			
Cold transient	20	5.20	126	3.6
	50	2.64	68	3.5
	75	1.30	17	2.7
	110	1.01	10	2.7
Hot stabilized	20	0.28	1.2	3.0
	50	0.24	1.5	2.9
	75	0.19	1.2	2.1
	110	0.23	1.9	2.3
Hot transient	20	0.35	4.8	3.1
	50	0.50	4.1	3.1
	75	0.45	5.2	2.5
	110	0.65	11.0	2.3
	6 vehicles without catalytic converters			
Cold transient	20	14.4	225	6.2
	50	9.97	162	6.8
	75	4.80	75	6.7
	110	3.60	42	7.1
Hot stabilized	20	3.70	32	6.9
	50	3.63	28	6.9
	75	3.69	23	5.7
	110	3.60	44	5.2
Hot transient	20	3.11	23	9.5
	50	3.06	20	9.0
	75	3.29	23	7.2
	110	3.77	49	6.9

a. Source: reference [32].

effects of air temperature on the average emissions rates of four vehicles with and six vehicles without catalytic converters. The effects of temperature on cold start HC and CO emissions are particularly marked in the vehicles with catalytic converters. During the cold transient phase of the FTP, these vehicles' average rate of HC emissions is higher by a factor of 4 and their average rate of CO emissions is higher by a factor of 3 at 20°F than they are at 75°F.

5.7 *Mathematical Models of the Relations between Motor Vehicle Emissions and Driving Conditions*

To construct transportation emissions inventories and to estimate the effects on emissions of changes in the design or operation of urban transportation systems, it is necessary to be able to convert information on driving conditions and other relevant variables into estimates of motor vehicle emissions. The conversion is carried out by means of mathematical models that relate emissions rates to driving conditions and the other variables.[10] Two types of models will be discussed in this section: microscale models and macroscale models. Microscale models relate a vehicle's instantaneous exhaust HC, CO, and NO_x emissions per unit time to its instantaneous speed and acceleration. Macroscale models relate a vehicle's total emissions or average emissions per unit distance traveled (including evaporative emissions) during an entire trip or a large part of a trip to the average speed of the vehicle during the trip or part of a trip. Microscale models are appropriate for use in situations in which detailed estimates of emissions on short sections of roadway are needed and detailed information on vehicles' speeds and accelerations can be obtained. An example of such a situation is the estimation of CO emissions and concentrations in the vicinity of a signalized intersection. Macroscale models are appropriate for estimating average motor vehicle emissions in geographical areas that are sufficiently large to contain many complete trips or substantial parts of trips. Examples of such areas are traffic corridors and entire urban regions. Typically, the detailed speed and acceleration data that are needed to operate microscale models cannot be obtained for such areas, whereas information on average trip speeds can be obtained from direct measurements or estimated with mathematical models. (Models for estimating speeds and accelerations are discussed in section 6.3.)

A microscale emissions model is developed from data obtained by measuring the emissions of vehicles that are driven in various driving modes on chassis dynamometers. Examples of driving modes are listed in table 5.8. The accelerations are constant in the listed acceleration and deceleration modes. Suppose that a driving mode has duration t_{mode} and constant acceleration a. (If the mode consists of a steady state cruise or idle, then $a = 0$.) Suppose also that at time $t \leq t_{mode}$ following the start of the mode, the speed is $s(t)$. Let $f_p(s, a)$ denote a vehicle's instantaneous pollutant p emissions rate, in units of mass per unit time, expressed as a function of speed and acceleration. Let E_p^{mode} denote the total mass of pollutant p emitted during the driving mode. Then E_p^{mode} is given by

$$E_p^{mode} = \int_0^{t_{mode}} f_p[s(t), a]\, dt. \tag{5.2}$$

To develop a microscale emissions model, values of E_p^{mode} for different driving modes are obtained from emissions measurements. The function f_p, which is not known a priori, is then estimated by means of regression analysis or other curve-fitting techniques. The fitted function f_p constitutes the microscale model of the relation between instantaneous emissions of pollutant p and instantaneous speed and acceleration.

An example of a microscale emissions model is the Automobile Exhaust Emission Modal Analysis Model that has been developed for the US Enironmental Protection Agency [33]. In this model, the function f_p is specified as

$$f_p(s, a) = \sum_{m=0}^{2} \sum_{n=0}^{2} b_p(m, n) s^m a^n [1 - h(a)] + \sum_{m=0}^{2} c_p(m) s^m h(a), \tag{5.3}$$

where

$$h(a) = \begin{cases} 1 - a, & 0 \leq a \leq 1.0 \text{ mi/hr/sec} \\ 1 + a/1.2, & -1.2 \leq a \leq 0 \text{ mi/hr/sec} \\ 0, & \text{otherwise.} \end{cases} \tag{5.4}$$

The quantities $b_p(m, n)$ and $c_p(m)$ denote coefficients whose values are estimated by fitting the function f_p to modal emissions data through equation (5.2). Separate sets of coefficients are estimated for vehicles of different model years. The values of the coefficients are updated periodically to include late-model vehicles and to incorporate the effects of vehicle age on emissions.[11] The Modal Analysis Model applies only to exhaust emissions from warmed-up automobiles. Start-up emissions (that is, excess emissions that are caused by cold and hot starts) and emissions from vehicles other than automobiles are not treated. Crankcase emissions could be treated by including them in the constant terms $b_p(0, 0)$ and $c_p(0)$. However, this usually is not done, as crankcase emissions from most automobiles currently in use are negligible.

A traffic stream usually includes vehicles from many different model years. Therefore, when equation (5.3) is used to estimate emissions from automobiles in a traffic stream, $f_p(s, a)$ usually is computed as a weighted average over the model years of automobiles in the stream. The weight for a given model year is equal to the proportion of automobiles from that model year in the traffic stream. Let ρ_i denote the density of model year i automobiles in the stream, ρ the total automobile traffic density, and $f_{pi}(s, a)$ the instantaneous emissions rate in units of mass per unit time for pollutant p and automobiles from model year i.[12] Then the weighted average emissions rate, $\bar{f}_p(s, a)$, is given by

$$\bar{f}_p(s, a) = \sum_i (\rho_i/\rho) f_{pi}(s, a). \tag{5.5}$$

The accuracy of the EPA's Modal Analysis Model has been investigated in a limited way by using it to estimate the average emissions of automobiles during the hot stabilized and hot transient phases of the FTP and comparing the resulting estimates with measured emissions [33]. The investigation was performed in the early 1970s and included vehicles only through model year 1971. The vehicles used in the investigation were different from those used to estimate the coefficients of the model. The percentage differences between the predicted and measured average emissions rates of all tested vehicles in the hot stabilized phase were 18% (CO), 9% (HC), and 44% (NO_x). In the hot transient phase, the differences were 15% (CO), 12% (HC), and 26% (NO_x) [33]. The accuracy of the Modal Analysis Model under driving conditions similar to those found, for example, near signalized intersections or other locations where traffic queueing occurs is unknown.

Microscale emissions models can be used in conjunction with traffic flow models to estimate the spatial distributions of emissions along roadways and to estimate emissions at locations such as signalized intersections where the mix of driving modes may be substantially different from the mix that occurs during entire trips. (Macroscale models can be used to estimate emissions from entire trips or large parts of trips and require considerably fewer input data than microscale models require.) In these applications, traffic volume data are obtained from exogenous sources, such as traffic measurements, and supplied as inputs to the traffic flow model. The traffic flow model then is used to generate the speed and acceleration data that are needed as inputs to the emissions model and the traffic density information that is needed to convert individual vehicles' emissions rates into aggregate emissions rates for an entire traffic stream.[13] The profile of emissions in the vicinity of a signalized intersection that is shown in figure 3.15 was obtained in this manner by using the EPA's Modal Analysis Model together with traffic density, speed, and acceleration data that were obtained as outputs from a simple hydrodynamic model of traffic flow [34]. Examples of the use of the Modal Analysis Model in conjunction with a more sophisticated traffic flow model to estimate emissions in the vicinity of signalized intersections are given in references [35–38].

The use of traffic measurements or a traffic flow model to generate the speed and acceleration data needed to operate a microscale emissions model normally entails considerable complexity and expense. Consequently, microscale emissions estimation is not feasible in most emissions inventory development activities and transportation planning studies. An alternative emissions estimation method, which sacrifices spatial detail but achieves a substantial reduction in input data requirements, uses the average speed of travel over an entire trip or a large part of a trip to represent the effects of variations in driving modes during the trip or part of a trip. Total emissions or the average mass-per-unit-distance emissions rate from the entire trip or part of the trip then are estimated as functions of the length and average speed of the trip or part of a trip and other relevant variables, such as air temperature, the occurrence of cold or hot starts, and the age, model year, and type of vehicle (e.g., gasoline-powered automobile, heavy-duty diesel) involved. This emissions estimation approach constitutes a macroscale model since it does not enable detailed estimates of variations in a vehicle's emissions rates during a trip to be developed.

There are two ways to formulate macroscale emissions models. One way is as follows. Let $E_p^{\text{trip}}(s, L, x)$ denote the total mass of pollutant p that is emitted during a trip of average speed s and length L, where x denotes all other relevant variables. Let $h_p(x)$ denote the mass of pollutant p hot soak evaporative emissions that occur at the end of the trip; $e_{1p}(s, x)$, the mass of start-up exhaust emissions of pollutant p that occurs at the beginning of the trip; and $e_{2p}(s, x)$, the pollutant p exhaust emissions rate, in units of mass per unit distance (e.g., g/mi), from hot stabilized operation during the trip. Finally, let $c_p(x)$ denote the pollutant p crankcase emissions rate in units of mass per unit distance. Start-up exhaust emissions of pollutant p, e_{1p}, are defined as the difference between total exhaust emissions of this pollutant on the trip and the exhaust emissions that would occur if the engine were warmed up and running at the beginning of the trip. Note that h_p and c_p are nonzero only for hydrocarbons and that e_{1p} depends on whether the engine is warm or cold when it is started. The total mass of pollutant p that is emitted during the trip is

$$E_p^{\text{trip}} = h_p(x) + Lc_p(x) + e_{1p}(s, x) + Le_{2p}(s, x). \quad (5.6)$$

The functions h_p, c_p, e_{1p}, and e_{2p} can be evaluated using data and procedures that will be described later in this section. These functions usually must be evaluated separately for each vehicle type, model year, and age. If, as is usually the case in practical applications, information on the ages and model years of the vehicles used for particular trips is not available, then for each vehicle type, E_p^{trip} can be computed as a weighted average over the ages and model years of vehicles in the region of concern. The weights equal the proportions of vehicles of each age and model year that are likely to be present in a traffic stream in the region of interest. Data and procedures for computing such weights are described in reference [5].

Equation (5.6) is an example of a macroscale emissions estimation model. The travel data required to use this model to estimate motor vehicle emissions in an urban region or part of a region consist of the number of trips that take place in the region or part of a region according to length and average speed and the proportion of trips that begin with cold starts. (All trips that do not begin with cold starts begin with hot starts.) These data usually can be obtained or estimated from standard urban transportation data sets and planning models. Thus, equation (5.6) provides a convenient method for estimating motor vehicle emissions in urban transportation planning studies. The equation can be adapted for use in computing emissions from parts of trips by dropping the term h_p if the parts in question do not include the hot soak period and dropping the term e_{1p} if the parts in question do not include the start-up period. When hot soak and start-up emissions are not included in the trip segments under consideration, it often is convenient to work in terms of the average emissions per unit distance traveled. This is given by $c_p(x) + e_{2p}(s, x)$. Examples of the use of standard transportation data to estimate cold start proportions are given in reference [39]. Example 5.1 below gives a simple illustration of the use of equation (5.6). Examples of the use of equation (5.6) in connection with standard transportation data and models to compute emissions are given in chapter 6 and references [40, 41].

Example 5.1—Use of Equation (5.6) to Compute Emissions

Suppose that shopping trips originating in a certain neighborhood go either to shopping location A or to shopping location B. Trips that go to location A have lengths of 3 mi and average speeds of 15 mi/hr. Trips that go to location B have lengths of 5 mi and average speeds of 25 mi/hr. Sixty percent of the trips begin with cold starts. Let the HC emissions functions in equation (5.6) be as follows (for simplicity, the variable x is not used in this example): $h_{HC} = 6.0$ g; $c_{HC} = 0.0$; $e_{1HC} = 11.0$ g for cold starts and 3.3 g for hot starts; $e_{2HC}(s) = 0.16 + 38.8/s$ g/mi. Assume that these emissions functions correspond to weighted averages over the ages and model years of vehicles that are used for shopping trips. Since 60% of trips begin with cold starts and the rest begin with hot starts, average start-up HC emissions per trip are $\bar{e}_{1HC} = 0.6(11.0) + 0.4(3.3) = 7.92$ g. Based on equation (5.6), average total HC emissions per trip to location A are

$$E_{HC}^{\text{trip A}} = 6.0 + 3(0.0) + 7.92 + 3(0.16 + 38.8/15) = 22.16 \text{ g.} \qquad (5.7a)$$

Average total HC emissions per trip to location B are

$$E_{HC}^{\text{trip B}} = 6.0 + 5(0.0) + 7.92 + 5(0.16 + 38.8/25) = 22.48 \text{ g.} \qquad (5.7b)$$

If there are 700 trips to location A and 300 trips to location B, then total emissions of HC from all of the trips are 22,556 g.

The second way of formulating a macroscale emissions model is as follows. Let α_c and α_h denote, respectively, the fractions of the total length of a trip that are driven during the first 505 sec following cold and hot starts. (Note that only one of these fractions can be nonzero for a single trip.) Let $e_{cp}(s, x)$ and $e_{hp}(s, x)$ denote, respectively, the average pollutant p exhaust emissions rates, in units of mass per unit distance, during the first 505 sec following cold and hot starts. As in equation (5.6), let $e_{2p}(s, x)$ denote the average pollutant p exhaust emissions rate, in units of mass per unit distance, during hot stabilized operation. Then, the total mass of pollutant p exhaust emissions from a trip of length L and average speed s, $E_p^{ex}(s, x)$, can be written as

$$E_p^{ex}(s, L, x) = L[\alpha_c e_{cp}(s, x) + \alpha_h e_{hp}(s, x) + (1 - \alpha_c - \alpha_h) e_{2p}(s, x)]. \qquad (5.8)$$

Total emissions from the trip, including crankcase and hot soak emissions, can be obtained by adding the quantity $h_p(x) + Lc_p(x)$ [as defined in equation (5.6)] to equation (5.8).

Equation (5.8) constitutes a macroscale model of exhaust emissions that is more closely related to standard exhaust emissions measurement procedures, notably the FTP, than the exhaust emissions terms in equation (5.6) are. Consequently, most data that the EPA publishes for computing motor vehicle exhaust emissions rates are based on equation (5.8). The published exhaust emissions data consist of formulas for computing e_{cp}, e_{hp}, and e_{2p} [5]. Equation (5.8) is somewhat less convenient than equation (5.6) for use in urban transportation planning studies, as equation (5.8) requires computation of the quantities a_c and a_h. However, the travel data that are needed to use the two equations are similar. The following example illustrates the use of equation (5.8) to compute emissions.

Example 5.2—Use of Equation (5.8) to Compute Emissions

In this example, equation (5.8) is used to compute HC emissions from the set of shopping trips that was described in example 5.1. Let the exhaust emissions functions for HC in equation (5.8) be given by

$$e_{cHC}(s) = 0.16 + 117.22/s \text{ g/mi}, \tag{5.9a}$$

$$e_{hHC}(s) = 0.16 + 62.32/s \text{ g/mi}, \tag{5.9b}$$

$$e_{2HC}(s) = 0.16 + 38.8/s \text{ g/mi}. \tag{5.9c}$$

As in example 5.1, the variable x is not used here. The emissions functions in equations (5.9) correspond to weighted averages over vehicle ages and model years. In the first 505 sec of travel, trips to location A travel 2.10 mi, or 70% of their total lengths. Since 60% of these trips begin with cold starts, it follows that $a_c = (0.60)(0.70) = 0.42$ and $a_h = (0.40)(0.70) = 0.28$. Therefore, based on equation (5.8), average exhaust HC emissions per trip to location A are

$$E_{HC}^{ex\,A} = (3)\,[(0.42)(0.16 + 117.22/15) + (0.28)(0.16 + 62.32/15) + (0.30)(0.16 + 38.8/15)] = 16.14 \text{ g}. \tag{5.10a}$$

Trips to location B travel 3.51 mi in their first 505 sec, or 70% of their total lengths. Therefore, as with trips to location A, $a_c = 0.42$ and $a_h = 0.28$. Exhaust HC emissions per trip to location B are

$$E_{HC}^{ex\,B} = (5)\,[(0.42)(0.16 + 117.22/25) + (0.28)(0.16 + 62.32/25) + (0.30)(0.16 + 38.8/25)] = 16.46 \text{ g}. \tag{5.10b}$$

If the functions e_{cp}, e_{hp}, and e_{2p} are known (e.g., from EPA publications), then e_{1p} in equation (5.6) can be computed. [Note that e_{2p} is the same in equations (5.6) and (5.8).] If a trip begins with a cold start, then e_{1p} is the difference between the total mass of emissions from a 505-sec trip that begins with a cold start and the total mass of emissions from a 505-sec trip that has an equal average speed and takes place under hot stabilized conditions. Similarly, if a trip begins with a hot start, then e_{1p} is the difference between the total mass of emissions from a 505-sec trip that begins with a hot start and the total mass of emissions from a 505-sec trip that has an equal average speed and takes place under hot stabilized conditions. Let $L_{505}(s)$ denote the distance traveled during the first 505 sec of a trip whose average speed is s. Then $e_{1p}(s, x)$ is given by

$$e_{1p}(s, x) = \begin{cases} L_{505}(s)\,[e_{cp}(s, x) - e_{2p}(s, x)] \text{ for a trip that begins with a cold start} \\ L_{505}(s)\,[e_{hp}(s, x) - e_{2p}(s, x)] \text{ for a trip that begins with a hot start.} \end{cases} \quad (5.11)$$

Estimates of evaporative and crankcase emissions for use in macroscale emissions models can be obtained from EPA publications. For example, MOBILE2, the most recent version of the emissions estimation computer program disseminated by the EPA, gives estimates of the combined diurnal, hot soak, and crankcase emissions that a vehicle produces during an entire day, divided by the distance that the vehicle travels during a day. The estimation formula is

$$r_p = c_p + (d_p + Th_p)/M, \quad (5.12)$$

where r_p is the combined evaporative and crankcase emissions rate expressed in units of mass of pollutant emitted per unit distance traveled, d_p is the mass of diurnal evaporative emissions, T is the number of trips the vehicle makes in a day, and M is the total distance traveled in a day. By evaluating r_p for three different combinations of values of T and M, equation (5.12) can be converted into a system of three simultaneous, linear equations in the three unknown quantities c_p, d_p, and h_p. These equations then can be solved for the values of c_p, d_p, and h_p. (Readers should note that MOBILE2 did not become available in time for use in preparing the motor vehicle emissions estimates presented in this book. The estimates presented here are based on the latest information that was available prior to the issuance of MOBILE2.)

The EPA's published emissions estimation formulas are based on a combination of measurements, modeling results, and simplifying assumptions. In principle, the emissions functions c_p, d_p, h_p, e_{cp}, and e_{2p} all can be estimated by direct measurement (in the case of evaporative and crankcase emissions) or (in the case of exhaust emissions) by fitting curves to emissions data that are obtained by driving vehicles on chassis dynamometers under many different driving conditions. In practice, estimates of evaporative and crankcase emissions are obtained in this way, using emissions measurements made during the FTP. However, generating the data needed to estimate exhaust emissions rates directly from emissions measurements made under many different driving conditions can entail a lengthy and costly measurement program. Therefore, the EPA's estimates of mass-per-unit-distance exhaust emissions rates incorporate several simplifying assumptions and procedures that reduce the number of measurements that are needed. One assumption is that for vehicles of a given type, age, and model year, the dependence of each of the functions e_{cp}, e_{hp}, and e_{2p} on air temperature T, speed s, and other variables x can be represented as a product form:[14]

$$e_{cp} = F_{cp}(T)\,G_{cp}(s)\,H_{cp}(x), \tag{5.13a}$$

$$e_{hp} = F_{hp}(T)\,G_{hp}(s)\,H_{hp}(x), \tag{5.13b}$$

$$e_{2p} = F_{2p}(T)\,G_{2p}(s)\,H_{2p}(x), \tag{5.13c}$$

where the F, G, and H are functions that must be determined. This assumption reduces measurement requirements because it implies that the dependence of the emissions functions e_{cp}, e_{hp}, and e_{2p} on T, s, and x can be inferred by measuring the T dependence for single s and x values, the s dependence for single T and x values, and the x dependence for single T and s values. It is not necessary to carry out a measurement program in which T, s, and x are changed simultaneously. A further reduction of measurement requirements is achieved by adopting a relatively simple definition of a cold start. For noncatalyst vehicles, a cold start is considered to occur if the engine is started after having been off for at least 4 hr. For catalyst vehicles, a cold start is considered to occur if the engine is started after having been off for at least 1 hr. This definition amounts to making the approximation that engine and catalyst cooling occur instantaneously at the end of the specified engine-off periods rather than gradually. Consequently, the definition avoids the need to measure emissions for a large number of different engine-off periods.

The functions F and H in equations (5.13) are estimated by fitting curves to data obtained directly from emissions measurements. However, the Modal Analysis Model [equation (5.3)] is used to generate the data from which the speed dependence functions G for gasoline-powered automobiles, gasoline-powered light trucks, and motorcycles are estimated. In this procedure, driving cycles representing trips at different average speeds are constructed from data obtained by driving instrumented vehicles in street traffic. Vehicles' exhaust emissions during these driving cycles then are estimated with the Modal Analysis Model rather than measured by driving vehicles through the cycles on dynamometers. The data from which the Modal Analysis Model is estimated do not include start-up emissions or emissions from light trucks or motorcycles. Therefore, it is assumed in the EPA procedure that the speed dependence of emissions during transient operation is similar to the speed dependence of emissions during hot stabilized operation and that the speed dependence of emissions from gasoline-powered light trucks and motorcycles is similar to that of automobiles.

There has not been a systematic evaluation of the accuracy of the estimates of motor vehicle emissions rates that are published by the EPA. However, the EPA estimates are the only readily available information on motor vehicle emissions rates, and they are in widespread use. All of the emissions computations performed in this book are based on the EPA estimates. Further information on the use of data published by the EPA to estimate motor vehicle emissions is available in reference [5].

5.8 *Conclusions*

Emissions controls for new motor vehicles have caused and will continue to cause substantial reductions in motor vehicles' emissions of air pollutants. Nonetheless, as discussed in chapter 2, motor vehicles in general and automobiles in particular are likely to remain important emissions sources in cities for the foreseeable future, and transportation-related air pollution will remain a problem in many cities. Therefore, it is useful to consider the possibility of reducing motor vehicle emissions through transportation and traffic management measures. The effects of such measures on emissions are discussed in chapter 6.

In chapter 6 it will be useful to bear in mind that several distinct components of motor vehicle emissions have now been identified. These are start-up (notably cold start) exhaust emissions, exhaust emissions from warmed-up (hot stabilized) vehicles, hot soak evaporations, and diurnal evaporations. (Crankcase emissions from most vehicles now in operation are negligible.) Within any given geographical area, the magnitudes of these emissions components are related in different ways to the traffic characteristics that are affected by transportation and traffic management measures. Transportation and traffic management measures affect motor vehicle trip volumes (i.e., total numbers of trips), vehicle miles traveled, and the quality of traffic flow (e.g., average speeds), among other characteristics. The size of the vehicle population also may be affected if changes in the transportation system cause people to alter the number of vehicles they own. Changes in the trip volume in an area will affect the magnitudes of start-up and hot soak emissions in the area. However, trip volume changes will affect the magnitudes of exhaust emissions from warmed-up vehicles only to the extent that they also affect aggregate vehicle miles traveled or traffic flow conditions in the area. Thus, for example, measures that mainly affect the frequencies of the shortest trips in an area may have large effects on the magnitudes of the start-up and hot soak components of emissions in the area while having relatively minor effects on the magnitude of the warmed-up running component. Measures that affect aggregate vehicle miles traveled or traffic flow characteristics in an area will affect the magnitude of the warmed-up running component of emissions but will not substantially affect the magnitudes of the start-up or hot soak components unless trip volumes are also affected. Finally, only measures that affect the number of hours per day that vehicles spend in an area will substantially affect the magnitude of diurnal evaporative emissions in the area. In an area the size of an entire metropolitan region, such measures consist mainly of changes in the size of the vehicle population.

References

1

Patterson, D. J., and Henein, N. A., *Emissions from Combustion Engines and Their Control*, Ann Arbor Science Publishers, Inc., Ann Arbor, MI, 1972.

2

Seinfeld, J. H., *Air Pollution: Physical and Chemical Fundamentals*, McGraw-Hill, New York, 1975.

3

US Environmental Protection Agency, *Control Techniques for Carbon Monoxide Emissions*, Report No. EPA-450/3-79-006, Research Triangle Park, NC, June 1979. NTIS Publication No. PB80-140510.

4

National Academy of Sciences, *Air Quality and Automobile Emission Control*, Vol. 4, Report by the Coordinating Committee on Air Quality Studies to the Committee on Public Works of the United States Senate, Serial No. 93-24, US Government Printing Office, September 1974.

5

US Environmental Protection Agency, *Mobile Source Emission Factors*, Report No. EPA-400/9-78-005, Washington, DC, March 1978. NTIS Publication No. PB 295672.

6

US Environmental Protection Agency, "Control of Air Pollution from New Motor Vehicles and New Motor Vehicle Engines: Certification and Test Procedures," *Code of Federal Regulations*, 40 CFR 86, revised as of 1 July 1980.

7

Calspan Corporation, *Automobile Exhaust Emission Surveillance: A Summary*, Report No. APTD-1544, US Environmental Protection Agency, Ann Arbor, MI, May 1973. NTIS Publication No. PB 220755.

8

Automotive Testing Laboratories, *A Study of Emissions from Passenger Cars in Six*

Cities, Report No. EPA-460/3-78-011a, US Environmental Protection Agency, Ann Arbor, MI, January 1979. NTIS Publication No. PB80-149800.

9

US Environmental Protection Agency, *The Cost of Clean Air and Water: Report to Congress,* Report No. EPA-230/3-79-001, Washington, DC, August 1979. NTIS Publication No. PB 300446.

10

General Motors Corporation, *Pocket Reference,* GM Technical Center, Warren MI, 1 February 1980.

11

Murrell, J. D., "Light Duty Automotive Fuel Economy . . . Trends Through 1979," SAE Paper 790225, Society of Automotive Engineers, Warrendale, PA, 1979.

12

Murrell, J. D., "Light Duty Automotive Fuel Economy . . . Trends Through 1978," SAE Paper 780036, Society of Automotive Engineers, Warrendale, PA, 1978.

13

US Environmental Protection Agency, *A Report on Automotive Fuel Economy,* Washington, DC, October 1973. NTIS Publication No. PB 258686.

14

Murrell, D., *Passenger Car Fuel Economy: EPA and Road,* Report No. EPA 460/3-80-010, US Environmental Protection Agency, Ann Arbor, MI, September 1980.

15

Lawrence, R., *Gasohol Test Program,* US Environmental Protection Agency, Ann Arbor, MI, December 1978. NTIS Publication No. PB 290569.

16

Gurney, M. D., Allsup, J. R., and Merlotti, C. L., "Gasohol: Laboratory and Fleet Test Evaluation," SAE Paper 800892, Society of Automotive Engineers, Warrendale, PA, 1980.

17

Naman, T. M., and Allsup, J. R., "Exhaust and Evaporative Emissions from Alcohol and Ether Fuel Blends," SAE Paper 800858, Society of Automotive Engineers, Warrendale, PA, 1980.

18

Stamper, K. R., "Evaporative Emissions from Vehicles Operating on Methanol/Gasoline Blends," SAE Paper 801360, Society of Automotive Engineers, Warrendale, PA, 1980.

19

Allsup, J. R., and Eccleston, D. B., *Ethanol/Gasoline Blends as Automotive Fuel,* Report No. BETC/RI-79/2, US Department of Energy, Bartlesville, OK, May 1979. NTIS Publication No. BETC/RI-79/2.

20

Allsup, J. R., *Experimental Results Using Methanol and Methanol/Gasoline Blends as Automotive Engine Fuel*, Report No. BERC/RI-76/15, US Department of Energy, Bartlesville, OK, January 1977. NTIS Publication No. BERC/RI-76/15.

21

Calspan Corporation, *An Evaluation of Restorative Maintenance on Exhaust Emissions of 1975-1976 Model Year In-Use Automobiles*, Report No. EPA-460/ 3-77-021, US Environmental Protection Agency, Ann Arbor, MI, December 1977. NTIS Publication No. PB 284031.

22

US Environmental Protection Agency, *Motor Vehicle Tampering Survey (1978)*, Washington, DC, November 1978.

23

US Environmental Protection Agency, *Control Strategies for In-Use Vehicles*, Washington, DC, November 1972.

24

Horowitz, J., "Inspection and Maintenance for Reducing Automobile Emissions: Effectiveness and Cost," *Journal of the Air Pollution Control Association*, Vol. 23, pp. 273–276, April 1973.

25

Meltzer, J., Hinton, M. G., Iura, T., Burke, A., Forrest, L., Smalley, W. M., and Augustine, F., *A Review of Control Strategies for In-Use Vehicles*, Report No. EPA-460/3-74-021, US Environmental Protection Agency, Ann Arbor, MI, December 1974. NTIS Publication No. PB 241768.

26

Kincannon, B. F., and Castaline, A. H., *Information Documents on Automobile Emissions Inspection and Maintenance Programs*, Report No. EPA-400/2-78-001, US Environmental Protection Agency, Washington, DC, February 1978. NTIS Publication No. PB 277776.

27

Cackette, T., Lorang, P., and Hughes, D., "The Need for Inspection and Maintenance for Current and Future Motor Vehicles," SAE Paper 790782, Society of Automotive Engineers, Warrendale, PA, 1979.

28

White J. T., Jones, G. T., and Niemczak, D. J., "Exhaust Emissions from In-Use Passenger Cars Equipped with Three-Way Catalysts," SAE Paper 800823, Society of Automotive Engineers, Warrendale, PA, 1980.

29

Becker, J. P., and Rutherford, J. A., "Analysis of Oregon's Inspection and Maintenance Program," Paper No. 79-7.3, presented at the 72nd annual meeting of the Air Pollution Control Association, June 1979.

30

Rutherford, J. A., and Waring, R. L., "Update on EPA's Study of the Oregon Inspection/Maintenance Program," Paper No. 80-1.2, presented at the 73rd annual meeting of the Air Pollution Control Association, June 1980.

31
Rutherford, J. A., *Automobile Exhaust Emission Surveillance—Analysis of the FY 1975 Program*, Report No. EPA-460/3-77-022, US Environmental Protection Agency, Ann Arbor, MI, December 1977. NTIS Publication No. PB 279535.

32
Eccleston, B. H., and Hurn, R. W., *Ambient Temperature and Vehicle Emissions*, Report No. EPA-460/3-74-028, US Environmental Protection Agency, Ann Arbor, MI, October 1974. NTIS Publication No. PB 247692.

33
Kunselman, P., McAdams, H. T., Domke, C. J., and Williams, M., *Automobile Exhaust Emission Modal Analysis Model*, Report No. EPA-460/3-74-005, US Environmental Protection Agency, Ann Arbor, MI, January 1974.

34
Edie, L. C., "Flow Theories," in D. C., Gazis, ed., *Traffic Science*, John Wiley & Sons, New York, 1974, pp. 1–108.

35
Cohen, S. L., Euler, G. W., Radelat, G., and Ross, P., "Energy and Emissions: The Simulation Approach," in *Transportation and Energy*, American Society of Civil Engineers, New York, 1978, pp. 193–215.

36
Cohen, S. L., and Euler, G., "Signal Cycle Length and Fuel Consumption and Emissions," *Transportation Research Record*, No. 667, pp. 41–48, 1978.

37
Cohen, S. L., "Use of Traffic Simulation in Analysis of Carbon Monoxide Pollution," *Transportion Research Record*, No. 648, pp. 74–76, 1977.

38
Lieberman, E. B., and Cohen, S., "New Technique for Evaluation of Urban Traffic Energy Consumption and Emissions," *Transportation Research Record*, No. 599, pp. 41–45, 1976.

39
Ellis, G. W., Camps, W. T., and Treadway, A., *The Determination of Vehicular Cold and Hot Operating Fractions for Estimating Highway Emissions*, US Department of Transportation, Washington, DC, September 1978.

40
Horowitz, J. L., and Pernela, L. M., "Analysis of Automobile Emissions According to Trip Type," *Transportation Research Record*, No. 492, pp. 1–8, 1974.

41
Horowitz, J. L., and Pernela, L. M., "Comparison of Automobile Emissions Based on Trip Type in Two Metropolitan Areas," *Transportation Research Record*, No. 580, pp. 13–21, 1976.

6 *Transportation and Traffic Management Measures*

A GENERAL CONCEPTUAL FRAMEWORK

6.1 *Introduction*

The term "transportation and traffic management measures" is used in this book to refer to actions whose objective is to change traffic volumes or traffic flow conditions and, thereby, motor vehicle emissions. Examples of such actions include synchronization of the signal lights on an arterial street, improvements in transit service quality, and changes in parking or gasoline prices. Over long periods of time, policies whose objective is to change urban travel patterns by changing urban land use patterns also can be included among transportation and traffic management measures.

The numbers of different transportation and traffic management measures that can be identified, ways in which these measures can be combined together, and conditions under which they can be implemented are vast. It is neither feasible nor desirable to present an exhaustive catalog of these measures here or to attempt to provide precise and generally applicable estimates of the effects of each measure or combination of measures on emissions or concentrations. Rather, the objectives of this chapter are to show how the problem of estimating the effects of transportation and traffic management measures on emissions can be dealt with in a systematic way, to identify factors that can have important influences on the emissions reduction effectiveness of different types of measures, and to provide estimates of the ranges of emissions and, where possible, concentration reductions that can be achieved by implementing various broad types of measures.

There have been few meaningful attempts to measure the air quality effects of transportation and traffic management measures, either directly through observations of air quality changes resulting from implementation of such measures or indirectly through observation of changes in appropriate surrogate variables. Therefore, the effects estimates that are presented here rely heavily on results derived from qualitative reasoning and mathematical models. Whenever possible, measurements of the effects of transportation and traffic management measures on air quality, emissions, and related variables are presented to corroborate these estimates.

Most of the mathematical models that are used in this chapter are relatively simple ones that can be implemented by hand or with the aid of a desk calculator. Although the elaborate, computer-based models that usually are used in transportation planning can treat a wider range of transportation and traffic management options, incorporate more detailed representations of the characteristics of these options, and provide more detailed indicators of the effects of implementing the options than simpler models can, the simpler models provide results that are quite satisfactory for many purposes. Moreover, the simpler models, which can be made to incorporate the same analytical concepts as advanced computer-based models do, are more accessible and better suited to clarifying relations among variables than the computer-based models are. Although the results of several computer-based analyses are presented here—partly for the purpose of corroborating results obtained with simpler methods—the reader is referred to sources cited in the text for detailed information on computer-based models.

The state of the art in transportation systems analysis is such that the effects of transportation and traffic management measures on small-scale spatial variations in CO emissions rates (e.g., variations along individual links of a roadway network) can be predicted only with great difficulty if they can be predicted at all. This is due partly to the complexity of currently available microscale models of traffic flow and partly to the inability of these models to predict the changes in traffic volumes that transportation and traffic management measures can cause. In order to minimize the complexity of the analytical methods that are used here, small-scale spatial variations in CO emissions and concentrations are not treated in this chapter. Rather, the discussion of the effects of transportation and traffic management measures on CO emissions deals mainly with average CO emissions rates along roadway links and in corridors. Link and corridor average CO emissions rates are approximately linearly related to spatially averaged CO concentrations on the links or in the corridors in question, but are not necessarily good predictors of CO concentrations at individual monitoring sites. Occasionally, reference will be made to total CO emissions from groups or entire networks of roadways. Such CO emissions totals contribute to general background CO concentrations, but are not closely related to CO concentrations along individual roadways. Techniques that may be useful for analyzing small-scale effects of transportation and traffic management measures on CO emissions rates were discussed briefly in section 5.7 and will be discussed briefly again in section 6.3.

HC and NO_x emissions are of concern primarily because of their contributions to regional O_3 concentrations. Because excessive O_3 concentrations in an urban region usually are caused by HC and NO_x emissions that take place over very large areas, possibly exceeding the size of the region, the effects of transportation and traffic management measures on HC and NO_x emissions are discussed in terms of total emissions from groups or entire networks of roadways.

This chapter is concerned mainly with measures for reducing emissions from motor vehicles that are used for personal transportation, notably automobiles.[1] The discussions of CO control measures in section 6.2 and traffic engineering measures in sections 6.3–6.5 also apply to traffic flows that include service and freight vehicles. There is no discussion of measures that are oriented primarily toward reducing emissions from service and freight vehicles. Although these vehicles can be significant sources of emissions in cities, there has been little research on transportation and traffic management measures for reducing their emissions, and little is known about such measures.

The chapter is divided into six parts for expositional clarity. Part A, consisting of this section and the next, presents a general conceptual framework for the rest of the chapter. Part B, consisting of sections 6.3–6.5, discusses traffic engineering measures for reducing motor vehicle emissions. Part C, which includes sections 6.6–6.9, discusses transit improvement and carpool incentive measures. Part D, consisting of sections 6.10–6.12, is concerned with traffic pricing and restraint measures. Part E discusses land use measures, and part F presents a general summary of the chapter.

6.2 *Inferences Based on Aggregate Characteristics of Urban Travel*

Important insights into the potential effectiveness of transportation and traffic management measures in improving air quality can be obtained by investigating the relative magnitudes of the emissions caused by trips with different purposes, origins and destinations, and times of day. Although such investigations cannot yield precise results, they can help greatly to establish realistic expectations concerning the effects of transportation and traffic management measures on emissions and concentrations of air pollutants. For example, it is possible to identify broad classes of trips that transportation and traffic management measures likely must affect if these measures are to be effective in reducing emissions and improving air quality. In addition, for some pollutants it is possible to develop rough upper bounds on the concentration reductions that transportation and traffic management measures can be expected to achieve.

Consider CO first. Significant reductions in excessive CO concentrations at a site usually can be achieved by reducing traffic-related emissions in the vicinity of the site during the hours that the concentrations are excessive. If the CO concentration tends to exceed the 1-hr CO air quality standard during several hours of the day, then correcting this condition requires reducing CO emissions during each of the hours in question. Examples of broad classes of transportation and traffic management measures that may achieve this result include improving traffic flow conditions along roadways where CO concentrations are excessive, diverting traffic from roadways where CO concentrations are relatively high to roadways where CO concentrations are relatively low, encouraging travelers to use carpools and transit instead of low-occupancy automobiles, and, possibly, encouraging travelers to change the times at which they travel from hours with relatively high CO concentrations to hours with relatively low CO concentrations.

Correcting violations of the 8-hr CO air quality standard requires achieving emissions reductions that are significant on 8-hr time scales. The reductions need not be spread evenly over the 8-hr periods in question, and it may be tempting in many cases to orient emissions reduction measures primarily toward peak period traffic, particularly peak period work trips. Some rough indications of the reductions in 8-hr average CO concentrations that can be achieved by reducing emissions only in peak periods can be obtained as follows. Peak period traffic (including truck traffic) usually accounts for roughly 40% of the total traffic volume on major urban roadways during an 8-hr period that includes a peak [1, 2].[2] Work trips usually account for 30–60% of total peak period traffic volumes in cities, although they undoubtedly account for larger proportions of traffic on some roadways [1, 2]. If the approximation is made that the CO concentrations and emissions rates along roadways are proportional to the traffic volumes on the roadways (this approximation is justified by the relations between traffic volumes and CO concentrations that are presented in section 3.2), then these statistics imply that all peak period traffic is responsible for roughly 40% of 8-hr average CO emissions and concentrations. They also imply that peak period work trip traffic typically is responsible for 12–24% (30–60% of 40%) of 8-hr average emissions and concentrations. Thus, for example, a 50% reduction in CO emissions from all peak period traffic would reduce 8-hr average CO concentrations by approximately 20%. A 50% reduction in CO emissions from peak period work trip traffic would reduce 8-hr average CO concentrations by 6–12%. These estimates indicate that relatively large reductions in peak period traffic volumes and emissions often are likely to produce relatively small reductions in 8-hr average CO emissions and concentrations.

Because the effects of peak period emissions reductions are likely to be diminished greatly when they are averaged over 8-hr periods, in most situations achieving large reductions in 8-hr average CO concentrations is likely to require reducing emissions during the entire 8-hr period of interest. (Possible exceptions to this statement include cases in which peak period emissions account for unusually large proportions of total 8-hr emissions, very large reductions in peak period emissions are possible, or meteorological conditions are such that off-peak emissions contribute little to 8-hr average concentrations.) Transportation and traffic management measures that improve traffic flow conditions on the appropriate roadways, divert traffic to alternative routes, or encourage the use of high-occupancy modes all may be effective in achieving the needed emissions reductions if the measures' effects on traffic can be maintained throughout the entire 8-hr period. However, measures that are designed mainly to shift travel from peak periods to other times of day are unlikely to have substantial effects on 8-hr average CO concentrations, particularly if the principal effect of these measures is to redistribute peak period traffic within the same 8-hr period in which excessive CO concentrations occur.

Achieving substantial reductions in O_3 concentrations in and near cities requires achieving reductions in HC and, possibly, NO_x emissions that are significant on at least a metropolitan regional scale.[3] Most of the transportation and traffic management measures that usually are considered for reducing regional motor vehicle emissions (e.g., transit improvements, carpool incentives, and various types of pricing and restraint measures) are oriented mainly toward reducing emissions from automobiles. As will be seen later in this chapter, the reductions in regional HC and NO_x emissions from automobiles that most of these measures can achieve are likely to be less than 15% for each pollutant. Given this information, a rough upper bound on the reductions in O_3 concentrations that can be achieved during the mid-1980s by automobile-oriented transportation and traffic management measures can be developed as follows. In the mid-1980s automobiles are expected to be responsible for roughly 15–40% of HC and NO_x emissions in cities, depending on the city [4–6]. Therefore, reductions of less than 15% in regional automobile emissions of HC and NO_x would produce reductions of less than 6% in total regional HC and NO_x emissions from all sources. The resulting reductions in O_3 concentrations would depend on factors such as existing emissions levels, the relative magnitudes of the HC and NO_x emissions reductions, and the spatial and temporal distributions of the emissions reductions. However, measures that reduce HC and NO_x emissions by less than 6% would be unlikely to achieve substantially larger (e.g., more than twice as large) percentage reductions in O_3 concentrations. Accordingly, it is likely that the reductions in regional maximum O_3 concentrations that can be achieved by all but the most stringent transportation and traffic management measures are less than roughly 12%. (For practical purposes, this rough bound can be rounded to 10%, and this will be done in the subsequent discussion.)

Further insight into the potential effects of transportation and traffic management measures on O_3 concentrations can be gained by examining relations between urban automobile emissions and travel patterns. Trips that have at least one end in the suburbs (i.e., outside of the principal city) account for 80–90% of automobile HC and NO_x emissions in urban areas, and trips with both ends in the suburbs account for 50–70% of automobile emissions of these pollutants [3, 7]. All work trips account for 30–40% of automobile HC and NO_x emissions, but work trips to the principal city account for only about 20% of total automobile HC and NO_x emissions, and peak period work trips to the principal city cause slightly more than 10% of automobile emissions of these pollutants [3, 7]. These statistics suggest that to be effective in reducing O_3 concentrations, transportation and traffic management measures almost certainly must reduce emissions from suburban travel, nonwork travel, or some combination of the two. Moreover, measures that affect only peak period work trips to the principal city are likely to have minimal effects on O_3 concentrations. For example, a 15% reduction in HC and NO_x emissions from such trips would reduce total HC and NO_x emissions from automobiles by less than 2% and would reduce total emissions of these pollutants from all sources by less than 1%. It is unlikely that emissions reductions of this magnitude, by themselves, would have perceptible effects on regional O_3 concentrations.

Annual average NO_2 concentrations tend to be associated with region-wide NO_x emissions. Hence, to be effective in reducing annual average NO_2 concentrations, transportation and traffic management measures must be effective in reducing region-wide NO_x emissions from motor vehicles. Reducing traffic-related NO_x emissions tends to be more difficult than reducing HC emissions is. Therefore, since automobiles will be responsible for at most 40% of NO_x emissions in and near cities during the mid-1980s, the reductions in annual average NO_2 concentrations that can be achieved through the implementation of most transportation and traffic management measures are likely to be less than 10%.

Short-term (e.g., 1-hr) average NO_2 concentrations currently are not regulated in the United States, and there has been little research into the effects that implementation of transportation and traffic management measures might have on these concentrations. However, given the tendency of high short-term average NO_2 concentrations to occur near heavily traveled roadways (among other places) as a result of chemical reactions between O_3 and the NO emitted by motor vehicles [see reaction (3.4)], it is likely that with one important exception, the effects of transportation and traffic management measures on short-term NO_2 concentrations near roadways would be similar to these measures' effects on CO concentrations for the same averaging time. The exception concerns measures whose main effects are to increase traffic speeds. Whereas such measures tend to reduce CO emissions rates per vehicle mile traveled, they tend to increase NO_x emissions rates (see figure 5.8). In addition, it is possible that measures that reduce regional O_3 concentrations may be helpful in reducing short-term NO_2 concentrations, owing to the importance of the O_3-NO reaction in producing high levels of NO_2 near roadways.

B TRAFFIC ENGINEERING MEASURES

Two broad classes of measures are discussed in this part. One class consists of traffic flow improvement measures and is discussed in section 6.3. The other class consists of vehicle-free zones and traffic cell systems and is discussed in section 6.4. Section 6.5 presents a summary of the discussion of traffic engineering measures. Although most of the discussion in this part deals explicitly only with automobile emissions, the main conclusions that are reached apply also to emissions from traffic streams that include trucks.

6.3 Measures to Improve Traffic Flow

The average HC and CO emissions rates of the vehicles in a traffic stream tend to decrease when the frequencies and durations of accelerations and idles decrease and when the average travel speed increases. Measures whose main purpose is to affect traffic flow conditions in these ways are called "traffic flow improvement measures." These measures, which usually are considered for implementation in connection with efforts to reduce congestion, include a wide variety of traffic engineering actions, such as:

- widening of roadways and intersections;
- installation of turn lanes and controls or restrictions on turning movements;
- on-street parking prohibitions;
- conversion of two-way streets into one-way streets;
- installation of reversible lanes on major arterial streets;
- improving signalization, including synchronizing signal lights;
- implementation of ramp metering on freeways.

Most of these measures are designed to improve traffic flow conditions and reduce congestion in relatively small areas, such as the central business districts of cities, or in heavily traveled corridors. Traffic flow improvement measures usually do not affect traffic flow conditions over entire urban areas. Consequently, the air quality improvement value of these measures consists mainly of their potential ability to achieve reductions in localized CO concentrations. Traffic flow improvement measures often increase NO_x emissions, as NO_x emissions rates from gasoline-powered motor vehicles tend to increase as average speeds increase.[4]

Increasing the average speed of traffic flow along a roadway can cause significant reductions in the HC and CO emissions rates of individual vehicles on the roadway. This is particularly true if the average speed is low prior to increasing it. For example, increasing the average speed of flow on an arterial street from 15 to 20 mi/hr would reduce the average rates of HC and CO emissions (expressed in units of g/veh mi) from warmed-up, gasoline-powered automobiles on the street by approximately 20% for each pollutant. The average rate of NO_x emissions from these vehicles would increase by roughly 10% [8]. Increasing the average speed of traffic flow on a freeway from 40 to 45 mi/hr would reduce the average HC and CO emissions rates (also expressed in units of g/veh mi) from warmed-up, gasoline-powered automobiles on the freeway by roughly 5% for each pollutant while increasing the average rate of NO_x emissions by a similar percentage [8].[5]

If traffic flow improvements are implemented on a roadway and the volume of traffic on the roadway does not change, then the percentage change in total emissions of a pollutant from vehicles on the roadway will equal the percentage change in the average g/veh mi emissions rate of the pollutant. However, improvements in traffic flow conditions frequently cause traffic volumes on the improved roadways to increase. This is because flow improvements tend to increase the convenience of travel on these roadways, thereby attracting traffic from competing routes and, possibly, from nonautomobile modes. In addition, flow improvements may draw increased numbers of trips to areas served by the improved roadways. These trips previously may have gone to other areas not served by the roadways, or they may not have been made at all prior to the flow improvements. If the flow improvements occur mainly during relatively short time periods (e.g., during peak periods), then traffic volumes may increase during these periods due to attraction of trips from other times of day.

Increases in traffic volumes decrease the HC and CO emissions reduction effectiveness of traffic flow improvement measures in two ways. First, given fixed g/veh mi emissions rates, increases in traffic volumes cause proportional increases in aggregate HC and CO emissions. Second, increases in traffic volumes cause average speeds to decrease, relative to the speeds that would have prevailed with constant volumes, owing to increased congestion. This causes g/veh mi HC and CO emissions rates to increase, thereby further increasing aggregate emissions. Increases in traffic volumes also cause aggregate NO_x emissions to increase, other things remaining unchanged, but the accompanying decreases in average speeds tend to decrease the g/veh mi NO_x emissions rates of gasoline-powered vehicles. Therefore, increases in traffic volumes may either accentuate or moderate the effects of traffic flow improvement measures on aggregate NO_x emissions, depending on which of these competing factors is dominant.

Because of the changes in traffic volumes that tend to accompany traffic flow improvement measures, these measures can have unexpected and counterproductive effects on aggregate HC and CO emissions along a roadway. For example, implementing traffic flow improvement measures along a roadway may cause aggregate emissions of these pollutants on the roadway to increase, even though the average speed of traffic on the roadway increases. In addition, implementing traffic flow improvement measures on one roadway may cause either increases or decreases in aggregate HC and CO emissions on other roadways. Several numerical examples that illustrate these effects are presented later in this section.

6.3.1 A Simple Predictive Model

Quantitative prediction of the effects of traffic flow improvement measures on emissions is, in general, a difficult task. There are at least two reasons for this. First, it is difficult to predict the effects of specific flow improvement measures on traffic flow conditions, even if the traffic volumes on the affected roadways do not change. Second, roadways are linked together into networks, and travelers can compare flow conditions in different parts of these networks when making travel decisions. Therefore, the traffic volume on a given roadway usually depends on traffic flow conditions on other roadways as well as on its own traffic flow conditions. Similarly, improving traffic flow conditions on one roadway usually will cause traffic volumes and flow conditions to change on other roadways as well as on the roadway where the flow improvement measures are implemented. As a result, predicting the effects of traffic flow improvement measures on traffic volumes, traffic flow conditions, and emissions usually requires predicting the outcome of a complex set of interactions among the roadways in the network.

Although prediction of the effects of traffic flow improvement measures on emissions is difficult, the principal elements of the relation between flow improvements and emissions can be represented in a relatively simple mathematical model. This model can be used directly to estimate the effects of traffic flow improvement measures in simple roadway networks, and it has the same basic components that are needed to make predictions for more complex networks. The components of the model are as follows.

1. A Demand Function

This specifies the relation between the demand for travel on each link of the roadway network and the traffic flow conditions on these links. The demand for travel on a link is expressed in terms of traffic volume (i.e., the average number of vehicles that pass over the link in a unit of time). It is assumed in this model that traffic flow conditions on a link can be described adequately by the average speed of traffic on the link. Therefore, the demand for travel on a link i can be written as

$$V_i = D_i(\mathbf{s}), \tag{6.1}$$

where V_i is the demanded traffic volume on the link, $\mathbf{s}$ is a vector of traffic speeds on all relevant links of the network, and D_i is the demand function. D_i describes the behavior of travelers who are faced with a set of traffic flow conditions $\mathbf{s}$. Specifically, it summarizes travelers' decisions whether to travel on link i under these conditions. D_i does not specify the traffic volume that link i actually can accommodate under the flow conditions $\mathbf{s}$.

2. *A Link Performance or Supply Function*

This describes the relation between the speed of traffic on a link, the volume of traffic that the link is capable of carrying at that speed, and the physical design and operation of the link. It is assumed in this model that the physical design and operation of a link can be described adequately by the link's physical or practical capacity.[6] It is also assumed that traffic flow improvement measures can be represented as changes in physical or practical capacity. Hence the relation between traffic speed, traffic volume, and physical design for a link i can be written as

$$s_i = P_i(V_i, c_i), \tag{6.2}$$

where s_i is the speed on link i, V_i is the volume on link i, c_i is the physical or practical capacity of link i, and P_i is the link i performance function. In the following discussion, c_i will be called, simply, the capacity of link i and can be interpreted as referring to either physical or practical capacity unless otherwise indicated.

3. *An Emissions Function*

This specifies the relation between the average emissions rates of individual vehicles on a link and traffic flow conditions on the link. Since traffic flow conditions are described by speeds in this model, the emissions function is a speed-emissions relation. This relation is different for different pollutants, but it is assumed to be the same for all roadway links. Average emissions rates are expressed in terms of mass of pollutant emitted per unit of distance traveled (e.g., g/veh mi). Thus, the average rate of emissions of pollutant p by vehicles on link i can be written as

$$e_{pi} = R_p(s_i), \tag{6.3}$$

where e_{pi} is the pollutant p emissions rate on link i, s_i is the speed on link i, and R_p is the emissions function for pollutant p.

Let E_{pi} denote the density (or, more precisely, the flux) of emissions of pollutant p on roadway link i. E_{pi} is expressed in units of mass per unit distance of roadway per unit of time (e.g., g/roadway mi/hr). Then

$$E_{pi} = V_i R_p(s_i), \tag{6.4}$$

where V_i is the actual traffic volume on link i. This volume must be consistent with travel demand at the prevailing speeds, and the speeds must be consistent with the traffic volumes and roadway capacities. Therefore, for every link i

$$V_i = D_i(\mathbf{s}) \tag{6.5}$$

and

$$s_i = P_i(V_i, c_i). \tag{6.6}$$

If the capacities c_i are given, then equations (6.5) and (6.6) determine the volumes V_i and speeds s_i. Hence, by virtue of equation (6.4), the capacities also determine the emissions densities on each link. Total emissions of a pollutant on a link can be obtained by multiplying the emissions density of the pollutant on the link by the length of the link.[7]

6.3.2 Applications of the Simple Model

The following three examples illustrate the application of the model of equations (6.4)–(6.6) and provide additional insight into the effects of traffic flow improvement measures on emissions. The first two examples deal only with CO emissions. CO is discussed in terms of emissions densities along roadways in order to emphasize the localized nature of CO emissions and concentrations. The examples show that implementing traffic flow improvement measures on a roadway can cause the CO emissions densities on that roadway and on nearby roadways either to increase or to decrease, depending on the circumstances in which the measures are implemented. The examples also show that the CO emissions density on a roadway does not necessarily decrease when congestion on the roadway is reduced. The third example treats HC and NO_x emissions, in addition to CO emissions. Since HC and NO_x usually are of interest because of their effects on regional O_3 concentrations, HC and NO_x are discussed mainly in terms of total emissions from all of the roadways being considered. The example shows that traffic flow improvement measures may decrease emissions of some pollutants while increasing emissions of others. The example also shows that even measures which cause trip volumes and lengths to increase may reduce emissions of some pollutants.

Example 6.1—Implementation of Traffic Flow Improvement Measures on One of Two Parallel Roadways

Assume that two locations A and B are connected by two parallel roadways, 1 and 2, as shown in figure 6.1a. The two roadways have equal lengths and carry traffic only in the direction A to B. Let V denote the total volume of traffic that flows from A to B, and let V_1 and V_2 denote the traffic volumes on roadways 1 and 2, respectively. V is assumed to be independent of the speeds of travel on the two roadways. V, V_1, and V_2 must satisfy

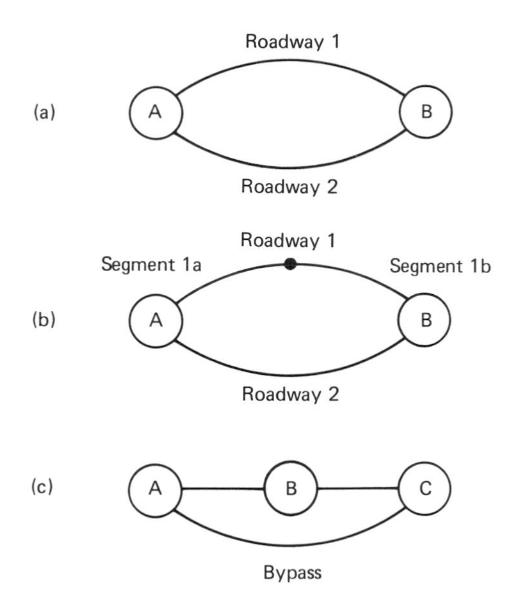

Figure 6.1 Examples of simple roadway networks: (*a*) examples 6.1 and 6.7; (*b*) example 6.2; (*c*) example 6.3.

$$V = V_1 + V_2. \tag{6.7}$$

It is assumed that travelers from A to B choose the roadway that minimizes their travel times. Since the lengths of roadways 1 and 2 are equal, this is equivalent to choosing the roadway with the higher speed. Hence, the demand function for roadway i ($i = 1$ or 2) is

$$D_i(s_1, s_2) = \begin{cases} 0 & \text{if} \quad s_i < s_j \\ V & \text{if} \quad s_i > s_j, \end{cases} \tag{6.8}$$

where $j = 1$ or 2, $j \neq i$, and s_i is the speed on roadway i. If $s_1 = s_2$, then the volumes on the two roadways have the values that are consistent with equation (6.7) and the roadways' performance functions. The performance functions are given by

$$P_i(V_i, c_i) = s_0/[1 + 0.9(V_i/c_i)^{4.5}], \qquad i = 1 \text{ or } 2, \tag{6.9}$$

where s_0 is the free-flow speed (assumed to be equal on both roadways) and c_i is the practical capacity of roadway i. In this example, s_0 is set at 23 mi/hr.[8] The performance functions and the demand function together imply that traffic distributes itself between the two roadways such that $s_1 = s_2$ and $V_1/c_1 = V_2/c_2$.

The speed-emissions relation for CO on the roadways is given by

$$e_{CO}(s) = 6.4 + 464/s, \tag{6.10}$$

where s is in mi/hr and e_{CO} is in g/veh mi. In the speed range 10–25 mi/hr, equation (6.10) is a good approximation to the predicted composite CO speed-emissions relation of warmed-up vehicles in the 1982 automobile fleet. The composite consists of a weighted average of the speed-emissions relations of the different model years of automobiles that are expected to comprise the 1982 fleet. The weights reflect the proportions of vehicles from each model year that are likely to be present in a typical traffic stream.

If the values of V, s_0, and the c's are given, then the demand function and equations (6.8) and (6.9) together determine the traffic volumes, speeds, and emissions densities on roadways 1 and 2. The volumes are given by

$$V_i = V\frac{c_i}{c_1 + c_2}, \qquad i = 1 \text{ or } 2. \tag{6.11}$$

The speeds are obtained by substituting these volumes into equation (6.9), and the emissions densities are obtained by substituting the speeds and volumes into equations (6.10) and (6.4).

Assume that c_2, the practical capacity of roadway 2, is fixed and that traffic flow improvement measures are implemented on roadway 1, thereby causing c_1, the practical capacity of roadway 1, to increase. Table 6.1 shows the CO emissions densities on roadways 1 and 2 as functions of c_1 for several different values of c_2 and the total traffic volume V. It can be seen from the table that increasing the practical capacity of roadway 1 causes the speed on both roadways to increase. Thus, congestion is reduced on both roadways. The speed increase on roadway 2 is caused by diversion of traffic away from that roadway. As a result of this diversion, increasing the practical capacity of roadway 1 causes the CO emissions density on roadway 2 to decrease. However, increasing c_1 may either increase or decrease the CO emissions density on roadway 1. When $V =$ 2,250 veh/hr and $c_2 = 750$ veh/hr, increasing c_1 causes the roadway 1 CO emissions density to decrease. When $V = 2{,}250$ veh/hr and $c_2 =$ 1,500 veh/hr, increasing c_1 causes the roadway 1 CO emissions density to increase. When $V = 1{,}500$ veh/hr and $c_2 = 750$ veh/hr, increasing c_1 decreases the CO emissions density on roadway 1 until c_1 reaches 1,500 veh/hr. Further increases in c_1 beyond this value cause the CO emissions density on roadway 1 to increase. Changes in c_1 tend to cause larger percentage changes in the roadway 2 CO emissions density than in the roadway 1 emissions density. For example, when $V = 2{,}250$ veh/hr and $c_2 = 750$ veh/hr, increasing c_1 from 1,500 to 2,250 veh/hr (roughly the equivalent of adding one lane of traffic to roadway 1) reduces the roadway 1 CO emissions density by roughly 20%, but it reduces the roadway 2 emissions density by nearly 50%. Similarly, when $V = 1{,}500$ veh/hr and $c_2 = 750$ veh/hr, increasing c_1 from 750 to 1,500 veh/hr reduces the CO emissions densities on roadways 1 and 2 by 12 and 65%, respectively. When $V = 2{,}250$ veh/hr and $c_2 = 1{,}500$ veh/hr, increasing c_1 from 750 to 1,500 veh/hr causes the roadway 1 CO emissions density to increase by 6%, whereas the roadway 2 CO emissions density decreases by nearly 50%.

Example 6.2—Implementation of Traffic Flow Improvement Measures on One Segment of a Two-Segment Roadway

Assume that roadway 1 in example 6.1 consists of two segments, 1a and 1b, as shown in figure 6.1b. If traffic flow improvement measures are implemented on segment 1a, thereby causing its practical capacity to increase, but the practical capacity of segment 1b remains unchanged, then the CO emissions density on segment 1b will tend to increase, even if the CO emissions density on segment 1a decreases. This is because increasing the practical capacity of segment 1a increases the effective practical capacity of route 1 as a whole and, therefore, causes traffic to divert to this route from route 2. However, the additional traffic volume of route 1 causes the speed on segment 1b to decrease. The CO emissions density on this segment increases as a consequence of the combined effects of increased traffic volume and decreased speed. An analogous conclusion would be reached if the practical capacity of segment 1b were increased but the practical capacity of segment 1a remained constant.

The effect of increasing the practical capacity of segment 1a but not of 1b can be illustrated quantitatively as follows. Assume, as in example 6.1, that all traffic flows from A to B and that travelers choose the route that minimizes their travel time. If the travel times are equal, then the volumes on the two roadways are selected to be consistent with equation (6.7) and the roadways' performance functions. Let route 1 and route 2 have equal lengths, and let segments 1a and 1b also have equal lengths. Let roadway 2 and each segment of roadway 1 have the performance function of equation (6.9) (with $i = 1\text{a}, 1\text{b}$, or 2). Note that $V_{1a} = V_{1b}$. Define c_{eff}, the effective practical capacity of roadway 1, as the practical capacity that causes the average speed of travel from A to B along route 1 to be given by

Table 6.1 Traffic flow characteristics and CO emissions densities on two parallel roadways (from example 6.1)

Total traffic volume (veh/hr)	Practical Capacity (veh/hr)		Traffic volume (veh/hr)		Speed on both roadways (mi/hr)	Average CO emissions rate of vehicles on both roadways (g/veh mi)	CO emissions densities (kg/roadway mi/hr)	
	Roadway 1	Roadway 2	Roadway 1	Roadway 2			Roadway 1	Roadway 2
1,500	750	750	750	750	12.1	44.7	33.5	33.5
	1,000	750	857	643	15.9	35.6	30.6	22.9
	1,250	750	938	562	18.4	31.5	29.6	17.7
	1,500	750	1,000	500	20.1	29.5	29.5	14.8
	1,750	750	1,050	450	21.1	28.4	29.8	12.8
	2,000	750	1,091	409	21.7	27.8	30.3	11.4
	2,250	750	1,125	375	22.1	27.4	30.8	10.3
	2,500	750	1,154	346	22.4	27.1	31.3	9.4
	2,750	750	1,179	321	22.6	27.0	31.8	8.7
	3,000	750	1,200	300	22.7	26.9	32.2	8.1
2,250	1,500	750	1,500	750	12.1	44.7	67.1	33.5
	1,750	750	1,575	675	14.7	37.9	59.6	25.6
	2,000	750	1,636	614	16.9	33.9	55.5	20.8
	2,250	750	1,688	562	18.4	31.5	53.2	17.7
	2,500	750	1,731	519	19.6	30.0	52.0	15.6
	2,750	750	1,768	482	20.5	29.1	51.4	14.0
	3,000	750	1,800	450	21.1	28.4	51.1	12.8
	750	1,500	750	1,500	12.1	44.7	33.5	67.1

1,000	1,500	900	1,350	14.7	37.9	34.1	51.1
1,250	1,500	1,023	1,227	16.9	33.9	34.7	41.6
1,500	1,500	1,125	1,125	18.4	31.5	35.5	35.5
1,750	1,500	1,212	1,038	19.6	30.0	36.4	31.2
2,000	1,500	1,286	964	20.5	29.1	37.4	28.0
2,250	1,500	1,350	900	21.1	28.4	38.3	25.6
2,500	1,500	1,406	844	21.5	27.9	39.3	23.6
2,750	1,500	1,456	794	21.9	27.6	40.2	21.9
3,000	1,500	1,500	750	22.1	27.4	41.1	20.5

$$s_{ave} = \frac{s_0}{1 + 0.9(V_1/c_{eff})^{4.5}}, \tag{6.12}$$

where V_1 is the traffic volume on route 1. Then c_{eff} is related to the individual segments' practical capacities, c_{1a} and c_{1b}, by

$$c_{eff} = [0.5\,(c_{1a}{}^{-4.5} + c_{1b}{}^{-4.5})]^{-1/4.5}. \tag{6.13}$$

Moreover, if V is the total volume of traffic flowing from A to B, then V_1 is given by

$$V_1 = V\frac{c_{eff}}{c_{eff} + c_2}, \tag{6.14}$$

and V_2 is obtained from V_1 and V by means of equation (6.7). Given V_1 and V_2, the speeds on segments 1a and 1b and on roadway 2 can be computed from the performance functions, and the CO emissions densities can be computed from equations (6.4) and (6.10).

Table 6.2 shows the relation between c_{1a} and the CO emissions densities on segments 1a and 1b and roadway 2 for the case $V = 1{,}350$ veh/hr, $c_{1b} = c_2 = 750$ veh/hr, and $s_0 = 23$ mi/hr. As expected, increases in c_{1a} cause decreases in the segment 1a and roadway 2 emissions densities and cause increases in the segment 1b emissions density. Also as expected, the speed on segment 1b decreases as c_{1a} increases. Because the fixed practical capacity of segment 1b limits the number of vehicles that divert to roadway 1 from roadway 2, increases in c_{1a} tend to cause larger percentage reductions in the segment 1a CO emissions density than in the roadway 2 density. For example, increasing c_{1a} from 750 to 1,500 veh/hr causes the segment 1a CO emissions density to decrease by 23%, whereas the roadway 2 emissions density decreases by 15%. The corresponding increase in the emissions density on segment 1b is 19%.

Example 6.3—Bypassing a Congested Area

Assume that a roadway connects three locations, A, B, and C, as shown in figure 6.1c. Consider travel in the direction A to C, and assume that the roadway carries both local traffic associated with trips that originate or terminate at B and through traffic associated with trips from A to C that do not stop at B. The distances from A to B and from B to C are each 1 mi.

Suppose that the roadway is congested and that a bypass route is opened to enable the through traffic to flow directly from A to C without passing through B. Let the length of the bypass route be 3 mi. If the volumes of traffic flowing from A to B, B to C, and A to C all stay constant, then

Table 6.2 Traffic flow characteristics and CO emissions densities on a two-segment roadway and a parallel one-segment roadway (from example 6.2)[a]

Practical capacity of segment 1a (veh/hr)	Traffic volume (veh/hr)		Speed (mi/hr)			CO emissions density (kg/roadway mi/hr)		
	Segments 1a and 1b	Roadway 2	Segment 1a	Segment 1b	Roadway 2	Segment 1a	Segment 1b	Roadway 2
750	675	675	14.7	14.7	14.7	25.6	25.6	25.6
1,000	709	641	19.3	13.6	15.9	21.6	28.8	22.8
1,250	720	630	21.4	13.1	16.3	20.2	30.0	22.0
1,500	724	626	22.2	13.0	16.4	19.7	30.4	21.7
1,750	725	625	22.6	13.0	16.5	19.5	30.6	21.6
2,000	726	624	22.8	12.9	16.5	19.4	30.7	21.5
2,250	726	624	22.9	12.9	16.5	19.4	30.7	21.5
2,500	727	623	22.9	12.9	16.5	19.4	30.7	21.5

a. Segment 1b and roadway 2 each have practical capacities of 750 veh/hr. The total traffic volume on roadways 1 and 2 combined is 1,350 veh/hr.

opening the bypass will cause some of the through traffic to use that roadway, thereby reducing congestion on the original (nonbypass) roadway and increasing the speeds of both the local and through traffic. The reduced traffic volume and increased average speed on the original roadway will reduce the CO emissions density on that roadway. Total NO_x emissions from all traffic will increase, owing to the increased length of the bypass route and the tendency of NO_x emissions rates to increase when speeds increase. However, the effect of the bypass on total HC emissions from all traffic is not immediately clear, as the speed increase associated with the bypass decreases HC emissions/VMT (vehicle mile traveled), whereas the added length of trips using the bypass increases total VMT. The following numerical example shows that the bypass can cause large reductions in HC emissions, even if it also causes a substantial increase in VMT.

Let the bypass and segments AB and BC of the original roadway each have capacities of 1,500 veh/hr. Assume that prior to opening the bypass, the total volume of traffic on each segment is 1,500 veh/hr, and that at least 200 veh/hr of this volume is due to through traffic from A to C. Let the performance functions of the segments AB and BC and of the bypass after it has opened be given by equation (6.9) with $s_0 = 23$ mi/hr. The predicted composite exhaust HC speed-emissions relation for warmed-up automobiles in the 1982 vehicle fleet is approximated well in the speed range 10–30 mi/hr by

$$e_{HC}(s) = 0.16 + 38.8/s, \tag{6.15}$$

where e_{HC} is in g/veh mi and s is in mi/hr. Evaporative HC emissions are not affected by the bypass and, therefore, are not considered in this example. The NO_x emissions rate for 1982 vehicles over the same speed range is given by

$$e_{NO}(s) = 1.1 + 0.04s. \tag{6.16}$$

The CO emissions rate is given by equation (6.10).

Equation (6.9) implies that before the bypass is opened, the average speed of traffic on segments AB and BC is 12.1 mi/hr. Therefore, the average emissions rates of automobiles on these segments are 44.7 g/veh mi (CO), 3.37 g/veh mi (HC), and 1.58 g/veh mi (NO_x). The corresponding emissions densities are [from equation (6.4)] 67.1 kg/roadway mi/hr (CO), 5.05 kg/roadway mi/hr (HC), and 2.38 kg/roadway mi/hr (NO_x). Total HC and NO_x emissions from all vehicles on the 2-mi-long roadway are 10.1 and 4.75 kg/hr, respectively. Total VMT on the roadway are 3,000 veh mi/hr.

Assume that after the bypass has opened, through traffic uses the roadway that minimizes the travel time from A to C. This implies that traffic distributes itself between the original and bypass roadways so as to make the travel time from A to C along the two roadways equal. (If the travel times were not equal, then some of the travelers on the roadway with the higher travel time would change routes.) From equation (6.9), the travel time from A to C on the original roadway is

$$T_o = (2/23)\,[1 + 0.9(V_o/1{,}500)^{4.5}] \quad \text{(hours)}, \tag{6.17}$$

where V_o is the traffic volume on either segment of the original roadway. The travel time on the bypass is

$$T_b = (3/23)\,[1 + 0.9(V_b/1{,}500)^{4.5}] \quad \text{(hours)}, \tag{6.18}$$

where V_b is the bypass traffic volume. Since the total volume of traffic on all roadways is assumed not to be affected by the opening of the bypass, V_o and V_b are related by

$$V_o + V_b = 1{,}500 \text{ (vehicles per hour)}. \tag{6.19}$$

Equations (6.17) and (6.18) imply that V_o and V_b also are related by

$$(V_o/1{,}500)^{4.5} = 5/9 + 1.5(V_b/1{,}500)^{4.5}. \tag{6.20}$$

This equation implies that

$$(V_o/1{,}500)^{4.5} \geq 5/9, \tag{6.21}$$

from which it follows that $V_o/1{,}500 \geq 0.88$ and $V_b/1{,}500 \leq 0.12$. However, if $V_b/1{,}500 \leq 0.12$, then the second term on the right-hand side of equation (6.20) is negligibly small. Therefore, V_o is approximately $(1{,}500)\ (5/9)^{1/4.5}$, or 1,316 veh/hr, and V_b is approximately 184 veh/hr. The speeds on the original roadway and the bypass are approximately 15.3 and 23.0 mi/hr, respectively. The emissions densities on the original roadway are 48.3 kg/roadway mi/hr (CO), 3.54 kg/roadway mi/hr (HC), and 2.25 kg/roadway mi/hr (NO_x). The densities on the bypass are 4.89 kg/roadway mi/hr (CO), 0.340 kg/roadway mi/hr (HC), and 0.372 kg/roadway mi/hr (NO_x). Total HC emissions on the two roadways, combined, are 8.10 kg/hr, total NO_x emissions are 5.62 kg/hr, and total VMT are 3,184 veh mi/hr. Comparing these results with the prebypass situation, it can be seen that the opening of the bypass has caused a 28% reduction in the CO emissions density on the original roadway, a 20% reduction in total HC emissions, an 18% increase in total NO_x emissions, and a 6% increase in total VMT.

The opening of the bypass and the accompanying reduction in congestion in the vicinity of locations A, B, and C may attract additional traffic to these locations, thereby raising the possibility that the ultimate effect of the bypass will be to increase total HC and NO_x emissions without significantly reducing the CO emissions density on the original roadway. However, equation (6.20) requires V_o to have a value close to 1,316 veh/hr over a wide range of V_b values. Therefore, moderate increases in the total volume of traffic that is attracted to B and C will be reflected mainly as increases in the volume of traffic on the bypass. Additional through traffic from A to C will use the bypass, and additional local traffic will divert nearly its own volume of through traffic to the bypass as long as sufficient through traffic continues to use the original roadway. As a result, moderate increases in the total volume of traffic will not increase the CO emissions density on the original roadway. Moreover, the 20% reduction in total HC emissions that accompanies the opening of the bypass when the traffic volume is constant enables additional traffic to be accommodated without increasing total HC emissions above their prebypass level. However, total NO_x emissions and the CO emissions density on the bypass will both increase as a result of increased traffic.

As an example of the effects of moderate increases in traffic volumes that accompany the opening of the bypass, suppose that this opening causes the traffic volumes associated with travel from A to B, B to C, and A to C to increase by 20% each, relative to their prebypass levels. Then equation (6.19) becomes

$$V_o + V_b = 1{,}800 \text{ (vehicles per hour).} \tag{6.22}$$

Assume that the through traffic volume exceeds 479 veh/hr after the increase (and exceeded 400 veh/hr before the increase). Then equation (6.22) and equation (6.20), together, imply that $V_o = 1{,}321$ veh/hr and $V_b = 479$ veh/hr. The CO emissions density on the original roadway is 48.6 kg/roadway mi/hr, which is only 1% above the level that existed without the increased traffic volume and is still 28% below the prebypass level. Total HC emissions are 9.81 kg/hr, or roughly 3% below the prebypass level. Total NO_x emissions are 7.41 kg/hr, or roughly 56% above the prebypass level. The CO emissions density on the bypass is 12.8 kg/roadway mi/hr, or roughly 2.6 times the value that prevails without the increase in the traffic volume. This large increase in the CO emissions density on the bypass is due to the large increase in the traffic volume on that roadway. Finally, total VMT are 4,079 veh mi/hr after the increase in traffic volume, or 36% above the prebypass level.

6.3.3 *Predicting Emissions on Real Roadways*

The foregoing examples show that the effects of traffic flow improvement measures on motor vehicle emissions are complex and not necessarily beneficial. Therefore, it would be useful to be able to forecast the effects of these measures on emissions from real roadway systems before the measures are implemented. Although a variety of techniques for making such forecasts are available, they all have important limitations that make them suitable only for rough estimation purposes. The most important techniques and their limitations are described in the following paragraphs.

In situations where traffic flow improvement measures can be represented easily as changes in capacities or other parameters of roadway performance functions, the main problem involved in forecasting the emissions effects of these measures is that of determining the traffic volumes and speeds on the roadway links of interest. This is equivalent to solving equations (6.5) and (6.6) for the traffic volumes V_i and speeds s_i in terms of the capacities c_i and any other performance function parameters. Problems of this type typically are solved by means of network equilibration procedures. Network equilibration refers to the process of determining mutually consistent link traffic volumes and speeds, taking into account the dependence of traffic volumes on speeds through the demand functions and the dependence of speeds on link volumes through the performance functions. Procedures are available that permit systematic treatment of route choice within the equilibration framework, and some procedures also include one or more of the demand dimensions travel frequency, destination choice, and mode choice in this framework [9, 11–17].[9] In terms of the simple model of equations (6.4)–(6.6), this corresponds to including at least route choice and possibly including other dimensions of travel demand in the demand functions D_i. In principle, all of the demand dimensions should be treated in the equilibration framework. However, this tends to be very difficult, and one or more demand dimensions usually is omitted from this framework in practice.

Current network equilibration procedures represent substantial improvements over earlier traffic assignment approaches, which could treat only route choice and required effects of congestion to be treated with arbitrary and, in some cases, incorrect methods. However, the ability of current equilibration procedures to deal with emissions estimation problems is limited in several important ways. First, as has already been noted, it usually is not feasible to include all relevant dimensions of travel demand in the equilibration framework. Moreover, most procedures that include travel frequency, destination, or mode choice in addition to route choice in this framework require the use of aggregate travel demand models (i.e., models that express travel demand in terms of flows between geographical subareas within a city). However, modeling the effects of most traffic flow improvement measures on choices of travel frequencies, destinations, and modes can best be done with disaggregate travel demand models (i.e., models that express travel demand in terms of the behavior of individual travelers or households). This is because the ability of disaggregate models to represent the effects of changes in transportation service levels on these dimensions of demand is considerably greater than that of aggregate models [18]. At present, methods that combine network equilibration procedures with disaggregate demand models are not available for routine use. (But see reference [19] for a discussion of an experimental approach to network equilibration with disaggregate models.) Therefore, in practice it usually is necessary to use relatively crude models to treat any dimensions of travel demand other than route choice that may be included in the equilibration framework. In general, it is necessary to use considerable caution when attempting to apply current equilibration procedures to situations in which substantial changes in dimensions of travel demand other than route choice are likely to occur.

A second limitation on the usefulness of current equilibration procedures for emissions estimation is that these procedures yield information only on average traffic flow conditions along each of the links in a roadway network. They provide no representation of variations in traffic densities or speeds along links and, hence, no representations of variations in emissions densities along links. This seriously limits the ability of current equilibration procedures to provide information that is useful for estimating the effects of traffic flow improvement measures on CO concentrations since, as was discussed in section 3.2, CO concentrations can be very sensitive to variations in emissions densities along links.

A third limitation of current equilibration procedures is that they can make large errors in estimating traffic volumes and speeds on individual network links. It is not unusual for even the best procedures to make errors of over 30% in link volumes and over 50% in speeds [20, 21]. Errors of these magnitudes can induce severe errors in emissions estimates. For example, if the true speed on a network link is 15 mi/hr but is estimated to be 7.5 mi/hr, then the CO emissions density from warmed-up automobiles on the link will be overestimated by 90%, even if the link volume is estimated correctly [8]. If the true speed is 15 mi/hr but is estimated to be 22.5 mi/hr, then the CO emissions density will be underestimated by approximately 25% [8]. The magnitudes of these errors exceed the magnitudes of the changes in emissions densities that most traffic flow improvement measures are likely to produce.

Many important traffic flow improvement measures, such as synchronization of signal lights, cannot be represented easily as changes in capacities or other parameters of roadway performance functions. The effects of such measures on emissions and other traffic flow characteristics must be estimated through the use of traffic flow simulation models. Several of these models have been adapted for use in emissions estimation. The model NETSIM simulates the motion of each vehicle in a city street network [22–27]. It is able to treat all major forms of traffic control, including stop and yield signs, fixed time traffic signals, and vehicle-actuated signals. Moreover, because it keeps track of the positions of all vehicles as they move through the network, it is able to simulate variations in emissions densities along network links. Thus, for example, NETSIM is able to simulate increases in emissions densities near the stop lines of signalized intersections. However, NETSIM has the considerable disadvantage of being unable to forecast changes in traffic volumes that occur as a result of the implementation of traffic flow improvement measures. As illustrated in examples 6.1–6.3, traffic volume changes can be extremely important in determining the effects of traffic flow improvement measures on emissions. Consequently, NETSIM is useful for forecasting these effects only if forecasts of changes in traffic volumes can be obtained from other sources.

The model INTRAS permits detailed simulation of traffic flows and emissions on freeways [28–31]. Like NETSIM, INTRAS requires traffic volume information to be provided as an input to the model.

The model TRANSYT6C simulates the flow of traffic on arterial streets and estimates (among other things) changes in emissions that are caused by implementing traffic flow improvement measures on these streets [32]. TRANSYT6C does not treat as broad a range of traffic flow improvement measures as NETSIM does, and it does not have NETSIM's ability to compute variations in emissions densities along network links. However, unlike NETSIM, TRANSYT6C incorporates a procedure for making a rough estimate of the change in traffic volume that occurs on an arterial street as a result of implementing traffic flow improvement measures on that street. TRANSYT6C also is able to design signal light synchronization plans that are optimal according to user-specified criteria that can include consideration of travel time, stop frequency, gasoline consumption, and emissions.

The results of using TRANSYT6C to estimate the effects of synchronizing the signal lights on sections of San Pablo Avenue in Berkeley, California, are shown in table 6.3. San Pablo Avenue is a two-way arterial street with three lanes of traffic in each direction. The tabulated results apply to a section of the street that is 2.75 mi long and has nine signalized intersections. The synchronization objective is to minimize passenger delay. The results show that signal light synchronization causes a 42% increase in vehicle miles traveled on San Pablo Avenue. This increase is due to diversion of vehicles to San Pablo Avenue from parallel routes. In spite of the large increase in traffic volume, the speed on San Pablo Avenue after synchronization of the signal lights is slightly higher than it was before synchronization. However, the increase in traffic volume causes emissions of CO, HC, and NO_x to increase by roughly 40% each. If the traffic volume did not increase, then emissions of these pollutants would decrease by approximately 3% each.

The model FREQ simulates traffic flow on freeways and estimates the effects of implementing freeway traffic management measures, such as ramp metering and priority treatment for high-occupancy vehicles [33, 34]. The model assigns traffic to the freeway under consideration and to a parallel arterial street. It also includes a simple mode choice submodel. These features enable FREQ to estimate changes in traffic volumes that are caused by the implementation of freeway traffic management measures. The outputs of the model include estimates of emissions rates on the freeway and parallel arterial street before and after the freeway traffic management measures are implemented. In addition, FREQ has the ability to design ramp metering plans that maximize any of several indicators of roadway performance. However, FREQ does not provide detailed information on the spatial distributions of freeway and arterial street emissions.

The results of using FREQ to estimate the effects of ramp metering on a section of the Eastshore Freeway (I-80) in the Berkeley-Oakland, California, area are shown in table 6.4. The tabulated results apply to northbound traffic on a 10-mi section of the freeway that has 10 on-ramps. The parallel arterial street is San Pablo Avenue. The ramp metering plan causes a small reduction in passenger hours of travel. However, it also causes increases in most categories of emissions. The increase in freeway CO emissions is caused by queueing at the metered ramps, the increase in freeway NO_x emissions is caused by an increase in the speed of freeway traffic, and the increase in emissions on the arterial street is caused by diversion of traffic from metered freeway ramps to the street.

Table 6.3 Effects of synchronizing the signal lights on an arterial street[a]

	Before synchronization	After synchronization if traffic volume is unchanged		After synchronization and with estimated changes in traffic volume	
		Level	Percent change[b]	Level	Percent change[b]
Traffic flow (veh mi/hr)	4,754	4,754	0	6,774	+43
Average speed (mi/hr)	18.9	19.4	+3	19.1	+1
CO emissions (kg/hr)	273	264	−3	382	+40
HC emissions (kg/hr)	26.6	25.9	−3	37.2	+40
NO_x emissions (kg/hr)	13.7	13.3	−3	19.5	+42

a. Source: reference [32]. Based on simulation studies of traffic flow on San Pablo Avenue in Berkeley, California. The simulation model is TRANSYT6C. The synchronization objective is minimization of passenger delay.
b. Percent change is computed relative to values before synchronization.

Table 6.4 Effects of ramp metering on an urban freeway[a]

Roadway	Percentage change					
	Passenger hours traveled	Vehicle miles traveled	CO emissions	HC emissions	NO_x emissions	Fuel consumption
Freeway	−6	−0.7	+0.4	−0.5	+9	+0.4
Parallel arterial street	+10	+2	+9	+7	0	+2
Total	−2	−0.3	+2[b]	+1	+8	+0.8

a. Source: reference [34]. Based on simulation studies using the model FREQ6PE. The ramp metering plan maximizes vehicle input to the freeway. The freeway is the Eastshore Freeway in the Berkeley-Oakland area. The parallel arterial street is San Pablo Avenue.
b. Total CO emissions contribute to background CO concentrations, but are not closely related to CO concentrations along individual roadways.

Network equilibration procedures and traffic flow simulation models are complementary in an important way. Network equilibration procedures provide relatively sophisticated methods for forecasting changes in traffic volumes that are caused by the implementation of traffic flow improvement measures, provided that the effects of these measures on roadways' performance functions are known. However, these effects must be determined outside of the equilibration framework. In contrast, traffic flow simulation models provide sophisticated treatments of the effects of traffic flow improvement measures on roadway performance, but they are not well equipped to forecast changes in traffic volumes that accompany the implementation of these measures. The complementary capabilities of equilibration procedures and traffic flow simulation suggest the possibility of combining the two approaches in a single model that incorporates the strengths of both. In such a model, the roadway performance functions that currently are used in equilibration would be replaced by a traffic flow simulation model. At least one prototype of a combined network equilibration and traffic flow simulation model has been developed [35, 36], but no such models are currently available for routine use as planning tools.

A final problem in estimating the emissions effects of traffic flow improvement measures is that of developing satisfactory functional relations between individual vehicle emissions rates and traffic flow variables. Inaccuracies in emissions estimation models constitute potentially important sources of error in computations of the emissions effects of traffic flow improvement measures. As discussed in section 5.7, the accuracy of currently available emissions estimation models is largely unknown.

6.4 *Vehicle-Free Zones and Traffic Cells*[10]

High CO concentrations usually are caused by emissions from traffic near the locations where the concentrations are measured. In areas that are small enough to permit access on foot, high CO concentrations usually can be reduced substantially by making the areas into vehicle-free zones. Table 6.5 shows reductions in CO concentrations that were measured in three cities when vehicle-free zones were established. The tabulated results pertain to CO concentrations in the restricted areas. The reductions in average CO concentrations measured during daytime hours range from 17 to 81%.

No measurements of the effects of vehicle-free zones on O_3 or annual average NO_2 concentrations have been reported. However, since the

Table 6.5 Changes in CO concentrations due to establishment of vehicle-free zones in three cities[a]

City	CO concentration (ppm)[b]		Averaging time
	Before	After	
Tokyo[c]			
Ginza	8.4	2.5	1 P.M.–6 P.M.
Shinjuku	7.8	1.9	11 A.M.–7 P.M.
Ikebukuro	7.3	3.6	12 noon–7 P.M.
Asakusa	2.4	2.0	10 A.M.–7 P.M.
Marseilles[d]	18.8	3.6	8 A.M.–6 P.M.
Vienna[e]	10.0	4.0	9 A.M.–7 P.M.
	10.8	3.8	2 P.M.–6:30 P.M.

a. Source: reference [37].
b. "Before" and "after" refer to before and after establishment of vehicle-free zones. All concentrations were measured inside the vehicle-free areas.
c. Measurements are available for four separate vehicle-free zones. The tabulated concentrations represent averages of measurements made at two to three locations within each zone. Measurements were made on 1 day before and 1 day after establishment of the zones.
d. The tabulated concentrations represent averages of measurements made at four locations in the vehicle-free zone. Measurements made on 24 days before and 10 days after establishment of the zone are included in these averages.
e. The tabulated concentrations represent averages of measurements made at one location on 10 days before and 23 days after establishment of the vehicle-free zone.

zones are necessarily small in size (typically much less than 0.5 mi^2), these effects undoubtedly are negligible. The effects of vehicle-free zones on short-term NO_2 concentrations are unknown. However, it is possible that in cities with high O_3 concentrations the effects could be large. In such cities, high short-term NO_2 concentrations often are caused by the reaction of O_3 with NO that is generated by nearby traffic [reaction (3.4)]. Since the establishment of vehicle-free zones eliminates this source of NO in the affected areas, it is likely that it often reduces short-term NO_2 concentrations in these areas to levels similar to the ones that exist at locations that are in the same general vicinities but are removed from large NO sources.

Simulation studies have suggested that establishment of vehicle-free zones may lead to increased traffic volumes and emissions on roadways that are close to the zones [38]. However, little is known about the extent to which this problem occurs in practice or the extent to which it can be alleviated through the implementation of traffic flow improvement measures on the affected roadways.

The central areas of cities often include many roadways near which CO concentrations are excessive. These areas usually are too large to be made into vehicle-free zones. However, if a bypass route is available or can be established, then it may be possible to achieve significant reductions in central area CO concentrations by diverting through traffic to the bypass. One way of effecting this diversion is to divide the central area into traffic cells, or zones, that are accessible to vehicles only from the bypass route. Direct vehicular travel between cells is discouraged or prevented by means of systems of one-way streets, restrictions on turns, or physical barriers.

Through traffic accounts for more than 50% of daily vehicle entries into the central areas of some US cities. Hence, the reductions in central area CO emissions and concentrations that can be achieved in US cities by establishing traffic cell systems are potentially large. However, the reductions in central area CO concentrations that occur as a result of establishing a traffic cell system will not necessarily be proportional to either the reduction in vehicle entries into the central area or to changes in simple indicators of the aggregate volume of central area traffic (e.g., vehicle miles traveled in the central area). For example, if most through trips occur at times or on streets with relatively low CO concentrations, then diversion of through traffic will have little effect on maximum CO concentrations. Conversely, if the highest CO concentrations occur near streets that carry large volumes of through traffic, then diversion of through trips away from these streets may produce percentage reductions in maximum CO concentrations that exceed the percentage reductions in traffic volumes, owing to the increases in average speeds that would occur. In addition, the establishment of a traffic cell system will change the spatial distribution of emissions from trips whose origins and destinations are both in the restricted area owing to the need to use the bypass road for travel between cells. This redistribution may cause CO emissions and concentrations on some streets in the restricted area to remain constant or even to increase following establishment of a traffic cell system, despite the elimination of through traffic from these streets. (See example 6.4 for an illustration of this effect.)

The diversion of through traffic to bypass roads that occurs when a traffic cell system is established is likely to increase the lengths of through trips. The lengths of local trips may also increase owing to the need to use bypass roads for travel between cells. This increased circuity of travel can cause aggregate HC and NO_x emissions to increase. If the bypass route consists of roadways that existed before establishment of the traffic cell system, then the CO emissions densities on those roadways are likely to

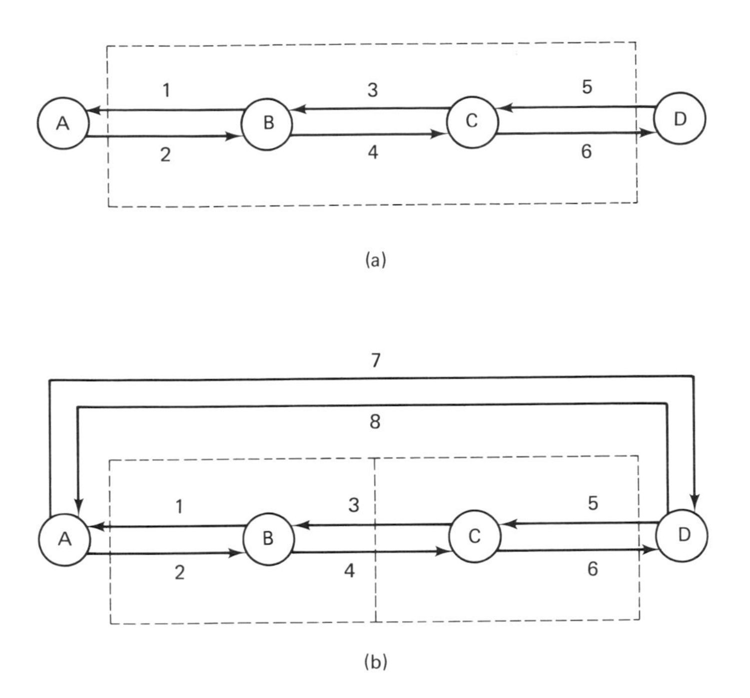

Figure 6.2 A simple traffic cell system: (*a*) before establishment of system (––– signifies boundary of central area); (*b*) after establishment of system (––– signifies boundaries of central area and traffic cells).

increase owing to the increased traffic volumes that diversion of through traffic will cause.

The following example illustrates the effects that establishment of a traffic cell system can have on CO emissions densities and aggregate HC and NO_x emissions. As in examples 6.1–6.3, CO is discussed here in terms of emissions densities in order to emphasize the localized nature of CO emissions and concentrations. HC and NO_x are discussed in terms of total emissions since these pollutants are of interest mainly because of their effects on regional O_3 concentrations.

Example 6.4—Traffic Cells

Refer to the simple street network that is shown in figure 6.2a. Let the area within the dashed lines, consisting of nodes B and C and roadway links 1–6, represent the central area of a city. Assume that each link is

0.5 mi long. Let T_{ij} $(i, j = A, B, C, D;\ i \neq j)$ denote the number of vehicle trips from node i to node j that occur in an hour, and let V_i $(i = 1, \ldots, 6)$ denote the traffic volume (in veh/hr) on link i. The link volumes are related to the trip interchange rates by

$$V_1 = V_{DA} + V_{CA} + V_{BA}, \tag{6.23a}$$

$$V_2 = V_{AB} + V_{AC} + V_{AD}, \tag{6.23b}$$

$$V_3 = V_{DA} + V_{DB} + V_{CA} + V_{CB}, \tag{6.23c}$$

$$V_4 = V_{AD} + V_{BD} + V_{AC} + V_{BC}, \tag{6.23d}$$

$$V_5 = V_{DA} + V_{DB} + V_{DC}, \tag{6.23e}$$

$$V_6 = V_{AD} + V_{BD} + V_{CD}. \tag{6.23f}$$

The number of vehicles entering the central area per hour, V_E, is

$$V_E = V_2 + V_5, \tag{6.24}$$

and the number of vehicles entering the central area per hour on through trips (i.e., trips from A to D or D to A), V_T, is

$$V_T = V_{AD} + V_{DA}. \tag{6.25}$$

The fraction of vehicle entries into the central area that is due to through trips is V_T/V_E.

Now assume that two traffic cells are established in the central area by severing links 3 and 4 and opening a 2.5-mi-long bypass road between nodes A and D. The traffic cells and the bypass are shown in figure 6.2b. Links 7 and 8 form the bypass. Note that trips between cells must use the bypass road. The traffic volumes on the links of the new network are

$$V_1 = V_{BA} + V_{BD} + V_{BC}, \tag{6.26a}$$

$$V_2 = V_{AB} + V_{CB} + V_{DB}, \tag{6.26b}$$

$$V_3 = 0, \tag{6.26c}$$

$$V_4 = 0, \tag{6.26d}$$

$$V_5 = V_{AC} + V_{BC} + V_{DC}, \tag{6.26e}$$

$$V_6 = V_{CD} + V_{CA} + V_{CB}, \tag{6.26f}$$

$$V_7 = V_{BD} + V_{BC} + V_{AD} + V_{AC}, \tag{6.26g}$$

$$V_8 = V_{CB} + V_{CA} + V_{DB} + V_{DA}. \tag{6.26h}$$

Let the performance functions of links 1–8 be given by equation (6.9) with $s_0 = 23$ mi/hr. Let links 1–6 have capacities of 1,500 veh/hr, and let links 7 and 8 have capacities of 2,250 veh/hr. Assume that the speed-emissions relations for vehicles on the eight links are given by equations (6.10), (6.15), and (6.16). Then, given fixed numerical values of the trip interchange rates T_{ij}, the changes in motor vehicle emissions along each link that are caused by implementation of the traffic cell system can be computed by combining equations (6.23), (6.26), and the speed-emissions relations with equations (6.4) and (6.6).

Table 6.6 shows the results of this computation for two cases. In both cases the fraction of vehicles entering the central area that is due to through trips is 0.50 before the establishment of the traffic cell system. However, in case I the volume of trips whose origins and destinations both are in the central area (i.e., trips from B to C and C to B) is higher and the volume of trips that have at least one end outside of the central area (i.e., trips that originate or terminate at A or D) is lower than it is in case II. In both cases the establishment of the traffic cell system reduces the CO emissions density on links 3 and 4 by 100%. However, aggregate HC and NO_x emissions increase by roughly 100–200%, depending on the case and the pollutant, owing mainly to the increased circuity of travel that the cell system causes. In case I, the establishment of the traffic cell system increases vehicle miles traveled by 190%, and in case II it increases vehicle miles traveled by 154%. In addition, in case I the CO emissions densities on links 1, 2, 5, and 6 increase by 24% each following establishment of the cell system, even though these links are in the central area and in traffic cells. However, in case II the cell system causes a 19% reduction in the CO emissions densities on the same links. These results illustrate the complexity of the effects that traffic cell systems can have on traffic flow and emissions.

Traffic cell systems have been implemented in several cities in Europe and Japan [39, 40]. Experience with these systems indicates that they can achieve substantial reductions in central area CO concentrations. In Gothenburg, Sweden, implementation of a central area traffic cell system is reported to have reduced half-hour average CO concentrations in the central area to less than 5 ppm from 60–70 ppm [40]. In Besançon, France, implementation of a central area traffic cell system together with traffic flow and transit improvement measures reduced 15-min average central area CO concentrations by 20–80%, depending on the location [40].

Table 6.6 Effects of establishing a simple traffic cell system[a]

Link	Traffic volume (veh/hr)		Speed (mi/hr)		CO emissions density (kg/mi/hr)		HC emissions (kg/hr)		NO_x emissions (kg/hr)	
	Before	After	Before	After	Before	After	Before	After	Before	After
Case I: $V_{AD} = V_{DA} = 450$ veh/hr; $V_{BC} = V_{CB} = 600$ veh/hr; $V_{AB} = V_{BA} = V_{AC} = V_{CA} = V_{BD} = V_{DB} = V_{DC} = V_{CD} = 225$ veh/hr										
1	900	1,050	21.1	19.5	25.6	31.7	0.90	1.13	0.87	0.99
2	900	1,050	21.1	19.5	25.6	31.7	0.90	1.13	0.87	0.99
3	1,500	0	12.1	23.0	67.1	0	2.52	0	1.19	0
4	1,500	0	12.1	23.0	67.1	0	2.52	0	1.19	0
5	900	1,050	21.1	19.5	25.6	31.7	0.90	1.13	0.87	0.99
6	900	1,050	21.1	19.5	25.6	31.7	0.90	1.13	0.87	0.99
7	0	1,500	23.0	20.1	0	44.3	0	7.84	0	7.14
8	0	1,500	23.0	20.1	0	44.3	0	7.84	0	7.14
Total HC and NO_x							8.64	20.2	5.86	18.24
Case II: $V_{AD} = V_{DA} = 550$ veh/hr; $V_{BC} = V_{CB} = 400$ veh/hr; $V_{AB} = V_{BA} = V_{AC} = V_{CA} = V_{BD} = V_{DB} = V_{DC} = V_{CD} = 275$ veh/hr										
1	1,100	950	18.8	20.6	34.2	27.5	1.22	0.97	1.02	0.91
2	1,100	950	18.8	20.6	34.2	27.5	1.22	0.97	1.02	0.91
3	1,500	0	12.1	23.0	67.1	0	2.52	0	1.19	0
4	1,500	0	12.1	23.0	67.1	0	2.52	0	1.19	0
5	1,100	950	18.8	20.6	34.2	27.5	1.22	0.97	1.02	0.91

6	1,100	950	18.8	20.6	34.2	27.5	1.22	0.97	1.02	0.91
7	0	1,500	23.0	20.1	0	44.3	0	7.84	0	7.14
8	0	1,500	23.0	20.1	0	44.3	0	7.84	0	7.14
Total HC and NO_x							9.92	19.56	6.46	17.92

a. Based on example 6.4 and figure 6.2. In case I there are 3,300 vehicle miles traveled (VMT) per hour before establishing the traffic cell system and 9,600 VMT/hr after. In case II there are 3,700 VMT/hr before and 9,400 after.

6.5 Summary of Discussion of Traffic Engineering Measures

The examples and simulation results that have been presented in sections 6.3 and 6.4 show that the effects of traffic engineering measures on emissions are highly complex and variable. The complexity and variability of these effects are due mainly to the changes in traffic volumes and flow patterns that implementation of traffic engineering measures can cause. The results of traffic flow simulation studies and measurements of the effects of implemented projects indicate that these measures are capable of causing emissions and, in the case of CO, concentration changes whose magnitudes span a range that includes increases of at least 40% and decreases of more than 80% on the affected roadways. Moreover, it cannot be assumed that traffic flow improvement measures that reduce congestion automatically will also reduce emissions. Several examples have been presented here in which traffic flow improvement measures increase the average speed of traffic on a roadway (thereby reducing congestion) but also increase the emissions density on the roadway.

Several analytical techniques are available for estimating the magnitudes of the emissions changes that specific types of traffic engineering measures would produce in specific situations. However, the state of the art in traffic flow analysis is such that these estimates of emissions changes are likely to be very rough. Often, it may be impossible to predict with confidence even whether implementation of a particular set of traffic engineering measures would increase or decrease emissions along a particular street. Therefore, in situations where the emissions and air quality effects of traffic engineering measures are important in determining the desirability of implementing the measures, it probably would be wise to implement the measures initially on an experimental basis and to observe the changes in emissions or air quality that accompany implementation. (Emissions observations must be carried out indirectly by measuring appropriate traffic flow variables and using these variables to estimate emissions.) The measures then can be modified if their observed effects on emissions or air quality are found to be inconsistent with the objectives that their implementation was intended to achieve.

C TRANSIT IMPROVEMENT AND CARPOOL INCENTIVE MEASURES

The objectives of this part are to develop bounds on the magnitudes of the emissions reductions that can be achieved by transit improvement and carpool incentive measures generally and to outline techniques for estimating quantitatively the effects of specific measures on emissions.

The part consists of four sections. Section 6.6 deals with the relation between the demand for travel by transit, carpool, and single-occupant automobile and the quality of service provided by these modes. Using mathematical models of travel demand and, to the extent available, other sources of information on the relation between travel behavior and service quality, implications of this relation for the emissions reduction effectiveness of transit improvement and carpool incentive measures are explored. Section 6.7 discusses some transit supply issues that have a major influence on the prospects for achieving large emissions reductions through transit improvements. Section 6.8 presents a numerical example that illustrates methods for quantitatively estimating the effects of transit service improvements in an arterial corridor. Section 6.9 discusses the usefulness of priority treatment for high-occupancy vehicles as an emissions reduction approach.[11]

6.6 *Demand Side Considerations*

Mode choice is the travel demand dimension that most strongly affects the emissions reduction effectiveness of transit improvement and carpool incentive measures. Travelers' mode choices are determined by the availabilities of modes (e.g., whether an automobile is available for a particular trip), by travelers' tastes (e.g., the relative values that travelers assign to such factors as travel time, travel cost, and service reliability), and by the quality of service offered by each mode in comparison to the available alternatives. A traveler will use the mode that he perceives as providing the highest quality of service, as measured in terms of his own tastes, among the modes that are available to him. The concept of service quality encompasses such factors as average travel time (referred to hereafter simply as travel time), travel cost, comfort, reliability, schedule flexibility and convenience, and safety, among others. Travel time often is subdivided into components, such as walk time, wait time, and in-vehicle time, that may be valued differently by travelers. The main purpose of most transit improvement and carpool incentive measures (carpool matching programs being possible exceptions) is to alter the service quality variables in ways that encourage the use of high-occupancy modes instead of low-occupancy ones. For example, the following measures all alter travel times or costs for one or more modes:

- changes in transit route structures and spacings;
- changes in transit stop spacings;
- changes in transit schedule frequencies or headways;
- use of express or skip stop transit service;
- priority treatment for transit vehicles and carpools on streets or freeways;

- preferential allocation of parking spaces for carpools;
- changes in transit fares;
- changes in parking prices.

Suppose that the relation between mode choice and the service variables can be approximated reasonably well with a mathematical model. Then the effects of transit improvements and carpool incentives on automobile use and emissions can be investigated in an abstract way by investigating the effects of varying the values of the service variables over ranges similar to those likely to occur in practice. This approach will be used in this section to place bounds on the emissions reduction effectiveness of transit improvements and carpool incentives. The approach enables these bounds to be developed without the need for detailed specification and analysis of large numbers of individual measures. To the extent possible, the results of the abstract analysis will be supplemented and corroborated with data obtained from actual experience with transit improvement and carpool incentive measures. The analysis is concerned mainly with emissions from trips between home and work. Nonwork trips are discussed briefly toward the end of the section.

6.6.1 A Model of Mode Choice

Quantitative forecasts of the effects on mode choice of changing the quality of service provided by one or more modes can be made with mathematical models called "mode choice models." These models describe the relation between travelers' mode choices, the quality of service provided by the available modes, and other relevant variables, such as incomes and automobile ownership levels. Among currently available mode choice models, disaggregate (i.e., individual choice) models with a mathematical form known as "multinomial logit" are the most useful for analyzing the effects of transit improvements and carpool incentives on mode choice. These models are based on a clearly defined principle of human behavior (that of utility maximization).[12] Subject to data availability, they are capable of treating virtually all of the modes that are likely to be available for most trips and of representing the effects on mode choice of changes in a wide variety of variables. Moreover, they can be developed statistically using relatively small data sets. The general structure of disaggregate, multinomial logit, mode choice models will now be summarized, since much of the following discussion involves their use. (See reference [18] for a detailed discussion of these models as well as a review and critique of urban travel demand models in general.)

Let an individual n face a choice among J different modes for a particular trip. Let X_{in} ($i = 1, 2, \ldots, J$) be a vector of known attributes of mode i and individual n that are relevant to individual n's choice of mode for the trip. For example, suppose that mode i is bus transit. Then the components of X_{in} might include the walk, wait, and in-vehicle travel times and the fare that will be required for individual n's trip if that trip takes place by bus. X_{in} also might include attributes of individual n, such as his household's income and the number of automobiles owned by the members of his household. The object of a disaggregate mode choice model is to forecast individual n's choice of mode in terms of the attributes X_{in}. Since the factors that determine mode choice are not completely known to analysts, the values of the vectors X_{in} (which represent known factors) cannot completely determine individual n's choice of mode. However, they can determine probabilities with which the various modes are chosen. In the multinomial logit formulation, the probability P_{in} that individual n chooses mode i is related to the known attributes of the modes and the individual in the following way:

$$P_{in} = [\exp V(X_{in})] \Big/ \sum_{j=1}^{J} \exp V(X_{jn}), \tag{6.27}$$

where V is a function that is called the utility function. This function relates individuals' preferences among modes to the known attributes of individuals and modes. Other things being equal, increasing the value of V for a particular individual and mode increases the likelihood that the individual will prefer that mode to the others. In practice, the function V is estimated statistically from observations of individuals' mode choices.

A disaggregate mode choice model forecasts the mode choices of individual travelers. In practice, however, one is usually concerned with the aggregate behavior of a large number of travelers (e.g., the travelers to work in a certain corridor), rather than with individual choices. Aggregate mode choice can be forecast with a disaggregate multinomial logit model by summing equation (6.27) over the relevant individuals (or over a random sample of these individuals). Thus, among N individual travelers, the percentage $\overline{P_{iN}}$ who use mode i (or the aggregate share of mode i among these travelers) is given by

$$\overline{P_{iN}} = \frac{1}{N} \sum_{n=1}^{N} P_{in}. \tag{6.28}$$

(See reference [41] for a more detailed discussion of methods for aggregating disaggregate logit models.)

The usual form of the utility function V in equation (6.27) is

$$V(X_{in}) = \sum_j a_j X_{jin}, \tag{6.29}$$

where X_{jin} is the jth component of the vector X_{in}, and a_j ($j = 1,2, \ldots$) is a set of numerical coefficients. Table 6.7 gives an example of the attributes (or variables) and coefficients of a disaggregate, multinomial logit, mode choice model whose utility function has this form. The subscript n is suppressed in the table since the definitions of the variables are the same for all individuals. The tabulated model pertains to trips between home and work. The modes treated by the model are drive alone, carpool, and transit. The values of the coefficients were estimated statistically using data on travel behavior in San Francisco, California [42, 43].

6.6.2 Preliminary Results Based on the Model

The model of table 6.7 can be used to investigate the effects of service quality changes on work trip mode choice and motor vehicle emissions. Table 6.8 shows examples of the model's predictions of work trip mode choice probabilities for a variety of values of the service quality variables. To preserve simplicity of presentation, the tabulated probabilities have been derived by applying the model to a single traveler, rather than by applying it to each member of a group of travelers and then aggregating the results by means of equation (6.28). The traveler has been assumed to have characteristics that are, roughly speaking, typical of urban workers from automobile-owning households, thereby enabling the tabulated probabilities to approximate the aggregate shares that would result from applying equation (6.28) to a group of such workers. Specifically, it has been assumed that there is one worker in the traveler's household and that the household's annual income is $25,000. In addition, it has been assumed that there is a probability of 0.50 that the household owns one automobile and a probability of 0.50 that it owns two.[13]

Referring to table 6.8, it can be seen that other things being equal, the probability of using the drive alone mode is higher and the probabilities of using carpools or transit are lower when the work place is outside of the central business district (CBD) than when it is inside. This presumably reflects the frustrations of driving in congested, CBD-oriented, peak period traffic, the greater ease of finding partners for carpooling to dense CBD locations than to less dense suburban ones, and, possibly, the difficulty of finding parking space (even at a high price) in the CBD. Despite the differences between mode choices for travel to CBD and non-CBD work

Table 6.7 Variables and utility function coefficients of a multinomial logit model of work trip mode choice[a]

Variable	Coefficient
Dummy variable equal to 1 for the drive alone mode, 0 otherwise	−3.385
Dummy variable equal to 1 for the carpool mode, 0 otherwise	−4.267
Round trip travel cost (in cents) divided by the annual income (in dollars) of the traveler's household	−21.43
Round trip in-vehicle travel time (minutes)	−0.0122
Round trip walk time (minutes)	−0.0335
Sum of round trip initial transit headways (in minutes) up to a maximum of 8 min if mode is transit, 0 otherwise[b]	−0.155
Excess over 8 min of the sum of initial round trip transit headways if mode is transit; 0 if mode is not transit or if the sum of round trip headways does not exceed 8 min[b]	−0.0107
One half the round trip sum of all transfer headways (in minutes) if the mode is transit, 0 otherwise	−0.0302
Dummy variable equal to 1 if the work place is in the central business district and the mode is drive alone, 0 otherwise	−1.067
Dummy variable equal to 1 if the work place is in the central business district and the mode is carpool, 0 otherwise	−0.347
Number of automobiles per worker in the traveler's household if the mode is drive alone, 0 otherwise	1.958
Number of automobiles per worker in the traveler's household if the mode is carpool, 0 otherwise	1.763
Number of automobiles per worker in the traveler's household if the mode is transit with auto access, 0 otherwise	1.389
Dummy variable equal to 1 if the mode is transit with auto access, 0 otherwise	−1.237
Dummy variable equal to 1 if the traveler is the primary worker in the household and the mode is drive alone, 0 otherwise	0.677
Number of workers in the traveler's household if the mode is carpool, 0 otherwise	0.327
Annual income (thousands of 1980 dollars) of traveler's household minus 2.4 times the number of household members, if mode is drive alone or carpool, 0 otherwise	0.00571

a. Source: references [42, 43]. The annual income coefficient has been scaled to pertain to 1980 dollars. The variables are evaluated for individual travelers and trips. The modes that are modeled are drive alone, carpool, and transit.

b. Initial headway is the headway on the first leg of a transit trip. Transfer headways are not included in the initial headway.

Table 6.8 Examples of the influence of service quality on work trip mode choice[a]

Case	Mode	Round trip cost (¢)	Round trip in-vehicle travel time (min)	Round trip walk time (min)	Work place in central business district (1, yes; 0, no)	Round trip transit headway (min)	Mode choice probability
1	Drive alone	150	40	2	1		0.48
	Carpool	80	45	2	1		0.21
	Transit	150	40	5	1	10	0.31
2	Drive alone	150	40	2	1		0.57
	Carpool	80	45	2	1		0.25
	Transit	150	60	15	1	30	0.18
3	Drive alone	150	40	2	1		0.59
	Carpool	80	45	2	1		0.26
	Transit	150	60	15	1	60	0.14
4	Drive alone	350	40	2	1		0.53
	Carpool	80	45	2	1		0.28
	Transit	150	60	15	1	30	0.20
5	Drive alone	150	40	10	1		0.50
	Carpool	80	45	2	1		0.29
	Transit	150	60	15	1	30	0.21
6	Drive alone	150	40	2	0		0.68
	Carpool	80	45	2	0		0.15
	Transit	150	40	5	0	10	0.17
7	Drive alone	150	40	2	0		0.75
	Carpool	80	45	2	0		0.16
	Transit	150	60	15	0	30	0.09

8	Drive alone	150	40	2	0		0.77
	Carpool	80	45	2	0		0.17
	Transit	150	60	15	0	60	0.07
9	Drive alone	350	40	2	0		0.72
	Carpool	80	45	2	0		0.18
	Transit	150	60	15	0	30	0.10
10	Drive alone	150	40	10	0		0.70
	Carpool	80	45	2	0		0.20
	Transit	150	60	15	0	30	0.11

a. Based on the mode choice model of table 6.7. It is assumed that the traveler is the principal and only worker in the household and that the household's annual income is $25,000. The household is assumed to own one car with probability 0.50 and two cars with probability 0.50. In addition, it is also assumed that no transfers are needed for transit trips and that there are two travelers in a carpool. Transit access is on foot. The sums of the drive alone, carpool, and transit probabilities do not necessarily equal 1.00 due to rounding.

places, transit service quality has a substantial effect on mode choice for both types of travel. For example, when transit service quality is relatively poor in comparison with drive alone service quality (cases 3 and 8 in table 6.8), the mode choice probabilities are 0.59 (drive alone), 0.26 (carpool), and 0.14 (transit) for CBD work places and 0.77 (drive alone), 0.17 (carpool), and 0.07 (transit) for non-CBD work places. However, when transit service quality is more nearly equal to that of the single-occupant automobile (cases 1 and 6), the mode choice probabilities are 0.48 (drive alone), 0.21 (carpool), and 0.31 (transit) for CBD work places and 0.68 (drive alone), 0.15 (carpool), and 0.17 (transit) for non-CBD work places. In other words, when transit service is good, the probability of driving alone is 12–19% less than it is when transit service is poor. Similarly, when transit service is good, the probability of using transit is more than double what it is when transit service is poor.

The CO and HC emissions of transit vehicles usually are very small in comparison to the emissions of the automobiles that transit replaces.[14] Consequently, if the average automobile trip length does not change greatly, the CO and HC emissions reductions resulting from transit service improvements are approximately proportional to the reductions in the numbers of automobile driver (or vehicle) trips that take place. The number of automobile driver trips, N_{AD}, that a group of travelers make is given by

$$N_{AD} = N_{DA} + N_{CP}/\mathrm{CPS}, \tag{6.30}$$

where N_{DA} and N_{CP}, respectively, are the numbers of drive alone and carpool person trips that the travelers make, and CPS is the size of carpools. To illustrate the reductions in CO and HC emissions that might result from improving transit service levels from those represented by cases 3 and 8 to those represented by cases 1 and 6, assume that the tabulated mode choice probabilities for these cases pertain to a large group of travelers. Then N_{DA} and N_{CP}, respectively, are proportional to the drive alone and carpool choice probabilities. Assume, also, that there are two persons per carpool and that transit service improvements do not cause a substantial change in the average length of automobile trips. Then, improving transit service quality from the conditions represented by cases 3 and 8 to those represented by cases 1 and 6 would reduce the number of automobile driver trips made by the affected travelers by 19 and 11%, respectively.[15] Motor vehicle CO and HC emissions from the affected trips would be reduced by approximately the same percentages. (In the following discussion, only percentage reductions in automobile driver trips will be stated. The corresponding percentage reductions in motor vehicle CO and HC emissions should be assumed to be approximately equal to the percentage reductions in automobile driver trips.)

The percentage reductions in NO_x emissions resulting from bus transit service improvements are likely to be considerably smaller than the percentage reductions in automobile driver trips, as diesel buses tend to have relatively high NO_x emission rates. In some cases, transit improvements that reduce CO and HC emissions may cause NO_x emissions to increase. More precise illustrations of the effects of transit service improvements on motor vehicle emissions from work trips are given later in this section and in section 6.8.

The foregoing examples indicate that high-quality transit service may be capable of attracting work trip travelers away from automobiles in sufficient numbers to achieve modest but potentially significant reductions in work trip motor vehicle emissions of CO and HC. However, cases 1 and 6 in table 6.8 represent very high levels of transit service quality. Although these levels can be achieved in high-volume corridors, for reasons that will be discussed in section 6.7 it is not likely that such high levels of transit service can be achieved over entire urban areas. The reductions in automobile driver trips and motor vehicle emissions that can be achieved by improving transit service decrease considerably when the quality of post improvement transit service decreases. For example, cases 2 and 7 in table 6.8 represent levels of transit service quality that are better than those represented by cases 3 and 8 but poorer than those represented by cases 1 and 6. Improving transit service quality from the conditions represented by cases 3 and 8 to those represented by cases 2 and 7 would reduce the number of automobile driver trips that the affected travelers make by 5 and 3%, respectively, compared to the reductions of 19 and 11% that would be achieved by improving transit service quality to the levels represented by cases 1 and 6.

Cases 2 and 7 probably represent higher levels of transit service quality than can be made available for most urban work trips (see section 6.7), although, as previously noted, transit service quality similar to that represented by cases 1 though 6 can be provided in high-volume corridors. Therefore, transit service improvements, by themselves, are probably better suited to reducing work trip motor vehicle emissions in high-volume corridors where existing transit service is poor than to reducing these emissions throughout entire urban areas.[16]

In deriving the foregoing results, it has been assumed that the transit system has sufficient capacity to carry all travelers who wish to use it.

This assumption may not be satisfied in some high-volume corridors. If the demand for transit use exceeds the available transit capacity, then reductions in automobile driver trips and emissions that are somewhat larger than those discussed so far may be achievable by implementing transit improvements that remove the capacity limitation. For example, in case 3 in table 6.8, suppose that the transit service can carry only half of the travelers that wish to use it. Assume that the travelers who are unable to use transit divide themselves among the drive alone and carpool modes in proportion to the tabulated probabilities of choosing those modes in case 3. Then the modal shares will be 0.65 (drive alone), 0.28 (carpool), and 0.07 (transit). Improving the quality of transit service to that shown in case 1 would provide sufficient capacity to accommodate demand, owing to the additional transit vehicles that would be needed to maintain the reduced headways. With a carpool size of two persons per vehicle, this improvement in transit service would cause a 26% reduction in automobile driver trips, compared to the 19% reduction that would be achieved if the case 3 transit ridership were not capacity limited.

The probability that work trip travelers will use the carpool mode is less sensitive to changes in travel times and costs than are the probabilities that these travelers will use the drive alone or transit modes. There is a variation of 0.11 among the drive alone choice probabilities for the tabulated cases with CBD work places and a variation of 0.09 among the drive alone choice probabilities for the cases with non-CBD work places. The transit choice probabilities vary by 0.17 and 0.10 for the two sets of cases. However, the carpool choice probabilities vary by only 0.08 and 0.05 for the CBD and non-CBD cases, respectively. The mode choice model does not indicate the reasons for this relative insensitivity of the carpool choice probability to changes in travel times and costs. However, it is likely that the insensitivity is due, at least in part, to the schedule inflexibility of carpooling, the difficulty of finding suitable carpool partners, and, possibly, aversion of some travelers to the socializing that is associated with carpooling.

Travel times and costs can be influenced more easily by transportation policy measures than can scheduling and social factors that affect carpooling. Therefore, the relative insensitivity of the carpool choice probability to changes in travel times and costs suggests that carpool incentives, by themselves, may not be capable of achieving large reductions in work trip automobile travel or emissions. This hypothesis is illustrated by cases 4, 5, 9, and 10, which represent examples of carpool incentive measures. In cases 4 and 9, carpools are given a cost advantage over single-occupant automobiles. In cases 5 and 10, carpools have an advantage in walk time, such as might be created by reserving desirable parking areas for carpools. In cases 2 and 7, carpool incentives are not in effect, but in other respects the service qualities of all modes are the same as they are in cases 4, 5, 9, and 10. Thus, cases 2 and 7 provide bases against which the effects of carpool incentives can be compared. If the size of carpools is assumed to be two persons, then relative to cases 2 and 7, the carpool incentives cause reductions in the numbers of automobile driver work trips made by the affected travelers of at most 6% in the cases with CBD work places and at most 4% in the cases with non-CBD work places. If the size of carpools is assumed to be three persons, which is larger than the average size of carpools in US cities is estimated to be [44], then the carpool incentives cause reductions in the numbers of automobile driver work trips made by the affected travelers of at most 8 and 5% for the CBD and non-CBD cases, respectively.[17]

The results that have been obtained so far, although they are very preliminary, suggest three hypotheses concerning the abilities of transit improvements and carpool incentives to reduce motor vehicle emissions from work trips. The first hypothesis is that transit improvements can cause moderate reductions in CO and HC emissions from work trips in high-volume corridors. The second hypothesis is that the ability of transit improvements to reduce emissions from all of the work trips in an entire urban region is probably quite limited. The third hypothesis is that the reductions in emissions that can be achieved through carpooling are likely to be small. In the following discussion, it will be seen that evidence from a variety of sources tends to support these hypotheses.

6.6.3 Corroborative Results

The results that have been obtained so far are based on a small number of very simple examples and a single model of mode choice. By themselves, these results provide only tentative indications of the reductions in work trip automobile travel and emissions that may be achievable through transit improvement and carpool incentive measures. However, evidence obtained from other sources tends to support these results. First, qualitatively similar results have been obtained in examples based on other mode choice models [45, 46]. Moreover, studies of psychological and scheduling factors that affect carpooling have confirmed that these factors can present serious barriers to carpooling for travelers who currently drive alone [47, 48]. Taken together, these two sets of results tend to support the hypotheses that transit improvements can cause moderate reductions in work trip motor vehicle emissions in high-volume corridors and that the emissions reduction effectiveness of carpooling is likely to be small.

To the extent that relevant data are available, operational experience with transit improvements and carpool incentives also tends to support these hypotheses. Two transit improvement projects in high-volume corridors that succeeded in achieving substantial increases in transit service quality and whose effects on transit usage and automobile travel have been relatively well documented are the Shirley Highway Express-Bus-on-Freeway demonstration project in the Washington, DC, area and the San Bernardino Freeway Express Busway project in the Los Angeles area. The bus service in both projects was oriented toward peak period work trips in the peak flow directions in their respective corridors.[18] Both projects made improvements in bus routes and schedules, and both included operation of express buses on specially reserved roadways (busways) next to otherwise congested freeways. Both projects achieved substantial reductions in transit travel times.

During the Shirley Highway project, the mode split to bus among travelers in the Shirley Highway corridor to whom the bus service was available (the bus market share) increased from 30% at the beginning of the project to 41% at the end of the project [49]. At the time of the project, approximately 20% of work trips in the entire Washington area used transit. It was estimated from surveys of Shirley corridor commuters and from traffic counts that the express bus service reduced the number of morning peak period, peak direction automobile trips in the Shirley Highway corridor by approximately 20%, relative to the number of trips that would have occurred without the bus service. In addition, it was estimated that emissions of HC, CO, and NO_x from morning peak period, peak direction automobile traffic were reduced by roughly 20% each, relative to the levels that would have occurred without the bus service [49]. Results for the afternoon peak period are not available.

During the San Bernardino Freeway project, the bus market share in the parts of the San Bernardino Freeway corridor that were best served by the project's buses increased from 4% at the beginning of the project to approximately 25% [50, 51]. It was estimated from traffic flow measurements that diversion of travelers from automobiles to buses caused emissions of HC and CO from combined morning and afternoon peak period, peak direction traffic to decrease by 13% for each pollutant, relative to the emissions levels that would have occurred without the bus service. Emissions of NO_x from the same traffic decreased by approximately 10% [51].

The effects of the Shirley Highway and San Bernardino Freeway projects on total motor vehicle HC and NO_x emissions in their respective metropolitan regions were not estimated but undoubtedly were negligible.

Experience with carpooling projects has been less encouraging than experience with the two transit projects has been. A review of the results of carpool matching and promotion programs in 19 cities found that these programs increased the numbers of urban regional work trips using carpools by less than 3% and decreased work trip vehicle miles traveled by at most 1% [52]. The opening of the Shirley Highway bus lane to carpools carrying four or more persons caused a reduction of approximately 1% in the volume of morning peak period, peak direction automobile traffic in the Shirley Highway corridor [49]. The opening of the San Bernardino Freeway bus lanes to carpools carrying three or more persons caused estimated reductions in emissions from morning and afternoon peak period, peak direction traffic of 7% each for CO, HC, and NO_x [51].

There have been several computer-based studies of the abilities of transit improvements and carpool incentives to achieve reductions in urban regional automobile emissions from work trips. These studies were based on the use of models similar to the one that was used to develop the examples in table 6.8. Like the examples, the computer-based studies involved varying the values of the service quality variables of the models in order to simulate the effects of transit improvements and carpool incentives. However, the computer-based studies predicted the mode choices of each worker in a large sample of workers and then aggregated the predictions using equation (6.28), rather than using the mode choice of a single "typical" worker to approximate the average mode choice of workers generally. As a result, the computer-based studies incorporated a more accurate treatment than the examples do of the effects of differences in the characteristics of the transportation options that are available to different workers (e.g., some workers have good transit service available to them, and others do not) and of variations in socioeconomic attributes (e.g., incomes) among travelers. In addition, many of the computer-based studies included automobile ownership and carpool size among the variables whose values were predicted. This enabled these studies to treat the possibilities that improvements in the quality of transit service may enable some households to reduce the numbers of automobiles that they own and that certain types of carpool incentives may encourage the formation of large carpools. Table 6.9 summarizes the results of several computer-based studies. In the case of studies that examined several different sets of transit improvement and carpool incentive measures, the measures that yielded the largest reductions in automobile use or

Table 6.9 Results of computer-based studies of the effects of transit improvements and carpool incentives on urban regional automobile emissions from work trips

City	Measure (see definitions in table notes)	Percentage reduction in work trip-related[a]					Reference
		Automobile driver trips	Automobile vehicle miles traveled	Automobile HC emissions	Automobile CO emissions[b]	Automobile NO_x emissions	
Washington, DC	Carpool incentives (A)		7				[53]
	Transit improvements and carpool incentives (B)		10				[53]
	Transit improvements (C)	7	8	8	8	8	[54]
	Transit improvements and carpool incentives (D)	11	9	10	10	9	[54]
Denver	Transit and carpool incentives (E)		4	3	3	4	[43]
	Carpool incentives (F)		4	3	3	4	[43]
Fort Worth	Transit and carpool incentives (E)		10	7	6	10	[43]
	Carpool incentives (F)		10	7	6	9	[43]
San Francisco	Transit and carpool incentives (E)		10	7	6	9	[43]
	Carpool incentives (F)		10	6	5	8	[43]
San Diego	Transit improvements (G)	5					[55]

a. Blanks signify information not available. Emissions reductions do not reflect any increases in emissions from transit vehicles that may occur as results of transit improvements. Definitions: carpool incentives (A)—carpool matching and promotion programs and preferential parking for carpoolers at all organizations with at least 100 employees. Vanpooling available to employees who live at least 10 mi from work and are employed by organizations with at least 500 employees; transit improvements and carpool incentives (B)—measures in A plus preferential treatment for high-occupancy vehicles on all major roadways, transit made available to all workers, and 20% reduction in all transit headways; transit improvements (C)—transit improved to point where total transit one-way travel time does not exceed total automobile one-way travel time by more than 19 min for any trip. One-way transit in-vehicle travel time does not exceed automobile in-vehicle travel time by more than 5 min. One-way transit out-of-vehicle travel time does not exceed automobile out-of-vehicle travel time by more than 14 min; transit improvements and carpool incentives (D)—transit improvements (C) plus carpool matching and promotion and preferential parking for carpools at all employment locations; transit and carpool incentives (E)—carpool matching and promotion programs at all organizations with at least 50 employees. Vanpooling available to employees who live at least 10 mi from work and are employed by organizations with at least 250 employees. Employers pay 50% of employees' transit fares; carpool incentives (F)—same as (D) but without transit fare subsidy; transit improvements (G)—reduce maximum headway to 10 min.

b. Total regional CO emissions contribute to general background CO concentrations but are not closely related to CO concentrations along individual roadways.

emissions have been included in the table. The tabulated results tend to support the hypothesis that transit improvements and carpool incentives are not likely to be capable of achieving large reductions in urban regional automobile emissions from work trips.

6.6.4 Discussion

None of the currently available evidence concerning the abilities of transit improvements and carpool incentives to achieve reductions in work trip automobile emissions is definitive. There have been few operational projects that have produced data useful for assessing the effects of transit improvement and carpool incentive measures on emissions, and the data obtained from these projects pertain to a small fraction of the measures that are, in principle, available. Moreover, the state of the art in travel behavior modeling is such that models can provide only rough estimates of the changes in automobile travel and emissions that transit improvement and carpool incentive measures would cause. Therefore, it is necessary to use considerable caution in attempting to draw general conclusions from the currently available results of operational projects and modeling studies.

To the extent that conclusions are justified, the available evidence suggests that transit improvements may be capable of achieving reductions in work trip motor vehicle emissions of at least 20% in individual high-volume corridors where existing transit service is poor. However, the reductions in urban regional emissions from work trips that can be achieved through transit improvements are likely to be less than 10%. The reductions in work trip emissions that can be achieved through carpool incentives also are likely to be less than 10%, both in individual corridors and in entire urban regions. Implementation of transit improvement and carpool incentive measures together can achieve somewhat greater emissions reductions than separate implementation of either type of measure can achieve. However, the available evidence suggests that even combinations of transit improvements and carpool incentives are unlikely to reduce regional emissions from work trips by substantially more than 10%. The ability of transit improvements to achieve regional emissions reductions is limited by the difficulty of making high-quality transit service available to large proportions of urban travelers. The reasons for this difficulty are explained in the next section. The ability of carpooling to reduce emissions is limited by the difficulty of achieving high levels of demand for the carpool mode.

Nonwork trips are not well suited to carpooling owing to their irregularity. There is little information available concerning the ability of transit improvements to reduce emissions from nonwork automobile travel. The results of a very limited set of modeling studies suggest that transit improvements would cause less than 5% reductions in nonwork automobile trips and vehicle miles traveled [46, 56]. The resulting emissions reductions also would be less than 5%. In addition, there is limited evidence to support the hypothesis that transit improvements and carpool incentives for work trips would cause small (1–2%) increases in nonwork automobile travel and emissions by making cars that formerly were used for work trips available for nonwork use [46, 53]. These emissions increases would tend to cancel at least part of the reductions in nonwork trip emissions that transit improvements for nonwork trips might otherwise cause. Therefore, in the remainder of this section it will be assumed that transit improvements would have negligible effects on emissions from nonwork automobile travel. However, it should be borne in mind that there have been too few systematic investigations of these effects to permit firm conclusions to be reached.

Work trips account for less than 40% of total automobile emissions of HC and NO_x in urban regions [3]. Therefore, the results presented in this section suggest that transit improvements and carpool incentives are unlikely to be capable of reducing regional automobile emissions of these pollutants by more than roughly 4% (i.e., 10% of 40%). However, transit improvements and carpool incentives appear to be capable of reducing morning peak period CO emissions by at least 20% in high-volume corridors where substantial improvements in transit service are possible. Provision of high-quality transit service appears to be the most important factor in achieving these emissions reductions. The reductions in afternoon peak period CO emissions that can be achieved by transit improvements and carpool incentives are likely to be less than the morning peak period reductions, as work trips account for a smaller proportion of afternoon than of morning peak period travel [2]. In addition, the reductions in peak period NO_x emissions that can be achieved through transit improvements are likely to be substantially less than the corresponding reductions in CO emissions owing to the relatively high NO_x emissions of diesel buses. (The effects of bus transit improvements on NO_x emissions are discussed further in section 6.8.)

The reductions in 8-hr average CO emissions that may be achievable in high-volume corridors through transit improvements and carpool incentives can be estimated as follows. Work trips typically account for approximately 35% of total travel in a daytime 8-hr period, and peak period work trips typically account for approximately 20% of 8-hr travel. The results that have been presented in this section suggest that transit improvements and carpool incentives may be capable of reducing work trip automobile emissions in high-volume corridors by 20–30%. These emissions reductions are most likely to be achievable during peak periods, but may also be achievable during hours adjoining peak periods. Accordingly, transit improvements and carpool incentives may be capable of reducing 8-hr average CO emissions in high-volume corridors by 5–10%. (In round numbers, this is the effect of reducing 20–35% of 8-hr average emissions by 20–30%.)

6.7 *Transit Supply Issues*

The foregoing discussion has dealt only with the demand side of the ability of transit improvements and carpool incentives to reduce automobile travel and emissions. In other words, the discussion has considered only the extent to which travelers would use transit and carpools if these modes were available and certain incentives in terms of service quality were provided for their use. However, as was indicated in the previous section, there are important supply side difficulties involved in making high-quality transit service available to large numbers of travelers over large parts of urban regions. These difficulties concern the relations among the cost of providing the service, the quality of the service, and the ridership levels that can be achieved. In addition, in some cities unusually large numbers of transit vehicles may be needed to carry large fractions of the trips that currently take place by automobile. Although the necessary vehicles and support facilities presumably could be acquired over sufficiently long periods of time, delays in acquiring them would prevent large numbers of automobile travelers from being diverted to transit quickly. As was discussed in the previous section, supply side difficulties are more important than demand side ones in limiting the emissions reductions that can be achieved through transit improvements. The purpose of this section is to explain these difficulties. Supply side difficulties are not present in the case of carpooling, as carpooling does not require increasing the number of vehicles that are operating in an area and is not likely to cause large increases in the costs of transportation.

6.7.1 A Note on Cost Comparisons

In the following discussion, the cost of automobile travel will be used as a standard for judging whether the cost of transit service is high or low. Roughly speaking, transit service whose average cost per person trip is considerably higher than that of automobile travel will be considered to be costly, possibly unacceptably so. The word "cost" will refer to the ordinary monetary costs of transportation, including the costs of automobile ownership to the extent that automobile ownership might be affected by transit improvements.[19] There are, of course, many other standards that might be used for judging whether transit costs are high or low. The cost of automobile travel is used here because the purpose of transit improvements in an emissions reduction program (as in many other programs) is to provide a substitute for automobile travel. Thus, the difference between the costs of transit and automobile travel provides an indicator of the cost (if transit is more expensive) or saving (if transit is less expensive) to society of the program.

6.7.2 Statement of the Problem

Using the foregoing standard for judging transit costs, the main supply side problem of providing high quality transit service to a relatively large percentage of the travelers in a city can be stated as follows: In order to keep the average cost per person trip of transit travel reasonably close to that of automobile travel, it is necessary to achieve a certain minimum level of transit vehicle occupancy. Given a level of transit service quality and the transit mode share that it would achieve, the necessary vehicle occupancy can be achieved only if the volume of person trips between the points served by transit is sufficiently large. The necessary person trip volume normally increases as the quality of transit service increases. Consequently, it is usually feasible to provide high-quality transit service only along heavily traveled routes. Conversely, it is particularly difficult to provide high-quality transit service in low-density residential areas, where many of the automobile trips that contribute to poor air quality originate and terminate, but where trip densities tend not to be sufficiently high to support high-quality service.

The following simple example illustrates the relations among transit service quality, transit costs, and person trip volumes.

Example 6.5—The Cost of Transit Service

Suppose that bus service is to be offered between two points A and B and that these two points are separated by a distance of L mi. Assume for simplicity that A and B are the only two points that will be served by the route, so that there are no intermediate stops. Let the cost of bus operation be $2.50/veh mi, and let the cost of automobile travel be $0.15/veh mi. Assume that parking is free. (Inclusion of nonzero parking costs would complicate the analysis in this example without altering the example's main conclusions.)

The cost per person trip of transit service can be related as follows to the quality of transit service and the total volume of person travel from A to B and B to A. Bus operation costs \$5.00L/veh round trip. If the schedule frequency on the bus route calls for f veh trips/hr, then the cost per hour of the bus service is \$5.00Lf. Assume that there are V_{AB} person trips/hr from A to B and V_{BA} person trips/hr from B to A. Let the transit mode shares among these trips, expressed as decimal fractions, be P_{AB} and P_{BA} for trips from A to B and B to A, respectively. Then there will be $P_{AB}V_{AB} + P_{BA}V_{BA}$ transit person trips/hr, and the average cost per transit person trip, C_T, can be written as

$$C_T = 5Lf/(P_{AB}V_{AB} + P_{BA}V_{BA}) \text{ (dollars)}. \tag{6.31}$$

If the average automobile occupancy is O_A persons/car, then the average cost/automobile person trip from A to B or B to A is \$0.15L/O_A. Denote this cost by C_A.

Note that the average transit vehicle occupancy is $(P_{AB}V_{AB} + P_{BA}V_{BA})/f$ persons/veh round trip. Thus, for example, to make the cost per person trip of transit travel equal to the cost per person trip of automobile travel, the average occupancy of transit vehicles must be $5O_A/0.15$ persons/round trip. If O_A is 1.25 persons/car, which is typical of work trip automobile occupancies in many cities, then the "break even" transit occupancy is 41.67 persons/round trip.

Suppose that it is desired to determine the minimum two-way person trip volume (transit passengers plus automobile drivers and passengers) that is needed to make the cost per transit person trip less than or equal to K times the cost per automobile person trip, where K is a positive constant. This volume can be obtained by solving the inequality

$$C_T \leq KC_A, \tag{6.32}$$

or

$$5Lf/(P_{AB}V_{AB} + P_{BA}V_{BA}) \leq 0.15KL/O_A \tag{6.33}$$

for the total volume $V_{AB} + V_{BA}$. If P_{AB} and P_{BA} are equal, then the solution to this inequality is

$$V_{AB} + V_{BA} \geq 5fO_A/0.15KP_{AB} = 5fO_A/0.15KP_{BA}. \tag{6.34}$$

If the mode shares are not equal in both directions, then the solution to the inequality is

$$V_{AB} + V_{BA} \geq 5fO_A(1 + r)/0.15K(P_{AB} + rP_{BA}), \tag{6.35}$$

where $r = V_{BA}/V_{AB}$ is the ratio of the volume of person trips in the off-peak direction of flow (assumed to be the direction B to A in this case) to the volume of person trips in the peak direction.

Suppose that the transit line is expected to serve mainly work trip travelers and that the levels of transit service quality that are under consideration correspond to cases 1, 2, and 3 in table 6.8 if one end of the line is in the CBD and to cases 6, 7, and 8 if neither end is in the CBD. Assume that these levels of transit service would produce transit mode shares that are equal to the mode choice probabilities in table 6.8. Note that the schedule frequencies corresponding to cases 1 and 6, 2 and 7, and 3 and 8, respectively, are 12, 4, and 2 vehicle round trips per hour. Let r be 0.4, and let the average automobile occupancy be 1.25. Let K be 1. Then the right-hand side of inequality (6.35) specifies the minimum two-way person trip volume ($V_{AB} + V_{BA}$) that is needed to make the average cost per transit person trip less than or equal to the average cost per automobile person trip. Table 6.10 shows this minimum volume for each of the cases that are under consideration. (Minimum volumes for other values of K can be obtained by dividing the tabulated volumes by the desired value of K.) The minimum volume is lower when the transit line has one end in the CBD than when it does not. This is because at any given level of transit service quality, the transit mode share is higher for travelers whose work places are in the CBD than for travelers whose work places are elsewhere (see table 6.8). However, in both the CBD-oriented and the non-CBD-oriented cases, the minimum person trip volumes that are needed to make the cost of transit travel less than or equal to the cost of automobile travel increase as the quality of transit service increases. If the quality of transit service is improved while the total person trip volume stays constant, then the level of transit use increases, but it does not increase sufficiently to pay for the increased cost of the improved service. Thus, the cost of high-quality transit service is competitive with the cost of travel by automobile only on routes where the total volume of person travel by all modes is relatively high.

6.7.3 Relations between Transit Service Quality, Ridership, and Cost in US Cities

Example 6.5 illustrates some of the supply side problems of providing high-quality transit service between two locations. The supply side problems of providing high-quality transit service to a relatively large percentage of the travelers in an entire city can be investigated with a

Table 6.10 Characteristics of transit service options for a route connecting two locations, A and B[a]

Case[b]		Transit schedule frequency (veh round trips/hr)	B is CBD?	Transit mode share in direction[c]		Minimum two-way person trip volume needed to make the average cost of transit service per transit person trip less than or equal to the average cost of automobile travel per automobile person trip
A to B	B to A			A to B	B to A	
1	6	12	Yes	0.31	0.17	1,852
2	7	4	Yes	0.18	0.09	1,080
3	8	2	Yes	0.14	0.07	694
6	6	12	No	0.17	0.17	2,941
7	7	4	No	0.09	0.09	1,852
8	8	2	No	0.07	0.07	1,190

a. Based on example 6.5. Automobile occupancy is assumed to be 1.25 persons per car. When B is the CBD, the volume of person travel from B to A is assumed to be 0.4 times the volume of person travel from A to B.
b. Cases correspond to the ones given in table 6.8.
c. Mode share is expressed as a decimal fraction.

simple model of urban bus transit service [57, 58]. The model involves designing hypothetical bus systems for the city and estimating the operating characteristics of these systems, including travel times, costs, and fleet sizes, as functions of transit ridership levels. Although this model does not produce information that is useful for the detailed design and evaluation of urban transit systems, it illustrates the potential difficulties of providing high-quality service to large numbers of urban travelers. The structure of the model is summarized here, and the results of applying it to two cities are presented. (A more detailed description of the model's structure is available in reference [58].)

To apply the model, the city under consideration is divided into a set of large districts. The districts are used to define transit service areas and the demand for transit trips. Since the model treats only transit supply and does not provide forecasts of ridership, the demand for transit trips is developed from exogenously specified, district-level person trip tables and exogenously specified transit mode split factors. The trip tables give the number of person trips per hour between each pair of districts according to trip purpose and time of day. The mode split factors specify the fraction of trips of each purpose that it is assumed will use transit if service is provided between their origin and destination districts or, in the case of intradistrict trips, within their districts of origin and destination. (The mode split factors do not represent forecasts of the demand for transit travel. They are parameters of the model that are used to establish supply side relations between transit mode split and other characteristics of the transit system.) Bus service is provided within and between districts where the volume of person trips exceeds an exogenously specified threshold. The set of all districts and pairs of districts within and between which transit service is provided is called the transit service area. Transit trips in the service area are assigned to bus routes on an idealized street network. Buses are assigned to the routes in sufficient quantities to both accommodate the demand for transit trips and achieve or exceed an exogenously specified minimum schedule frequency. Given the route assignments and schedule frequencies, average transit travel time is computed from estimates of average walk and wait times, in-vehicle distances, and bus speeds. Average transit cost per trip is computed from estimates of the purchase price of buses and auxiliary facilities (yards, shops, and stations) and estimates of the costs of operating buses. The average travel time and cost per trip that would be incurred if all transit trips were carried in automobiles also are computed.

Through repeated runs of the model using different values of the transit mode split factors, threshold trip volume for providing bus service within or between districts, and minimum schedule frequency, supply side relations among total transit ridership, the transit mode share in the areas served by transit (i.e., the transit market share), travel times and costs by transit and automobile, and transit fleet sizes can be developed. These relations are supply side, in contrast to demand side, because the transit mode share in the area served by transit does not constitute a prediction of the mode share that would be achieved if transit service with the specified travel times and costs were provided. Rather, it defines a mode share that must be achieved in order to achieve specified values of total transit ridership, average transit travel time per trip, and average transit cost per trip.

The foregoing model has been used to estimate the characteristics of bus transit systems that would be capable of carrying relatively large proportions of total person trips in the Los Angeles and Washington, DC, areas. The person trip tables used with the model were obtained from the Los Angeles and Washington area transportation studies.[20] It was assumed that transit service would be provided only between the hours of 6 A.M. and 8 P.M. Approximately 90% of daily person trips in the Los Angeles and Washington areas take place during this time period.

The characteristics of several transit options for each city are summarized in table 6.11. This table shows the transit mode shares that must be achieved in the transit service area to enable various percentages of total, regional person trips to be carried by transit with the specified differences between average transit and automobile travel times and costs. The table also shows the bus fleet sizes that are needed. The most prominent features of the tabulated results are the high transit mode shares that must be achieved in the transit service area in order to achieve the specified ridership levels and average travel times and costs. The required mode shares vary from 31 to 67%, depending on the city and the option, and apply to all-day, all-purpose travel. In US cities, transit mode shares in this range usually occur only for work trips in high-volume, CBD-oriented corridors.

To achieve the transit ridership levels shown in table 6.11 with mode shares below the tabulated values, it would be necessary to accept substantial increases in transit travel times, transit travel costs, or both. This is because reducing the mode share while keeping total ridership constant would require extending transit service to relatively low volume routes. (The structure of the transit supply model is such that transit service is extended to routes roughly in descending order of their volumes.) Hence, to maintain low transit travel times it would be

Table 6.11 Characteristics of Los Angeles, California, and Washington, DC, area transit options[a]

City	Ridership expressed as percentages of total daily and 6 A.M.–8 P.M. person trips that use transit		Differences (transit-auto) between average one-way travel times and costs of transit and automobile trips		Transit mode share that must be achieved in transit service area	Required size of bus fleet
	Daily	6 A.M.–8 P.M.	Time (min)	Cost (¢)		
Los Angeles	9	10	17	0	38	4,000
	9	10	17	15	31	4,500
	18	20	17	0	45	9,500
	18	20	17	15	38	9,000
	26	30	17	0	52	14,000
	26	30	17	15	46	15,500
Washington	12	13	17	15	50	1,200
	18	20	17	0	64	2,000
	18	20	17	15	52	3,000
	28	30	17	0	67	3,100
	28	30	17	15	55	5,000

a. Source: reference [58] and unpublished results obtained from applying the transit supply model described in reference [58]. Cost differences are in 1980 cents.

necessary to accept low vehicle occupancies and high average costs per trip. Conversely, to achieve the high vehicle occupancies that are needed to maintain low average costs per transit trip, it would be necessary to decrease service frequencies and increase route spacings, thereby increasing wait and walk times for transit users. As an example of the effects on transit travel times and costs of achieving transit mode shares below those shown in table 6.11, suppose in the case of Los Angeles that it is desired to carry 18% of daily person trips on transit. Assume that transit can attract only 25% of the person trips in its service area (which is still very high by US standards), instead of the 38–45% shown in the table. Then, to enable transit to carry 18% of total daily person trips and maintain an average cost per transit trip that is approximately equal to the average cost per automobile trip, it would be necessary to allow the average transit travel time to exceed the average automobile travel time by roughly 25 min. To enable transit to carry the same percentage of daily trips and maintain a 17-min difference between the average transit and automobile travel times per trip, it would be necessary to accept an average cost per transit trip that is approximately $1.50 more than the average cost per automobile trip.

In Washington, if transit carried 18% of daily trips, maintained a 17-min per trip difference between the average transit and automobile travel times, and achieved a mode share of 25% in its service area (in contrast to the share of 52–64% that is shown in table 6.11), the average cost of transit travel would exceed the average cost of automobile travel by approximately $2.50/trip.

Results analogous to those in table 6.11 are not available for cities other than Los Angeles and Washington. However, there are reasons for concluding tentatively that in most other US cities, transit mode shares similar to or higher than those in table 6.11 would be needed to enable transit to achieve the ridership levels and travel times and costs shown in the table. The mode share that is needed to achieve the tabulated travel times and costs depends mainly on the trip density in the transit service area. A lower mode share is needed when the trip density is high than when it is low. Although data on trip densities in US cities are not readily available, data on population densities, which often are useful indicators of trip densities, are available. In 1970, which is the census year closest to the years in which the data used to develop the results for table 6.11 were acquired, Los Angeles and Washington were among the 12 densest metropolitan areas in the United States and among the 9 densest outside of the New York City area [59]. Moreover, the suburbs of Los Angeles and Washington had higher densities than did the suburbs of any other large US cities, and Washington had one of the 5 highest central city densities in the United States [59]. Judging from these statistics, it seems likely that in most US cities, and not only in Los Angeles and Washington, high-quality, high-ridership transit service could be provided at an average cost per transit trip that is near to the average cost of automobile travel only if transit mode shares similar to or greater than those shown in table 6.11 could be achieved.

6.7.4 Discussion

The foregoing results imply that the feasibility of providing high-quality transit service that carries a relatively large proportion of total person trips in an urban region depends critically on the transit mode share that can be achieved in the area that transit serves. If a sufficiently high mode share cannot be achieved, then to provide high-quality, high-ridership service, it is necessary to tolerate an average cost per transit trip that is considerably greater than the average cost of an automobile trip. The limited evidence that is available suggests that in the foreseeable future, all-day transit mode shares as high as those in table 6.11 probably will not be achievable over sufficiently large areas of US cities to produce the transit ridership levels that are shown in the table.[21] Although mode shares similar to those in table 6.11 currently are achieved for work trips in some high-density, CBD-oriented corridors, these trips do not account for sufficiently large proportions of total travel in urban regions to produce the tabulated ridership levels. It is highly questionable whether transit service whose travel times are similar to those shown in table 6.11 could achieve such high transit mode shares for non-CBD work trips in the absence of severe restraints on automobile travel. It is also highly questionable whether this transit service could attract sufficiently many nonwork trips to permit attainment of the all-day mode shares shown in table 6.11.

The transit mode share that the transit service in table 6.11 might be capable of achieving for non-CBD work trips can be investigated with the mode choice model of table 6.7. Table 6.12 shows values of the model's service quality variables that correspond approximately to the average level of transit service quality in table 6.11. The model's predictions of the resulting transit mode choice probabilities for non-CBD workers from households with various socioeconomic characteristics are shown in table 6.13. The model predicts that non-CBD workers from households in which there is at least one automobile per worker would have transit mode choice probabilities of less than 0.25. Non-CBD workers from automobile-

Table 6.12 Average values of service quality variables for the transit service of Table 6.11

Variable	Value for		
	Drive alone	Carpool	Transit
Round trip travel cost (¢)	150	75	150
Round trip in-vehicle travel time (min)	32	37	38
Round trip walk time (min)	2	2	19
Sum of initial round trip transit headways (min) up to 8 min			8
Excess over 8 min of sum of initial transit headways			2
One half round trip sum of transfer headways (min)			7

Table 6.13 Transit mode choice probabilities for non-CBD workers with the values of the service quality variables shown in table 6.12[a]

Household characteristics				
Automobiles per worker	Primary worker[b]	Number of workers	Annual income ($)	Transit choice probability
1	1	2	10,000	0.16
1	0	2	10,000	0.23
1	1	1	10,000	0.17
0.5	1	2	10,000	0.32
0.5	0	2	10,000	0.43
0.33	1	3	10,000	0.37
0.33	0	3	10,000	0.46
0	1	1	10,000	0.85
1	1	2	30,000	0.15
1	0	2	30,000	0.21
1	1	1	30,000	0.16
0.5	1	2	30,000	0.31
0.5	0	2	30,000	0.41
0.33	1	3	30,000	0.36
0.33	0	3	30,000	0.45
0	1	1	30,000	0.85

a. Based on the mode choice model of table 6.7 and a four-member household.
b. The value of the primary worker variable is 1 if the traveler is the primary worker in the household and 0 otherwise.

owning households in which there is less than one automobile per worker would have transit mode choice probabilities of roughly 0.30–0.45, depending on the household characteristics. Only workers from non-automobile owning households would have transit mode choice probabilities exceeding 0.50. Similar results have been obtained in examples based on other mode choice models [45, 46]. Although these results are rough, since they do not take account of the deviations in transit service quality from the average that would occur in real transit systems, they are, nonetheless, informative. They suggest that transit service similar in quality to that of tables 6.11 and 6.12 would achieve a transit mode share of 30% or more for non-CBD work trips only if the service persuaded relatively large numbers of households with non-CBD workers to own fewer automobiles than they have workers. If the numbers of automobiles owned by these households were comparable to the numbers of workers in the households, then the transit mode share for non-CBD work trips would be well below 30%.

Average automobile ownership levels in most US cities currently are in the vicinity of one automobile per worker. Although data on automobile ownership levels among households with non-CBD workers (as distinct from all households) are not readily available, it is likely that these ownership levels are higher than the urban regional averages. Thus, it is likely that substantial reductions in automobile ownership levels would be needed to achieve non-CBD transit mode shares of 30% or more, given levels of transit service quality similar to those in table 6.11 and 6.12. Although it is difficult to predict how automobile ownership levels might be affected by the provision of transit service similar to that shown in the tables, it would be unwise to assume that provision of such service automatically would produce the necessary reductions in ownership.

The computer-based studies that were discussed in the previous section provide additional evidence concerning the likelihood of achieving transit mode shares as high as those shown in table 6.11. Although these studies did not use transit service that corresponds exactly to the service in tables 6.11 and 6.12, the two Washington studies investigated the effects of significantly improving the quality of transit service that is available to non-CBD workers [53, 54]. Neither study obtained work trip transit mode shares as high as 30%.

The prospects for achieving high transit mode shares for work trips could be enhanced by designing the transit system to provide travel times that are shorter than those shown in tables 6.11 and 6.12. However, this would almost certainly cause a large increase in the mode share needed to maintain any specified average cost per transit trip, thereby defeating the purpose of the reduction in travel times. For example, if transit headways were reduced by 50%, then the transit mode share needed to maintain a constant average cost per transit trip would roughly double, but it is highly unlikely that the mode share that is actually realized would double.

Work trip mode shares similar to those shown in table 6.11 undoubtedly could be achieved if automobile use were made sufficiently costly or were otherwise severely restrained. However, it is likely that very high automobile use costs or stringent restraints would be needed, indicating that large numbers of travelers would find using the transit service to be highly undesirable. Table 6.14 shows estimates of work trip transit mode shares that were produced in several computer-based modeling studies of the effects of automobile pricing and restraint measures on travel in the Los Angeles and Washington areas. These studies included increasing the average price per automobile driver round trip to work by roughly $1.25–4.00 (1980 dollars), depending on the study. None of the studies achieved estimated mode shares for their respective cities that are as high as those

Table 6.14 Estimates of work trip transit mode shares that would be achieved by implementation of combinations of transit improvement, pricing, and restraint measures in Los Angeles, California, and Washington, DC

City	Measures[a]	Work trip transit mode share (%)	Reference
Los Angeles	Transit improvements and 140% increase in gasoline price	29	[60]
Washington	Transit improvements, carpool incentives, double gasoline price, reduce CBD parking supply, minimum parking price of $2.80 per day in CBD and $1.40 per day elsewhere	39	[53]
Washington	Transit improvements and $5.50 per day minimum parking price	24	[54]
Washington	Transit improvements and triple gasoline price	22	[54]

a. Prices are in 1980 dollars and have been inflated from the years of the cited studies as necessary.

shown in table 6.11. Thus, if the results of the studies are reasonably accurate, increases in the costs of automobile travel that are greater than those represented in table 6.14 would be needed to achieve transit mode shares as high as those in table 6.11.

The transit options listed in table 6.11 require improbably high transit mode shares for nonwork trips to be achieved, even if large numbers of work trips can be attracted to transit. Table 6.15 shows the nonwork trip transit mode shares in the transit service area that are required by each of the options in table 6.11 for several different values of the transit mode share for work trips. The work trip mode shares have been set at very high levels in order to obtain indications of the minimum nonwork trip mode shares that are required. Depending on the city, the transit option, and the transit mode share for work trips, nonwork trip mode shares ranging from 19 to at least 75% are needed. Nonwork trip mode shares near the lower end of this range are satisfactory only if virtually all work trips in the transit service area use transit. As noted in the previous section, information on the ability of transit to attract nonwork travelers is extremely limited. However, the available evidence suggests that one should not be optimistic about the prospects for achieving the nonwork trip mode shares shown in table 6.15 with transit service similar to that shown in table 6.11 [46, 56]. Moreover, this evidence suggests that policies to restrain or increase the cost of nonwork automobile travel would not cause large increases in transit use for nonwork trips. Rather, it is likely that the main effect of these policies would be to reduce nonwork trip frequencies and trip lengths in the affected areas.

The foregoing difficulties would not be eased by orienting the transit service primarily toward work trips, even though it is easier to attract work trips to transit than to attract other trips. If the transit service operated all day, then it could not achieve the high average vehicle occupancy needed to maintain a low average cost per trip owing to the relatively low volume of work trips during off-peak hours. This is illustrated in table 6.15, which shows that even if the transit systems described in the table attracted all of the work trips in their respective service areas, substantial proportions of nonwork trips still would have to use transit in order to maintain a 17-min per trip difference between average transit and automobile travel times and an average cost per transit trip that is close to the average cost per automobile trip. If transit served only peak period work trips, then with any given transit service area and transit mode share in the service area, the number of trips over which the fixed costs of transit service could be spread would be considerably lower than it would be if the service operated all day and served trips of all purposes. Consequently, a higher

Table 6.15 Nonwork trip transit mode shares that are required for the Los Angeles, California, and Washington, DC, area transit options

City	Ridership expressed as percentage of total daily person trips that use transit	Differences (transit-auto) between average one-way travel times and costs of transit and automobile trips		Nonwork trip transit mode share (%) that must be achieved in the transit service area if the work trip mode share (%) is		
		Time (min)	Cost (¢)	50	75	100
Los Angeles	9	17	0	36	31	27
	9	17	15	28	23	19
	18	17	0	44	30	35
	18	17	15	36	31	27
	26	17	0	52	48	44
	26	17	15	45	41	36
Washington	12	17	15	50	38	25
	18	17	0	71	59	46
	18	17	15	53	41	28
	28	17	0	75	63	51
	28	17	15	57	45	33

transit mode share among the trips for which service is offered would be needed to maintain given values of the average transit travel time and cost per trip than would be needed if the service operated all day and carried trips of all purposes. Indeed, in the case of Los Angeles, the results obtained from the previously described transit supply model indicate that if the transit service areas used in constructing table 6.11 were held constant but service was provided only for peak period work trips, it would not be possible to achieve transit travel times and costs similar to the tabulated ones. In the case of Washington, mode shares approximately half again as large as those shown in table 6.11 would be needed to achieve the tabulated travel times and costs if the service areas were held constant and service was provided only for peak period work trips.

In summary, it is possible to provide transit service that carries a relatively large proportion of the total person trips in an urban region and that has travel times and costs per trip that are reasonably close to automobile travel times and costs only if a sufficiently high transit mode share can be achieved in the transit service area. Although the effects that widely available, high-quality transit service would have on urban travel behavior are far from certain, the available evidence suggests that the necessary mode shares probably are not achievable in US cities in the near future. Consequently, it is likely that if a transit system offered sufficiently high quality service over a sufficiently large area to attract a large proportion of the travelers in its urban region to transit, then the cost of providing this service would be considerably higher than the cost of accommodating the transit riders in automobiles. Providing the service would cause a substantial increase in the cost of transportation in the affected city. It is this costliness of high-quality transit service, more than any other factor, that limits the reductions in urban regional motor vehicle emissions that can be achieved by diverting automobile travelers to transit.

A possible additional obstacle to providing high-quality, high-ridership transit service in large cities is the number of transit vehicles that would be required. This problem is illustrated by the transit options for Los Angeles that are shown in table 6.11. At present, there are approximately 4,000 transit buses operating in the Los Angeles area. Thus, implementing the tabulated transit options that are capable of carrying 18–26% of total daily person trips in the Los Angeles area would require increasing the size of the existing fleet by a factor of roughly 2.25–5. The delays associated with acquiring the necessary additional vehicles and support facilities would prevent large numbers of automobile travelers from being diverted to transit quickly.

The foregoing discussion has been concerned with a single service concept: namely, fixed route, fixed schedule bus service. However, the conclusions that have been reached apply to other service concepts that are potentially capable of providing high-quality service to large proportions of urban travelers. Rail transit systems can provide high-quality service, but they tend to have high fixed costs. Therefore, they can achieve an average cost per trip that is near that of the automobile only in the highest-density, usually CBD-oriented, corridors [61]. This makes rail systems unsuitable for carrying large proportions of total person trips in urban regions. Jitneys and other forms of paratransit can provide high-quality service, but they use low-capacity vehicles. Consequently, they can achieve average costs per trip that are close to the cost of automobile travel only if their drivers are paid at wage rates that are substantially below those of unionized bus drivers in large cities [61, 62]. Although lower-than-union wages are paid in current paratransit operations, it cannot be assumed that this practice necessarily would continue if paratransit became a major form of mass transportation in large cities [63].

6.8 Analysis of Transit Improvements in a Corridor

The discussion in sections 6.6 and 6.7 has indicated that the potential for achieving emissions reductions through transit improvements is greatest in high-volume corridors where current transit service is poor. Therefore, it is useful to examine in greater detail ways in which the effects of transit improvement measures on transit service quality, travel behavior, motor vehicle emissions, and transportation costs in corridors can be estimated. The purpose of this section is to present an example of the estimation of these effects. Although simplifying assumptions are made in the example in order to minimize its mathematical complexity, the example is similar to a more elaborate analysis in terms of its conceptual structure and the main qualitative features of its results. The main simplifying assumptions will be identified as they arise.

Example 6.6—Transit Improvements in an Arterial Corridor

An arterial corridor connects residential areas of a city with the CBD. The corridor is L mi long and W mi wide. The outer L_R miles are in residential areas, and the inner L_C miles are in the CBD, with $L_R + L_C = L$ (see figure 6.3a). A bus route operates along the arterial street. Buses operate with headways of H min and make S stops/mi. The service is oriented toward peak period work trips, and the following analysis deals only with these trips. During the morning peak, trips originate in the residential area and terminate in the CBD. The opposite flow pattern occurs during the

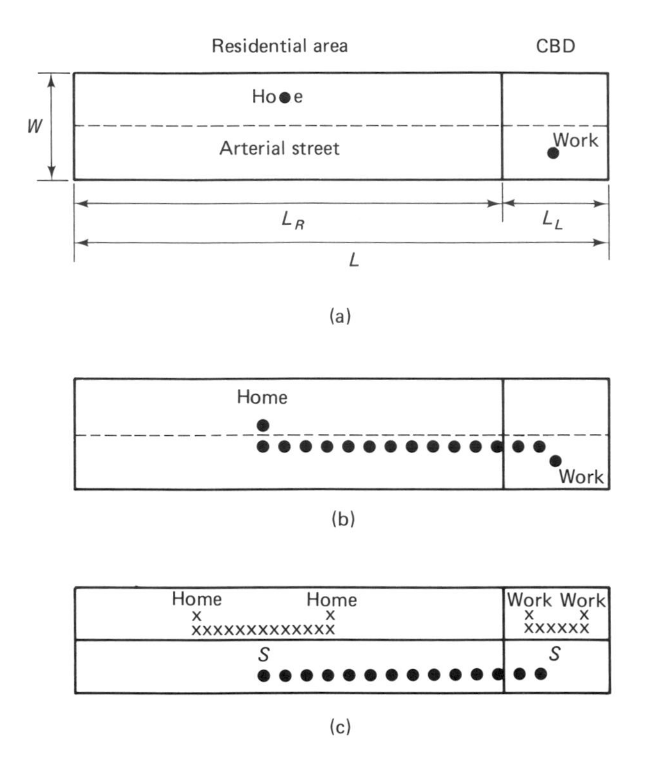

Figure 6.3 Diagram of arterial corridor for example 6.6: (*a*) dimensions and locations of home and work of typical traveler; (*b*) route of a drive alone trip from home to work; (*c*) routes of typical transit trips from home to work (S, bus stop; × × ×, walking; . . . , riding in bus).

afternoon peak. It is assumed for simplicity that the flows of work trips in the two periods are symmetrical, so that only one period needs to be analyzed explicitly. In addition, it is assumed that the flow of work trips in the off-peak direction (e.g., from the CBD to the residential area during the morning peak) is negligible.

In this example, the effects of transit headways and stop frequencies on work trip mode choice, motor vehicle emissions due to work trips, and the costs of transportation for work trips will be investigated. The investigation will make use of the disaggregate mode choice model of table 6.7. To simplify the presentation, the analysis will be based on modeling the travel behavior of a single "typical" traveler whose behavior will be considered to represent the average behavior of all travelers in the corridor.

The traveler's residence is located midway along the residential part of the corridor and at a distance of $W/4$ mi from the arterial street. The traveler's work place is located midway along the CBD part of the corridor and also is a distance of $W/4$ mi from the arterial street (see figure 6.3a). The personal characteristics of the traveler are the same as those used in table 6.8. There is one worker in the traveler's household, and the household's annual income is $25,000/yr. The household owns one car with probability 0.50 and two cars with probability 0.50.

In a more sophisticated analysis, the travel behavior of many corridor travelers would be modeled. The modeled travelers would be chosen so that the distribution of their personal characteristics and of the locations of their residences and work places approximate those of all travelers in the corridor. The modeled behavior of the individual travelers then would be aggregated using methods similar to those discussed in connection with equation (6.28).

The only aspect of travel behavior that will be modeled in this example is mode choice. (A more sophisticated analysis might also consider the effects of transit improvements on automobile ownership and, possibly, on choice of residential location.) Therefore, the first step of the analysis is to determine the values of the service quality variables in the mode choice model of table 6.7. This must be done for each of the three available modes: drive alone, carpool, and transit.

Drive Alone

The typical traveler's drive alone trip from home to work consists of driving from home to the arterial street, a distance of $W/4$ mi; driving along the arterial street to the center of the CBD, a distance of $L_R/2 + L_C/2 = L/2$ mi; and driving from the arterial street to the work place, a distance of $W/4$ mi (see the dotted line in figure 6.3b). The trip from work to home reverses this sequence. Thus, the total round trip distance traveled by automobile is

$$D_{DA} = 2\left(\frac{W}{4} + \frac{L}{2} + \frac{W}{4}\right) = W + L \text{ (miles)}. \tag{6.36}$$

Let V_A be the speed of peak period traffic flow in the corridor.[22] This speed and all other variables relating to traffic flow conditions and transportation levels of service will be assumed not to vary within peak periods. Then $IVTT_{DA}$, the round trip in-vehicle travel time for the drive alone mode, is D_{DA}/V_A. Expressed in terms of W and L, this is

$$IVTT_{DA} = (W + L)/V_A. \tag{6.37}$$

Since the mode choice model requires in-vehicle travel time to be expressed in units of minutes, V_A is expressed in units of mi/min. Assume that the round trip walk time for the drive alone mode is

$$WT_{DA} = 2 \text{ (minutes)}. \tag{6.38}$$

The costs to the traveler of driving alone include mileage-related costs (e.g., costs of gasoline and maintenance) and parking costs. Let the mileage-related cost be c_M ¢/mi, and let the price of parking be c_P ¢/day. Then the round trip cost to the traveler of a drive alone trip is

$$C_{DA} = c_M (W + L) + c_P \text{ (cents)}. \tag{6.39}$$

Carpool

Carpool travel tends to involve longer trip lengths and travel times than drive alone travel does owing to the need to collect and distribute carpool passengers. Assume that carpool trip lengths and travel times exceed the corresponding drive alone trip lengths and travel times by 1 mi and 5 min, respectively, per round trip. Then, the average carpool round trip distance is

$$D_{CP} = (W + L) + 1 \text{ (miles)}. \tag{6.40}$$

The average round trip in-vehicle travel time for carpool travelers is

$$IVTT_{CP} = (W + L)/V_A + 5 \text{ (minutes)}. \tag{6.41}$$

The walk time for carpoolers is assumed to be the same as that for drive alone travelers. Hence, the carpool walk time is given by

$$WT_{CP} = 2 \text{ (minutes)}. \tag{6.42}$$

The total cost to the travelers of a carpool vehicle round trip is $c_M D_{CP} + c_P$. Assume that carpoolers divide this cost equally among themselves and that the average size of a carpool is CPS. Then, the cost to each carpool member of a round trip by carpool is

$$C_{CP} = c_M \frac{W + L + 1}{\text{CPS}} + \frac{c_P}{\text{CPS}}. \tag{6.43}$$

Transit

To compute the transit walk distance correctly, it is necessary to take account of the fact that most travelers must walk along the arterial street to reach a bus stop. Therefore, for the purpose of computing the transit walk distance, it will be assumed that the typical traveler lives and works at randomly selected locations along lines that are parallel to the arterial street at a distance of $W/4$ mi from it. The lines extend for a distance of half the stop spacing on each side of a bus stop (see figure 6.3c).

To get from home to work by transit, the typical traveler must walk from home to the arterial street where the bus route is, a distance of $W/4$ mi; walk along the arterial street to the bus stop, an average distance of 1/4 the stop spacing; ride the bus to the CBD, an average distance of $L/2$ mi; walk along the arterial street from the CBD bus stop to the street on which the work place is located, an average distance of 1/4 the stop spacing; and walk from the arterial street to the work place, a distance of $W/4$ mi. This pattern is reversed for travel from work to home.

The round trip in-vehicle travel distance for the typical transit traveler is L mi. It is assumed that between stops buses move at the same speed V_A as other traffic moves. However, buses lose time at stops due to deceleration, boarding and alighting of passengers, and acceleration. If there are S stops/mi and buses are delayed t_S min/stop, then the delay per mile is St_S, and the delay on an L-mi trip is LSt_S min. Hence, the round trip in-vehicle travel time for the typical transit traveler is

$$\text{IVTT}_{\text{TR}} = L/V_A + LSt_S \text{ (minutes).} \tag{6.44}$$

If there are S stops/mi along the bus route, then the stop spacing is $1/S$ mi. Hence, the round trip walk distance for the typical transit traveler is

$$D_{\text{WTR}} = 2\left(\frac{W}{4} + \frac{1}{4S} + \frac{1}{4S} + \frac{W}{4}\right) = W + 1/S \text{ (miles).} \tag{6.45}$$

Let the average walking speed be V_W (expressed in units of mi/min). Then the round trip walk time for the typical transit traveler is

$$\text{WT}_{\text{TR}} = \frac{W + 1/S}{V_W} \text{ (minutes).} \tag{6.46}$$

The cost to travelers of transit travel is equal to the fare. Let the one-way fare, in cents, be F. Then the round trip cost for transit travelers is

$$C_{\text{TR}} = 2F \text{ (cents).} \tag{6.47}$$

Transit travel also entails waiting for the bus. However, in the mode choice model of table 6.7, the effects of waiting are subsumed in the headway variables. Therefore, it is not necessary to estimate waiting time for transit travelers.

Table 6.16 shows the values of the quantities L, W, H, V_A, c_M, c_P, CPS, t_S, S, V_W, and F that will be used in the remainder of this example, as well as the values of several other parameters that will be used later in the example. Table 6.17 shows the typical traveler's mode choice probabilities for several combinations of H and S values. These probabilities were computed using the mode choice model of table 6.7. In the subsequent analysis, it will be assumed that the tabulated mode choice probabilities represent average values for corridor travelers as a group.

A notable feature of the choice probabilities in table 6.17 is the relation between the transit stop frequency and the transit choice probability. Other things being equal, increasing the number of transit stops per mile reduces the walk time for transit travelers and, therefore, should increase the probability that travelers will choose transit. However, in table 6.17 increasing the number of stops per mile from 2 to 4, which reduces round trip walk time by 5 min, causes only a slight increase in the transit choice probability, given a constant headway. Increasing the number of stops per mile from 4 to 8, which reduces round trip walk time by 2.5 min, has virtually no effect on the transit choice probability. This independence of the transit choice probability from the transit stop frequency is caused by an interaction between the stop frequency and the transit in-vehicle travel time. Increasing the stop frequency increases the in-vehicle travel time by increasing the time that buses lose at stops [see equation (6.44)]. In the cases shown in table 6.17, the effects of reductions in walk times caused by increases in stop frequencies are almost exactly compensated by the effects of increases in in-vehicle travel times. Hence, increasing the stop frequency does not substantially increase the probability that travelers choose transit.

Having determined the mode choice probabilities of the typical corridor traveler, the next step of the analysis is to determine the volumes of person trips on each mode. These volumes are not necessarily proportional to the mode choice probabilities, even with the assumption that the typical traveler probabilities represent the average behavior of all travelers in the corridor, because the transit system may not have the passenger capacity needed to carry all of the travelers who wish to use transit. If the transit headway is H min, then the number of buses arriving at the end of a bus route per hour is $60/H$. Let k denote the passenger capacity of a transit vehicle (assumed in this example to be 70 passengers). Then the passenger capacity of the transit system in the corridor, K, is

$$K = 60k/H \text{ (passengers per hour).} \tag{6.48}$$

Table 6.16 Parameter values for example 6.6

Parameter	Symbol	Value
Length of corridor	L	6 mi
Width of corridor	W	0.5 mi
One-way transit headway	H	5, 15, 30 min
Speed of traffic flow	V_A	0.25 mi/min (15 mi/hr)
Cost per mile of operating an automobile	c_M	15¢/mi
Price to traveler of parking	c_P	100¢/day
Average carpool size	CPS	2.3
Bus stops per route mile	S	2, 3, 8/mi
Bus delay per stop[a]	t_S	0.33 min (20 sec)
Walking speed	V_W	0.05 mi/min (3 mi/hr)
One-way transit fare	F	75¢
Passenger capacity of a bus	k	70 passengers
Full cost of parking an automobile	C_P	$2.00/day
Cost per vehicle hour of transit service	C_B	$30.00/hr
Minimum transit layover time per round trip	t_L	15 min

a. Strictly speaking, t_S depends on transit ridership, since increases in ridership increase the numbers of passengers desiring to board or alight at each stop and reduce the likelihood that stops can be skipped. However, to avoid complicating the analysis, t_S is treated as a constant parameter in this example and is assigned a representative value.

Table 6.17 Mode choice probabilities for example 6.6[a]

One-way transit headway (min)	Transit stops per mile	Choice probability for		
		Drive alone	Carpool	Transit
5	2	0.53	0.24	0.23
15	2	0.55	0.25	0.19
30	2	0.58	0.27	0.15
5	4	0.52	0.24	0.24
15	4	0.54	0.25	0.21
30	4	0.57	0.26	0.17
5	8	0.52	0.24	0.24
15	8	0.54	0.25	0.21
30	8	0.57	0.26	0.17

a. Based on mode choice model of table 6.7. Probabilities do not necessarily sum to 1.00 due to rounding.

Suppose that there are N person trips/hr in the corridor during peak periods. (Fluctuations in travel volumes within peak periods will not be treated here.) Let P_{DA}, P_{CP}, and P_{TR} denote, respectively, the drive alone, carpool, and transit choice probabilities. Then, if transit capacity is sufficient to accommodate all travelers who wish to use that mode, the numbers of persons per hour using each of the modes are

$$N_{DA} = NP_{DA} \text{ (trips per hour for drive alone),} \quad (6.49a)$$

$$N_{CP} = NP_{CP} \text{ (trips per hour for carpool),} \quad (6.49b)$$

and

$$N_{TR} = NP_{TR} \text{ (trips per hour for transit).} \quad (6.49c)$$

The transit system will have sufficient capacity to accommodate the demand for transit trips if $NP_{TR} \leq K$. Equations (6.49) are applicable when this condition is satisfied.

A rigorous treatment of the effects of insufficient transit capacity on mode choice would require treating variations in the degree of bus crowding along the bus route and, possibly, including crowding explicitly as a variable of the mode choice model. This would severely complicate the mathematics of this example. To prevent this from happening, the following simple procedure will be used to determine the passenger volumes on each mode when transit demand exceeds transit capacity:

If the demand for transit trips exceeds the passenger capacity of the transit system, set the volume of transit trips equal to the transit passenger capacity. Assign the remaining trips to the drive alone and carpool modes in proportion to the choice probabilities for these modes.

This procedure leads to the following equations for passenger volumes when transit travel demand exceeds transit passenger capacity:

$$N_{DA} = (N - K)P_{DA}/(P_{DA} + P_{CP}) \text{ (trips per hour);} \quad (6.50a)$$

$$N_{CP} = (N - K)P_{CP}/(P_{DA} + P_{CP}) \text{ (trips per hour);} \quad (6.50b)$$

$$N_{TR} = K \text{ (trips per hour).} \quad (6.50c)$$

These equations apply if $NP_{TR} > K$.

The passenger volumes on the three modes in the corridor are given by equations (6.49) if transit demand does not exceed transit passenger capacity (i.e., if $NP_{TR} \leq K$), and by equations (6.50) otherwise. Table 6.18 shows the transit passenger capacities and the modal person trip volumes in the example corridor for the headways and stop frequencies used in table 6.17 and three different levels of the total person trip volume N.

Table 6.18 Transit capacities and person trip volumes by mode for example 6.6

One-way transit headway (min)	Transit stops per mile	Transit capacity (passengers/hr)	Modal volumes (person trips/hr) for three values of total volume (person trips/hr) in corridor[a] Total = 500			Total = 1,000			Total = 1,500		
			Drive alone	Carpool	Transit	Drive alone	Carpool	Transit	Drive alone	Carpool	Transit
5	2	840	266	121	113	533	241	226	799	362	339
15	2	280	277	126	97	554	252	194	839	381	280(CR)
30	2	140	291	133	76	591	269	140(CR)	934	426	140(CR)
5	4	840	260	118	122	520	235	245	780	353	367
15	4	280	271	123	106	542	246	212	839	381	280(CR)
30	4	140	286	130	84	591	269	140(CR)	934	426	140(CR)
5	8	840	261	118	121	522	236	242	782	354	364
15	8	280	272	123	105	544	247	210	839	381	280(CR)
30	8	140	287	131	83	591	269	140(CR)	934	426	140(CR)

a. Modal volumes do not necessarily sum to corridor totals due to rounding.

Cases in which the transit passenger volume is limited by the transit capacity are marked CR.

The person trip volumes for the drive alone and carpool modes can be used to compute the numbers of automobile vehicle trips per hour and vehicle miles traveled (VMT) per hour in the corridor. These quantities are needed to compute automobile emissions and the total costs of automobile travel. The number of automobile vehicle trips for the drive alone mode is equal to the number of drive alone person trips. The number of carpool vehicle trips is equal to the number of carpool person trips divided by the carpool size. Therefore, the total number of automobile vehicle trips per hour, T_A, is

$$T_A = N_{DA} + N_{CP}/\text{CPS} \text{ (trips per hour).} \tag{6.51}$$

The number of automobile VMT per hour for the drive alone and carpool modes is equal to the numbers of vehicle trips by these modes multiplied by the average one-way trip lengths. The one-way trip lengths are $(W + L)/2$ mi for the drive alone mode and $(W + L)/2 + 0.5$ mi for carpools. Denote the number of automobile VMT per hour by M_A. Then

$$M_A = N_{DA}\frac{W + L}{2} + \frac{N_{CP}}{\text{CPS}}\left(\frac{W + L}{2} + 0.5\right) \text{ (vehicle miles per hour).} \tag{6.52}$$

The number of transit vehicles that must operate in the corridor to maintain the desired headway and the number of transit VMT per hour can be determined from the corridor length and transit operating policies. Recall that traffic flow and transportation service variables are assumed not to vary within peak periods. Then, assuming that buses operate on a round trip basis, the number of transit VMT per hour, M_{TR}, is

$$M_{TR} = (2)(60)L/H = 120L/H \text{ (vehicle miles per hour),} \tag{6.53}$$

where the factor of 60 converts H from minutes to hours. The required number of buses can be computed from the schedule frequency (the inverse of the headway) and the time required for a vehicle round trip. If R is the time required for a round trip, then a single vehicle can maintain a schedule frequency of $1/R$ trips per unit time. B vehicles can maintain a schedule frequency of B/R trips per unit time. Therefore, if it is desired to maintain a schedule frequency of f trips per unit time, fR vehicles are needed. If this number is fractional, it is rounded up to the nearest integer. Noting that $f = 1/H$, the number of vehicles B that are needed to maintain a headway of H min is

$$B = R/H \text{ (vehicles),} \tag{6.54}$$

where R and H are both expressed in minutes, and fractional B values are rounded upward to the nearest integer.

The length of a bus round trip is $2L$ mi. Therefore, the time in motion required for the round trip is $2L/V_A$ min. In addition, the bus must make stops during a round trip. Since the volume of travel in the off-peak direction of flow in the example corridor is negligible, it will be assumed that buses make no stops during trips in the off-peak direction. Hence, the number of stops per round trip is LS, and the total time per round trip that is spent at stops is LSt_S. Finally, it is likely that buses will pause at the ends of one-way trips in order to maintain established schedules and to permit drivers to rest. Assume that this layover time is t_L min/round trip. Then the round trip travel time for a bus is given by

$$R = 2L/V_A + LSt_S + t_L \text{ (minutes)}. \tag{6.55}$$

Table 6.19 shows the numbers of automobile vehicle trips per hour, automobile and transit VMT per hour, and buses corresponding to each of the combinations of headways, stop frequencies, and total person trip volumes in table 6.18. The bus layover time, t_L, has been set at 15 min/round trip or, if this value does not yield an integer-valued bus fleet size B, at the smallest value above 15 min that permits B to be an integer.[23]

The emissions associated with the various levels of transit service in the example corridor can be computed from the data in table 6.19. Let e_{p1} denote the trip end-related component of automobile emissions of pollutant p, expressed in units of kg/trip. This component includes start-up and hot soak emissions. Let e_{p2} denote the hot stabilized exhaust component of autombile emissions of pollutant p, expressed in units of kg/VMT. Let e_{pB} denote the rate of bus emissions of pollutant p, also expressed in units of kg/VMT. Finally, let E_{pA} and E_{pB}, respectively, denote total emissions of pollutant p per hour by automobiles and buses operating in the corridor. Then E_{pA} and E_{pB} are given by

$$E_{pA} = e_{p1}T_A + e_{p2}M_A \text{ (kilograms per hour)}, \tag{6.56a}$$

$$E_{pB} = e_{pB}M_{TR} \text{ (kilograms per hour)}. \tag{6.56b}$$

Total pollutant p emissions per hour from all vehicles in the corridor, E_p, are given by

$$E_p = E_{pA} + E_{pB} \text{ (kilograms per hour)}. \tag{6.56c}$$

The costs associated with the various levels of transit service in the corridor also can be computed from the data in table 6.19. Recall that the mileage-related cost of operating an automobile in the corridor is c_M ¢/mi,

Table 6.19 Vehicle miles traveled per hour, vehicle trips per hour, and transit fleet size for example 6.6

Transit				Automobile VMT and vehicle trips per hour for three values of total person trip volume (trips/hr) in corridor[a]					
				500 trips/hr		1,000 trips/hr		1,500 trips/hr	
One-way headway (min)	Stops per mile	VMT	Fleet size	VMT	Vehicle trips	VMT	Vehicle trips	VMT	Vehicle trips
5	2	144	14	1,060	319	2,120	637	3,190	956
15	2	48	5	1,100	332	2,210	663	3,350	1,000
30	2	24	3	1,160	349	2,360	708	3,730	1,120
5	4	144	15	1,040	311	2,070	622	3,110	933
15	4	48	5	1,080	325	2,160	649	3,350	1,000
30	4	24	3	1,140	343	2,360	708	3,730	1,120
5	8	144	16	1,040	312	2,080	624	3,120	936
15	8	48	6	1,080	325	2,170	651	3,350	1,000
30	8	24	3	1,140	343	2,360	708	3,730	1,120

a. Results have been rounded to three significant figures.

or $c_M/100$ \$/mi. Denote the full daily cost of parking an automobile in the CBD by C_P. C_P may differ from the cost to a driver of parking a car, c_P, owing to parking subsidies that many urban travelers receive.[24] Let C_P have units of \$/day. For the purposes of this discussion, it can be assumed that the cost of parking a car is divided equally between the home-to-work and work-to-home trips. Therefore, the total cost per hour of automobile travel in the example corridor, TC_A, is given by

$$TC_A = c_M M_A/100 + C_P T_A/2 \text{ (dollars per hour)}. \quad (6.57a)$$

Let the cost of operating a bus be C_B \$/hr. Then the total cost of transit operations in the corridor, TC_{TR}, is given by

$$TC_{TR} = C_B B \text{ (dollars per hour)}. \quad (6.57b)$$

The total cost per hour of all transportation services in the corridor (within the context of this example, these services include only automobiles and buses), TC, is

$$TC = TC_A + TC_{TR} \text{ (dollars per hour)}. \quad (6.58)$$

The transit profit or deficit from operations in the corridor is equal to the difference between fare collections and the cost of operating the transit service. Let p_{TR} denote this profit (if the difference is positive) or deficit (if the difference is negative) in units of \$/hr. Then p_{TR} is given by

$$p_{TR} = N_{TR}F/100 - TC_{TR} \text{ (dollars per hour)}. \quad (6.59)$$

The factor of 100 in the first term on the right converts the fare F from units of ¢/trip to \$/trip.

Table 6.20 shows estimates of average 1982 automobile and bus emissions rates for the traffic speed in the example corridor. Table 6.21 shows total hourly emissions from automobiles, buses, and all vehicles in the corridor for the combinations of headways, stop frequencies, and total person trip volumes in tables 6.18 and 6.19.[25]

Table 6.20 Automobile and bus emissions rates for example 6.6

	CO	HC	NO_x
Automobile hot stabilized exhaust emissions (g/mi)	37	2.8	1.7
Automobile starting and hot soak emissions (g/trip)	149	17	3.7
Bus emissions (g/mi)	51.3	7.70	29.2

a. Source: references [8, 64]. It has been assumed that all trips begin with cold starts. Based on a 1982 mix of vehicle ages and model years.

Table 6.21 Total hourly motor vehicle emissions in the corridor of example 6.6

Transit headway (min)	Transit stops per mile	Emissions (kg/hr) from: Buses CO	Buses HC	Buses NO_x	Automobiles CO	Automobiles HC	Automobiles NO_x	All vehicles[a] CO	All vehicles[a] HC	All vehicles[a] NO_x
		Person trip volume = 500 trip/hr								
5	2	7.39	1.11	4.20	86.8	8.39	2.98	94.2	9.50	7.19
15	2	2.46	0.37	1.40	90.3	8.73	3.11	92.7	9.10	4.51
30	2	1.23	0.18	0.70	94.9	9.18	3.26	96.1	9.36	3.96
5	4	7.39	1.11	4.20	84.7	8.19	2.91	92.1	9.30	7.12
15	4	2.46	0.37	1.40	88.4	8.55	3.04	90.9	8.92	4.44
30	4	1.23	0.18	0.70	93.3	9.02	3.21	94.5	9.21	3.91
5	8	7.39	1.11	4.20	85.0	8.22	2.92	92.4	9.33	7.13
15	8	2.46	0.37	1.40	88.6	8.57	3.05	91.1	8.94	4.45
30	8	1.23	0.18	0.70	93.5	9.04	3.22	94.7	9.22	3.92
		Person trip volume = 1,000 trips/hr								
5	2	7.39	1.11	4.20	174	16.8	5.97	181	17.9	10.2
15	2	2.46	0.37	1.40	181	17.5	6.21	183	17.8	7.61
30	2	1.23	0.18	0.70	193	18.6	6.63	194	18.8	7.33
5	4	7.39	1.11	4.20	170	16.4	5.83	177	17.5	10.0
15	4	2.46	0.37	1.40	177	17.1	6.08	179	17.5	7.48
30	4	1.23	0.18	0.70	193	18.6	6.63	194	18.8	7.33
5	8	7.39	1.11	4.20	170	16.4	5.85	177	17.5	10.0

15	8	2.46	0.37	1.40	177	17.1	6.10	180	17.5	7.50
30	8	1.23	0.18	0.70	193	18.6	6.63	194	18.8	7.33
		Person trip volume = 1,500 trips/hr								
5	2	7.39	1.11	4.20	260	25.2	8.95	268	26.3	13.2
15	2	2.46	0.37	1.40	274	26.5	9.41	276	26.8	10.8
30	2	1.23	0.18	0.70	305	29.5	10.5	306	29.7	11.2
5	4	7.39	1.11	4.20	254	24.6	8.74	262	25.7	12.9
15	4	2.46	0.37	1.40	274	26.5	9.41	276	26.8	10.8
30	4	1.23	0.18	0.70	305	29.5	10.5	306	29.7	11.2
5	8	7.39	1.11	4.20	255	24.7	8.77	262	25.8	13.0
15	8	2.46	0.37	1.40	274	26.5	9.41	276	26.8	10.8
30	8	1.23	0.18	0.70	305	29.5	10.5	306	29.7	11.2

a. May not equal sum of bus and automobile emissions due to rounding.

It can be seen from table 6.21 that variations in the transit stop frequency have little effect on emissions in the example corridor. Among the tabulated cases, changing the transit stop frequency while keeping the headway constant causes changes in total emissions from all vehicles of slightly more than 2%, at most. Thus, changing the transit stop frequency is not an effective emissions control measure in the corridor.

Reducing the transit headway can cause moderately large CO and HC emissions reductions in the example corridor when the person trip volume is high. For example, when the person trip volume is 1,500 trips/hr, reducing the transit headway from 30 to 5 min reduces CO and HC emissions by 11–14%, depending on the pollutant and the stop frequency. Reducing the transit headway is much less effective in reducing NO_x emissions. When the person trip volume is 1,500 trips/hr, reducing the transit headway from 30 to 15 min reduces NO_x emissions by roughly 4%, compared to reductions of 10% for CO and HC. Reducing the headway from 15 to 5 min causes NO_x emissions to increase by approximately 20%, whereas CO and HC emissions decrease by up to 5 and 4%, respectively.

Headway reductions are less effective at reducing CO and HC emissions when the person trip volume in the corridor is low than when it is high. Table 6.21 shows that when the total person trip volume is 500 trips/hr, reducing the headway from 30 to 15 min reduces CO and HC emissions by only 4 and 3%, respectively, compared to the reductions of 10% for each pollutant that occur when the person trip volume is 1,500 trips/hr. Reducing the headway from 15 to 5 min when the person volume is 500 trips/hr causes CO and HC emissions to increase. This is because the additional buses that are needed to reduce the headway from 15 to 5 min do not attract enough travelers out of automobiles to compensate for the buses' emissions. When the person trip volume is 500 trips/hr, reducing the headway from 30 to 15 min causes NO_x emissions to increase by 14%. Reducing the headway from 15 to 5 min causes NO_x emissions to increase by 60%.

The effects of headway reductions on emissions are different at different person trip volumes mainly because transit ridership is capacity limited when the headway is long and the person trip volume is high (see table 6.18). Reducing the headway eases this constraint. Therefore, when the person trip volume is high, persons switching from automobile modes to transit include travelers who desired to use transit before the service was improved but were crowded off, as well as travelers who are attracted by the reduced waiting times that headway reduction provides. When the person trip volume in the corridor is low, transit ridership is not capacity limited. Therefore, only travelers who are attracted to transit by reduced waiting times switch from automobile modes to transit when the headway is reduced.

Table 6.22 shows cost information for the example corridor. It has been assumed that C_B has the value $30.00/hr and that C_P has the value $2.00/day. When the person trip volume in the corridor is 1,500 trips/hr, reducing the transit headway from 30 to 15 min reduces the total cost of transportation in the corridor and also is profitable for the transit system. When the person trip volume is 1,000 trips/hr, reducing the headway from 30 to 15 min may either increase or decrease the total cost of transportation in the corridor, depending on the transit stop frequency. However, the headway reduction always is unprofitable for the transit company in the tabulated cases. Thus, the tabulated results show that transit improvements may reduce total transportation costs, even if they cause a deficit for the transit system. When the person trip volume in the corridor is 500 trips/hr, reducing the transit headway from 30 to 15 min increases both the total cost of transportation in the corridor and the transit deficit.

Reducing the headway from 15 to 5 min increases both the total cost of transportation in the corridor and the transit deficit at all of the tabulated person trip volumes. However, the cost and deficit increases are larger when the person trip volume is low than when it is high.

In summary, two types of transit improvement measures have been examined in this example: changes in transit stop frequencies and reductions in transit headways. Changes in stop frequencies have relatively small effects on emissions in the cases that have been examined. Reductions in transit headways are capable of producing moderately large reductions in CO and HC emissions when the person trip volume in the example corridor is high. Headway reductions are less effective at reducing CO and HC emissions when the person trip volume is low. In addition, headway reductions are less costly when the person trip volume is high than when it is low. If the person trip volume is sufficiently high, then headway reductions can reduce the total cost of transportation in the corridor and be profitable for the bus system. The results of this example illustrate the conclusion reached in sections 6.6 and 6.7 that transit improvements are likely to be more useful for reducing emissions in high volume corridors where current transit service is poor than for reducing emissions in low volume corridors.

Table 6.22 Hourly costs of transportation services in the corridor of example 6.6

		500 trips/hr				1,000 trips/hr				1,500 trips/hr			
Transit headway (min)	Stops per mile	Transit cost ($)	Transit profit (+) or deficit (−) ($)	Automobile cost ($)	Total cost ($)	Transit cost ($)	Transit profit (+) or deficit (−) ($)	Automobile cost ($)	Total cost ($)	Transit cost ($)	Transit profit (+) or deficit (−) ($)	Automobile cost ($)	Total cost ($)
5	2	420	−335	478	898	420	−250	956	1,380	420	−166	1,430	1,850
15	2	150	−77	497	647	150	−4	995	1,140	150	+60	1,510	1,660
30	2	90	−33	523	613	90	+15	1,060	1,150	90	+15	1,680	1,770
5	4	450	−358	467	917	450	−266	934	1,380	450	−175	1,400	1,850
15	4	150	−71	487	637	150	+9	974	1,120	150	+60	1,510	1,660
30	4	90	−27	514	604	90	+15	1,060	1,150	90	+15	1,680	1,770
5	8	480	−389	468	948	480	−298	936	1,420	480	−207	1,400	1,880
15	8	180	−101	488	668	180	−23	976	1,160	180	+30	1,510	1,690
30	8	90	−28	515	605	90	+15	1,060	1,150	90	+15	1,680	1,770

6.9 *Priority Treatment for High-Occupancy Vehicles*

Priority treatment for high-occupancy vehicles consists of allocating roadway facilities preferentially to buses and, in some cases, carpools. The usual objective of implementing priority treatment for high-occupancy vehicles is to provide a speed advantage for these vehicles in comparison to nonpriority vehicles and, thereby, to encourage the use of high occupancy modes. In addition, increasing the average speed of buses on a route may reduce the number of vehicles needed to maintain a given bus schedule frequency on the route [see equations (6.54) and (6.55)], thereby reducing the cost of providing bus service on the route.

One way of achieving priority treatment for high-occupancy vehicles is to reserve one or more roadway lanes for the exclusive use of these vehicles. For example, the curb lane in the normal direction of traffic flow (normal flow curb lane) or the median lane of an arterial street can be reserved for the exclusive use of buses. On one-way streets, the left-hand curb lane can be reserved for buses that move in the direction opposite to that of normal traffic flow (contraflow curb lane). In some cases, it may be desirable to reserve an entire street for the exclusive use of buses. On freeways, normal flow or contraflow median lanes can be reserved for the exclusive use of priority vehicles. Priority treatment also can be provided through the use of busways. These consist of specially constructed, high-speed roadways for the exclusive use of buses and, possibly, carpools. Busways usually are constructed in freeway rights-of-way but are physically separated from the lanes used by nonpriority traffic.

Priority treatment also can be accomplished by signal preemption. On surface streets, buses can be equipped with devices that enable them to change signal lights from red to green and to extend green periods when approaching intersections. On freeways where entry is controlled by signals (ramp meters), special lanes can be used to permit high-occupancy vehicles to bypass the signals and any queues that have formed.

Priority treatment has been observed to increase bus speeds by a factor of 2 or more on freeways and by up to 50% on arterial streets [65]. However, priority treatment does not always increase the speeds of buses. For example, if priority treatment causes a large number of buses to be confined to a single lane, then the resulting congestion in the lane can reduce buses' speeds.

The immediate objective of priority treatment for high-occupancy vehicles is to alter the speeds of flow of priority and nonpriority traffic. Thus, priority treatment operates through its effects on traffic flow conditions.

As with the traffic flow improvement measures that were discussed in section 6.3, the effects of priority treatment on emissions depend on a complex set of interactions among traffic-flow conditions and travel demand factors. The two most important travel demand factors are route choice and mode choice. For example, suppose that a normal flow lane on an arterial street is removed from use by general traffic and reserved for buses, thereby increasing bus speeds relative to the speeds of non-priority vehicles. This will induce some automobile travelers to switch to transit. In this respect, establishment of the bus lane tends to reduce traffic volumes and emissions on the street where the reserved lane is established and, possibly, on parallel streets as well. However, the establishment of the bus lane reduces the roadway capacity that is available for use by nonpriority traffic, which tends to increase the degree of congestion of nonpriority traffic on the street where the lane is established and to induce some automobile travelers to divert to parallel routes. In this respect the establishment of the bus lane tends to increase emissions, both on the street where the lane is established and on parallel routes. The net effect of the bus lane, therefore, can be either to increase or decrease emissions on the streets that are involved, depending on whether mode switching or congestion effects are dominant.

The following example illustrates the interactions among traffic flow conditions, travel demand factors, and emissions that occur with priority treatment for high occupancy vehicles.

Example 6.7—Priority Treatment for Buses on an Arterial Street
Two parallel arterial streets connect two points A and B that are 2 mi apart (see figure 6.1a). Roadway 1 has three lanes of traffic in each direction, and roadway 2 has two lanes of traffic in each direction. The total volume of automobile traffic flowing from A to B on both roadways, combined, is 3,200 veh/hr. The total volume of automobile traffic flowing from B to A on both roadways, combined, is 1,400 veh/hr. In addition, roadway 1 carries 50 buses/hr in each direction of flow. There are no buses on roadway 2.

To estimate the effects of bus priority treatment on traffic flow conditions, it is necessary to define performance functions for the two roadways. For the purpose of defining and evaluating these performance functions, each bus will be counted as three automobiles. Thus, if V_{auto} and V_{bus}, respectively, are the automobile and bus volumes in a particular lane of a roadway, the effective volume in that lane, V_{eff}, is

$$V_{eff} = V_{auto} + 3V_{bus}. \qquad (6.60)$$

The performance function of each lane on each roadway is assumed to be

$$P(V_{eff}, c) = \frac{s_0}{1 + 0.9(V_{eff}/c)^{4.5}}, \qquad (6.61)$$

where c is the practical capacity of a lane and s_0 is the free flow speed. In this example, it is assumed that $s_0 = 23$ mi/hr and $c = 800$ veh/hr. It also is assumed that traffic distributes itself among the available lanes on a roadway so as to achieve equal effective traffic volumes in these lanes.

Automobile travelers are assumed to choose the roadway that minimizes their travel time. As in example 6.1, this implies that the automobile speeds on the two roadways are equal for each direction. However, the speed of travel in the direction AB is not necessarily equal to the speed of travel in the direction BA. When buses are in motion, they are assumed to move at the speed given by the performance function of equation (6.61).

The speed-emissions relations for automobiles in the example are given by equations (6.10), (6.15), and (6.16). Emissions rates for buses are assumed to be as in table 6.20.[26]

Assume, initially, that there is no priority treatment for buses, so that buses and automobiles are distributed over all lanes of roadway 1. Then for each direction of travel, all lanes of both roadways must have the same effective traffic volumes. Let $V^*_{Li}(D)$ denote the effective volume per lane on roadway i ($i = 1$ or 2) in direction D (D = AB or BA). Then for each D

$$V^*_{L1}(D) = V^*_{L2}(D). \qquad (6.62)$$

The total effective volume in all lanes in a given direction must equal the total automobile volume in that direction plus three times the bus volume. Let $L_i(D)$ denote the number of lanes on roadway i in direction D. In addition, let $V_a(D)$ and $V_b(D)$ denote, respectively, the total automobile and bus volumes in direction D. Then for each D

$$\sum_{i=1}^{2} L_i(D) V^*_{Li}(D) = V_a(D) + 3V_b(D). \qquad (6.63)$$

Equations (6.62) and (6.63) can be solved to obtain

$$V^*_{Li}(D) = [V_a(D) + 3V_b(D)] / \sum_{j=1}^{2} L_j(D), \qquad i = 1 \text{ or } 2. \qquad (6.64)$$

These effective volumes per lane can be substituted into equation (6.61) to obtain the speeds in each direction of the two roadways.

The actual (as opposed to effective) traffic volumes on each roadway in each direction can be obtained from equation (6.64) in the following way. The total effective volume for a particular direction and roadway is equal to the effective volume per lane for that direction multiplied by the number of lanes in that direction on the roadway in question. Denote the total effective volume in direction D of roadway i by $V_i^*(D)$. Then

$$V_i^*(D) = L_i(D)\,[V_a(D) + 3V_b(D)] / \sum_{j=1}^{2} L_j(D), \qquad i = 1 \text{ or } 2. \tag{6.65}$$

There are no buses on roadway 2. Therefore, the automobile volumes in the two directions on that roadway are equal to the effective volumes. The automobile volumes on roadway 1 can be obtained by subtracting the roadway 2 volumes from the total automobile volumes $V_a(D)$. The bus volumes on roadway 1 are equal to $V_b(D)$ since all buses operate on roadway 1. Denote the automobile volume in direction D of roadway i by $V_{ai}(D)$, and denote the bus volume by $V_{bi}(D)$. Then

$$V_{ai}(D) = [L_i(D)\,V_a(D) + 3(-1)^i L_2(D)\,V_b(D)] / \sum_{j=1}^{2} L_j(D), \qquad i = 1 \text{ or } 2, \tag{6.66a}$$

$$V_{bi}(D) = \begin{cases} V_b(D) & \text{if } i = 1 \\ 0 & \text{if } i = 2. \end{cases} \tag{6.66b}$$

Automobile emissions rates per vehicle mile traveled on roadways 1 and 2 can be obtained by substituting the speeds from equation (6.61) into the speed-emissions relations of equations (6.10), (6.15), and (6.16). Bus emissions rates per VMT are as given in table 6.20. The emissions densities (emissions per roadway mile per hour) on the two roadways can be obtained by substituting the emissions rates per VMT and traffic volumes from equations (6.66) into equation (6.4). The emissions densities are first computed separately for automobiles and buses and then added to obtain the total density of emissions from all vehicles. Total emissions on each roadway are computed by multiplying the roadways' emissions densities by their lengths. As in examples 6.1–6.4, CO will be discussed mainly in terms of emissions densities in order to emphasize the localized nature of CO concentrations. HC and NO_x will be discussed mainly in terms of total emissions, since these pollutants are of interest primarily because of their effects on O_3 formation.

The top section of table 6.23 shows the traffic volumes, speeds, and travel times per mile on the two example roadways when there is no priority treatment for buses. The top section of table 6.24 shows the

Table 6.23 Effects of a reserved lane for buses on traffic volumes, speeds, and travel times on two arterial streets (from example 6.7)

Direction	Percentage of direction AB automobile travelers switching to bus	Automobile volumes (veh/hr)		Speeds (mi/hr)		Time (min/mi)[a]	
		Roadway 1	Roadway 2	Automobile	Bus[a]	Automobile	Bus
	Without reserved lane						
AB	0	1,860	1,340	16.4	16.4	3.67	3.67
BA	0	780	620	22.7	22.7	2.64	2.64
	With reserved lane and two direction AB automobile lanes on roadway 1 (figure 6.4a)						
AB	0	1,600	1,600	12.1	23.0	4.96	2.61
AB	5	1,520	1,520	13.4	23.0	4.47	2.61
AB	10	1,440	1,440	14.7	23.0	4.07	2.61
AB	15	1,360	1,360	16.0	23.0	3.74	2.61
AB	20	1,280	1,280	17.3	23.0	3.47	2.61
BA	Any	Same as without reserved lane					
	With reserved lane and three direction AB automobile lanes on roadway 1 (figure 6.4b)						
AB	0	1,920	1,280	17.3	23.0	3.47	2.61
AB	5	1,824	1,216	18.2	23.0	3.29	2.61
AB	10	1,728	1,152	19.1	23.0	3.14	2.61
AB	15	1,632	1,088	19.9	23.0	3.02	2.61
AB	20	1,536	1,024	20.5	23.0	2.92	2.61
BA	Any	625	775	22.2	22.2	2.70	2.70

a. Travel time per mile is the inverse of the speed. Bus speeds and travel times refer to in-motion speeds and times only. The effects of delays at bus stops are not included.

emissions densities on each roadway and total emissions on the two roadways combined.

Now suppose that the curb lane of roadway 1 in the direction AB is reserved for buses, thereby reducing the number of automobile lanes in this direction from three to two (see figure 6.4a). Assume that a fraction δ of automobile travelers from A to B changes to bus as a result of the reserved lane. Thus, the total volume of automobile traffic from A to B is now $(1 - \delta) V_a(AB)$ veh/hr. The reserved lane does not affect travel in the direction B to A. The traffic volumes in the direction A to B following implementation of the reserved lane can be computed by making the following modifications to equations (6.64) and (6.66). To compute the volumes in the automobile lanes, set the total volume of automobile traffic in the direction AB equal to $(1 - \delta) V_a(AB)$. Set $V_b(AB) = 0$ since buses now use the bus lane and not the automobile lanes, and set $L_1(AB) = 2$ since one lane of roadway 1 in the direction AB is no longer available to automobiles. These modifications enable the automobile traffic volumes to be computed from equations (6.64) and (6.66). The traffic volume in the bus lane is $V_b(AB)$, and the effective volume in this lane for use in equation (6.61) is $3V_b(AB)$.

Once the traffic volumes in the direction AB have been computed, the automobile and bus speeds in this direction can be computed from equation (6.61). Note, however, that the bus speed obtained from this equation does not include the effects of delays at bus stops. Given the traffic volumes and speeds, emissions rates and densities are computed using equations (6.4), (6.10), (6.15), and (6.16) and table 6.20, as was done without the reserved lane in effect.

The traffic volumes, speeds, and travel times per mile on the example roadways with the reserved lane in effect are shown in the middle section of table 6.23 for a range of δ values. Emissions are shown in the middle section of table 6.24. Depending on the mode shift δ, the reserved lane reduces direction AB bus travel times by 0.84–2.35 min/mi, relative to automobile travel times.[27] The reserved lane diverts direction AB automobile traffic from roadway 1 to roadway 2 and reduces direction AB automobile speeds if the proportion of direction AB automobile travelers switching to transit is less than approximately 17%. (This value can be obtained from table 6.23 by interpolation.) This is because the reservation of the lane reduces the direction AB roadway capacity that is available to automobiles. The CO emissions density on roadway 1 decreases following implementation of the reserved lane if the proportion of direction AB automobile travelers switching to transit exceeds roughly 4%. (This value

Table 6.24 Effects of a reserved lane for buses on emissions on two arterial streets (from example 6.7)

Percentage of direction AB automobile travelers switching to bus	Roadway 1 emissions density (kg/mi/hr)			Roadway 2 emissions density (kg/mi/hr)			Total emissions on both roadways (kg/hr)[a]		
	CO	HC	NO_x	CO	HC	NO_x	CO[b]	HC	NO_x
	Without reserved lane								
0	90.6	6.93	25.1	63.1	4.54	16.6	307	22.9	83.4
	With reserved lane and two direction AB automobile lanes on roadway 1 (figure 6.4a)								
0	97.6	7.61	20.4	88.2	6.54	15.8	372	28.3	72.4
5	88.4	6.87	20.7	78.9	5.80	16.1	335	25.3	73.6
10	80.6	6.25	21.0	71.2	5.18	16.4	304	22.9	74.8
15	74.1	5.74	21.1	64.7	4.67	16.5	278	20.8	75.2
20	68.6	5.31	21.2	59.2	4.24	16.6	256	19.1	75.6
	With reserved lane and three direction AB automobile lanes on roadway 1 (figure 6.4b)								
0	86.0	6.57	24.6	63.7	4.56	18.0	299	22.3	85.2
5	80.3	6.13	24.5	59.9	4.26	17.9	280	20.8	84.8
10	75.3	5.75	24.3	56.5	4.01	17.8	264	19.5	84.2
15	70.8	5.41	23.9	53.5	3.78	17.6	249	18.4	83.0
20	66.7	5.11	23.5	50.8	3.58	17.3	235	17.4	81.6

a. Based on roadways that are 2 mi long.
b. Total CO emissions contribute to background CO concentrations, but are not closely related to CO concentrations along individual roadways.

can be obtained from table 6.24 by interpolation.) However, a mode switch of over 16% is needed to achieve a reduction in the roadway 2 CO emissions density owing to the diversion of automobile traffic from roadway 1 to roadway 2, and a mode switch of more than 10% is needed to reduce total HC emissions. With lower mode switches, the reserved lane increases CO emissions on roadway 2 and total HC emissions. The reserved lane either reduces NO_x emissions or leaves them unchanged on both roadways under all of the tabulated conditions.

In the configuration of figure 6.4a, the adverse effects of the reserved lane on roadway 2 CO and total HC emissions are caused by the reduction in the roadway capacity that is available to automobiles in the major direction of flow. Since the direction BA traffic volumes on roadways 1 and 2 are less than half of the direction AB volumes, it may be possible to mitigate these adverse effects by removing a lane from the BA direction in order to create the direction AB bus lane. The resulting configuration is shown in

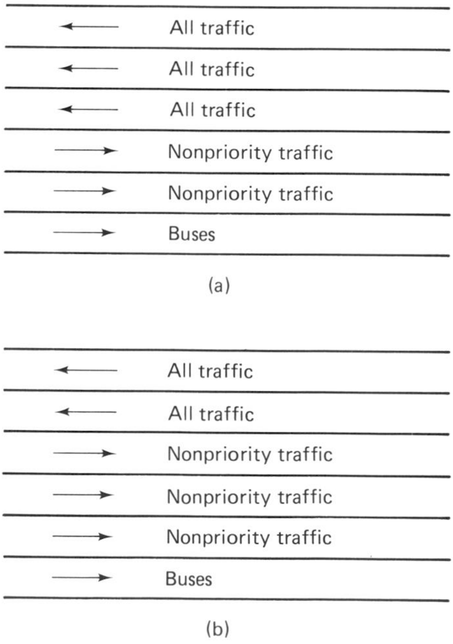

← All traffic
← All traffic
← All traffic
→ Nonpriority traffic
→ Nonpriority traffic
→ Buses

(a)

← All traffic
← All traffic
→ Nonpriority traffic
→ Nonpriority traffic
→ Nonpriority traffic
→ Buses

(b)

Figure 6.4 Two configurations of a reserved lane for buses on an arterial street. Arrows signify direction of traffic flow. (*a*) Lane removed from same direction traffic. (*b*) Lane removed from opposite direction traffic.

figure 6.4b. The direction BA traffic volumes in this configuration can be computed from equations (6.64) and (6.66) by setting $L_1(\mathrm{BA}) = 2$ to reflect the removal of a lane from the direction BA. The direction AB automobile volumes can be computed from the same equations by replacing $V_a(\mathrm{AB})$ with $(1 - \delta)V_a(\mathrm{AB})$ to account for the effects of mode switching, setting $V_b(\mathrm{AB}) = 0$ in the automobile lanes, and setting $L_1(\mathrm{AB}) = 3$ for automobile travel. The actual traffic volume in the bus lane is $V_b(\mathrm{AB})$, and the effective volume is $3V_b(\mathrm{AB})$. Once the volumes are determined, speeds and emissions can be computed by following the same procedures that were used when there was no bus lane and when the figure 6.4a configuration was in effect.

The traffic volumes, speeds, and travel times per mile with the bus lane configuration in figure 6.4b in effect are shown in the bottom section of table 6.23. Emissions are shown in the bottom section of table 6.24. In the configuration of figure 6.4b, the reserved lane reduces direction AB bus travel times by 0.31–0.86 min/mi relative to automobile travel times, depending on the proportion of automobile travelers switching to transit as a result of the lane. This time savings is roughly one third of the savings that was achieved in the configuration of figure 6.4b. However, the roadway 1 CO emissions density and total HC emissions on both roadways, combined, are reduced in the figure 6.4b configuration, regardless of the proportion of direction AB automobile travelers switching to transit. The roadway 2 CO emissions density is reduced if the mode switch proportion exceeds a value that is less than 1%. Thus, as anticipated, the configuration of figure 6.4b is more attractive in terms of its effects on CO and HC emissions than the configuration of figure 6.4a is. However, in the configuration of figure 6.4b, the reserved lane causes the NO_x emissions density on roadway 2 to increase, and it causes total NO_x emissions to increase if the mode switch proportion is less than approximately 13%.

The foregoing example illustrates the potential complexity of the effects of bus priority treatment on emissions. The TRANSYT6C [32] and FREQ [33, 34] models enable the effects of priority treatment to be estimated for traffic flow conditions that are more realistic than those used in the example. Table 6.25 shows the results of using FREQ to estimate the effects of ramp metering with priority entry for high-occupancy vehicles on a section of the Eastshore Freeway (I-80) in the Berkeley-Oakland area. The tabulated results include the changes in emissions that occur on the freeway and on a parallel arterial street (San Pablo Avenue) as a result of the priority treatment plan. The plan causes CO and HC emissions on

Table 6.25 Effects of ramp metering on an urban freeway with priority entry for vehicles with two or more occupants (percentage changes)[a]

Roadway	Passenger hours traveled	Vehicle miles traveled	CO emissions	HC emissions	NO_x emissions	Fuel consumption
Freeway	−13	−1	−4	−4	+9	−1
Parallel arterial street	+9	+3	+11	+9	0	+3
Total	−7	−0.4	−1[b]	−2	+9	0

a. Source: reference [34]. Based on simulation studies using the model FREQ6PE. The ramp metering plan maximizes passenger input to the freeway and allows vehicles with two or more occupants to bypass the ramp meters without delay. The freeway is the Eastshore Freeway in the Berkeley-Oakland area. The parallel arterial street is San Pablo Avenue.

b. Total CO emissions contribute to background CO concentrations, but are not closely related to CO concentrations along individual roadways.

the freeway and total HC emissions to decrease. However, CO and HC emissions on San Pablo Avenue increase, owing to diversion of traffic from the freeway to that street. In addition, NO_x emissions on the freeway and total NO_x emissions increase due to the priority treatment plan, but NO_x emissions on San Pablo Avenue do not change. These results further illustrate the potential complexity of the effects on emissions of priority treatment for high-occupancy vehicles.

D PRICING AND RESTRAINT MEASURES

6.10 Pricing Measures

Automobile use and emissions can be reduced by increasing the price that travelers must pay for using automobiles. Means of increasing this price include

- removing parking subsidies and increasing parking prices (e.g., through the imposition of taxes on parking facilities);
- increasing the price of gasoline through the use of gasoline taxes;
- area licensing—motorists who wish to operate automobiles in specified areas of a city or at specified times of day must purchase special licenses and display them on their vehicles;
- vehicle metering—vehicles are equipped with tamper-proof odometers, and a fixed tax per mile driven is assessed, based on the odometer reading.

These pricing mechanisms are not equivalent in terms of their effects upon travel and emissions or in terms of their ease or difficulty of implementation. The method of increasing parking prices is able to differentiate among geographical areas, times of day, and, at least approximately, trip purposes. For example, parking prices can be set at high levels in areas that have high CO concentrations or that have good transit accessibility, and prices can be set at lower levels in areas that have low CO concentrations or poor transit accessibility. However, parking prices do not affect trips through an area. This diminishes the effectiveness of parking prices as means of reducing automobile use and emissions in areas where through trips represent large proportions of total travel. In addition, parking pricing measures can be difficult to implement. In many areas, large numbers of parking facilities are privately owned and are reserved primarily for use by persons conducting business with the organizations that own or lease them. There may be no readily available means of forcing these organizations to increase the prices that they charge their employees or customers for parking or to begin charging for parking that currently is free.

Gasoline taxes provide a means of increasing the price of automobile travel over entire urban regions. However, gasoline taxes are not suitable for differentiating among different geographical subareas in a region, different times of day, and different trip purposes. A potential disadvantage of gasoline taxes as emissions control measures is that their effects on automobile use can be avoided to some extent through the purchase of high-fuel economy vehicles. However, high-fuel economy and low-fuel economy automobiles that comply with the current US emissions standards do not have substantially different emissions rates. Therefore, a gasoline tax that is intended as an emissions reduction measure may have to be increased periodically as the fuel economy of the automobile fleet improves. Very high gasoline taxes can be required to achieve modest increases in the cost of automobile travel. For example, the average one-way length of home-to-work trips in US cities is approximately 10 mi [66]. If the average fuel economy of automobiles is assumed to be 20 mi/gal, then a gasoline tax of $1.00/gal would be needed to increase the average daily cost of automobile travel for work trips by $1.00/veh round trip. Gasoline pricing measures are easy to implement, as mechanisms for imposing and collecting gasoline taxes are in place and operating.

Area licenses, like parking price measures, can differentiate among geographical areas and times of day. Unlike parking price measures, area licenses affect through trips and, therefore, are likely to be more effective than parking prices in discouraging automobile use in areas where there are large numbers of through trips. They are likely to be particularly effective in discouraging automobile use and reducing emissions in small geographical areas, such as the CBDs of cities, that individuals can choose to avoid traveling to by automobile. They also may be effective in discouraging regional travel during limited hours of the day. However, area licenses may not be as effective in discouraging all-day travel throughout entire urban regions as they are in discouraging travel in geographical subareas or during limited hours. This is because many individuals who could avoid automobile travel to certain places or during certain hours undoubtedly would find it difficult or impossible to completely avoid automobile travel on most days. Therefore, these individuals would have to purchase licenses for automobile travel on most days if a region-wide licensing plan were in effect. However, once the decision to own a license is made, the license provides no further disincentive to automobile use.

It is possible, though far from certain, that area licensing plans would be cumbersome and expensive to enforce, since they would require enforcement officers to observe whether vehicles have valid licenses and, possibly, to stop moving vehicles that do not have licenses.

Vehicle metering is a relatively exotic approach to increasing the price of automobile travel. Its main advantage is that it both provides a means of charging for all automobile travel in an urban (or, possibly, larger) region and it cannot be avoided through the purchase of fuel-efficient vehicles. The metering approach has many implementation problems, including those of installing and maintaining meters on large numbers of vehicles and the possibility that the public might perceive metering as an invasion of privacy. In addition, it would be necessary to find legal and convenient ways to disable, temporarily, the meters of automobiles that are taken on trips outside of a region in which a metering plan is in effect and to ensure that the meters are reactivated when the vehicles return to the region.

None of the foregoing pricing procedures enables the prices of individual automobile trips to be related closely to the emissions that the trips produce. Each procedure either ignores factors that are important in determining emissions or allows prices to vary substantially in response to factors that are independent of emissions. For example, parking prices, in addition to not affecting trips through an area, are independent of trip lengths and cannot easily be made to depend on the design, age, or mechanical condition of a vehicle. Prices of area licenses are independent of the distance driven in the licensed area and of whether vehicles are in warm or cold operating condition in the area. Although the price of a license conceivably could be made to depend on the design and age of the licensed vehicle, it could not reasonably be related to the vehicle's mechanical condition. Gasoline taxes cause the price of a trip to depend on the fuel economy of the vehicle that is used, among other factors. However, as has already been noted, automobiles with good fuel economy and automobiles with poor fuel economy do not have substantially different emissions rates. Finally, odometer-based vehicle metering would not be sensitive to hot and cold starts, the speed of travel, or a vehicle's mechanical condition.

There is a long-standing economic justification for increasing the price of automobile use. Many automobile travelers—particularly peak period work trip travelers—receive large explicit or implicit subsidies by not paying the full costs of the parking facilities that they use, the social costs represented by the congestion delays that they impose on other travelers and by the air pollution that they cause, and, possibly, the full costs of roadway facilities that are constructed to handle peak period traffic volumes and are underutilized during off-peak periods. Accordingly, it is

Table 6.26 Results of modeling studies of the effects of pricing policies on urban regional automobile emissions from work trips

City	Policy	Percentage reduction in[a]					Reference
		Automobile driver trips	Automobile vehicle miles traveled	Automobile emissions			
				CO[b]	HC	NO_x	
Denver	Increase parking price by $2.85/day[c]		1.9	1.7	1.7	1.9	[43]
	Double gasoline price		0.7	0.6	0.6	0.8	[43]
Fort Worth	Increase parking price by $2.85/day[c]		1.9	1.9	2.0	2.0	[43]
	Double gasoline price		1.1	1.0	1.0	1.2	[43]
San Francisco	Increase parking price by $2.85/day[c]		3.7	3.9	3.8	3.7	[43]
	Double gasoline price		3.2	2.2	2.6	3.4	[43]
Los Angeles	Increase parking price by $1.50/day[c]	22	14				[60]
	Increase gasoline price by 140% and improve transit service	28	43				[60]
San Diego	Increase parking price by $3.20/day and improve transit service[c]	28					[55]
Washington, DC	Transit improvements, carpool incentives, double gasoline price, and set minimum		19				[53]

	parking price at $2.80/day in the CBD and $1.40/day elsewhere[c]						
	Transit improvements, carpool incentives, and triple gasoline price	12	13	12	12	12	[54]
	Transit improvements, carpool incentives, and minimum parking price of $5.50/day[c]	15	13	14	14	14	[54]

a. Blanks signify information not available.
b. Total regional CO emissions contribute to general background CO concentrations, but are not closely related to CO concentrations along individual roadways.
c. Prices are in 1980 dollars and have been inflated from the years of the cited studies as necessary.

often argued on economic grounds that the price of automobile travel should be increased in order to remove these subsidies (see, e.g., references [67, 68]). Although subsidy removal provides a qualitative, conceptual justification for increasing the price of automobile travel as part of pollution control programs, it does not provide a basis for quantitative determination of appropriate price levels for such programs. This is because the magnitudes of pollution-related subsidies to automobile travel are unknown (see section 4.8). It cannot be assumed that pricing plans that remove parking or congestion-related subsidies necessarily will be effective in controlling air pollution. The magnitudes of parking and congestion subsidies depend on factors such as the values of land and travel time, which are not directly related to pollutant concentrations. Pollutant concentrations depend on factors such as weather conditions and vehicles' emissions rates, which are not directly related to parking or congestion costs. (Examples of congestion pricing measures that are not effective in controlling air pollution and of pollution pricing measures that do not remove congestion-related subsidies are given in reference [69].)

Table 6.26 shows estimates of the effects that increases in parking and gasoline prices would have on regional automobile use for and emissions from work trips in several US cities. Tables 6.27 and 6.28 show analogous information for nonwork trips and all trips, respectively. Table 6.29 shows estimates of the effects that parking pricing measures and area licensing would have in a 3.5-mi^2 area in the center of Boston. All of the tabulated estimates are based on the results of travel demand modeling studies, as pricing measures aimed at reducing automobile use or emissions have not been implemented in any US cities.[28] Since, in practice, it is likely that attempts to implement pricing measures often would be accompanied by transit improvements and carpool incentives, many of the tabulated estimates pertain to combinations of pricing, transit improvement, and carpool incentive measures. However, the emissions estimates do not include emissions from transit vehicles.[29]

The price increases in tables 6.26, 6.27, and 6.28 amount to roughly $1.00–4.00/day for automobile driver work trips and $0.00–1.50/round trip for automobile driver nonwork trips (in 1980 dollars), depending on the policy being considered. Most of the tabulated results (with the notable exception of those for Los Angeles) suggest that price increases in these ranges would produce reductions in total regional automobile emissions from all trips of roughly 6–15% if transit service were improved and carpool incentives were provided in connection with the price increases. Larger price increases would produce larger emissions re-

Table 6.27 Results of modeling studies of the effects of pricing policies on urban regional automobile emissions from nonwork trips

		Percentage reduction in[a]					
		Automobile driver	Automobile vehicle miles	Automobile emissions			
City	Policy	trips	traveled	CO[b]	HC	NO_x	Reference
Denver	Double gasoline price	2.0	16	8.5	9.4	18	[43]
Fort Worth	Double gasoline price	1.8	19	9.4	11	21	[43]
San Francisco	Double gasoline price	3.0	23	11	13	26	[43]
Los Angeles	Increase parking price by $1.50/trip[c]	21	17				[60]
	Increase gasoline price by 140% and improve transit service	31	29				[60]
San Diego	Increase parking price by $0.30/hr and improve transit service[c]	11					[55]
Washington, DC	Double gasoline price		3.3				[53]
	Triple gasoline price	3.8	10	5.4	3.7	7.9	[70]

a. Blanks signify information not available.
b. Total regional CO emissions contribute to general background CO concentrations, but are not closely related to CO concentrations along individual roadways.
c. Prices are in 1980 dollars and have been inflated from the years of the cited studies as necessary.

Table 6.28 Results of modeling studies of the effects of pricing policies on urban regional automobile emissions from all trips

City	Policy for		Percentage reduction in[a]					Reference
			Automobile driver trips	Automobile vehicle miles traveled	Automobile emissions			
	Work trips	Nonwork trips			CO[b]	HC	NO_x	
Denver	Increase parking price by $2.85/day[c]	None		0.7	0.8	0.8	0.6	[43]
	Double gasoline price	Double gasoline price		10	4.9	5.4	12	[43]
Fort Worth	Increase parking price by $2.85/day[c]	None		0.9	1.0	1.1	0.9	[43]
	Double gasoline price	Double gasoline price		10	4.9	5.7	12	[43]
San Francisco	Increase parking price by $2.85/day[c]	None		1.6	2.2	1.8	1.5	[43]
	Double gasoline price	Double gasoline price		14	5.9	8	17	[43]
Los Angeles	Increase parking price by $1.50/day[c]	Increase parking price by $1.50/trip[c]	21	15				[60]
	Increase gasoline price by 140% and improve transit service	Increase gasoline price by 140% and improve transit service	30	36				[60]
San Diego	Increase parking price by $3.20/day and improve transit service[c]	Increase parking price by $0.30/hr and improve transit service[c]	14	9				[55]

Washington, DC	Transit improvements, carpool incentives, double gasoline price, and set minimum parking price at $2.80/day in the CBD and $1.40/day elsewhere[c]	Double gasoline price		9.4				[53]
	Transit improvements, carpool incentives, and triple gasoline price	Triple gasoline price	5.8	11	8.4	7.4	9.7	[54, 70]
	Transit improvements, carpool incentives, and minimum parking price of $5.50/day[c]	None	3.2	6.3	6.1	6.0	6.2	[54, 70]

a. Blanks signify information not available.
b. Total regional CO emissions contribute to general background CO concentrations, but are not closely related to CO concentrations along individual roadways.
c. Prices are in 1980 dollars and have been inflated from the years of the cited studies as necessary.

Table 6.29 Results of modeling studies of the effects of pricing polices on automobile travel and CO emissions in the central area of Boston, Massachusetts[a]

		Percentage reductions in				
Policy	Peak or off-peak	Automobiles entering central area	Automobiles traveling to central area	Automobiles traveling through central area	Automobile vehicle miles traveled in central area	Automobile CO emissions in central area
Increase parking price by $2.85/day	Peak	15–44	28–89	0	16–39	26–63
Area license costing $2.85/day required for travel on surface streets	Peak	20–51	28–89	7–9	22–53	30–68
Increase parking price by $1.40/trip	Off-peak	17–32	33–63	0	16–31	16–39
Area license costing $1.40/day required for travel on surface streets	Off-peak	21–36	33–63	8	26–40	26–44

a. Source: reference [72]. Prices have been inflated to 1980 dollars. Ranges of effects reflect uncertainties in the results of the models that were used.

ductions. Not surprisingly, pricing measures that affect both work and nonwork trips tend to achieve larger emissions reductions than do measures that affect only work trips.

The Boston central area estimates (table 6.29) suggest that pricing measures may be capable of achieving relatively large reductions in CO emissions in the downtown areas of large cities with well-developed transit systems. Area licensing appears to be particularly useful in this respect, owing to its ability to divert through trips away from the controlled area.

Tables 6.26–6.29 also illustrate the considerable uncertainty that is associated with current forecasts of the effects of pricing measures on automobile emissions and travel. In Tables 6.26–6.28, the differences between the trip and VMT reduction estimates for Los Angeles and those for other cities are due mainly to differences in the price sensitivities of the travel demand models that were used in forming the various estimates, rather than to differences in the policies that were considered or in the characteristics of the cities. Similarly, the upper and lower limits of the ranges of the estimates shown in table 6.29 correspond to results that were obtained with two different sets of demand models. Thus, although the available evidence supports the qualitative conclusion that increases in the price of automobile travel would reduce automobile use and emissions, it is not possible at present to forecast accurately the magnitudes of the emissions reductions that would be produced by specific changes in transportation prices.

An area licensing plan has been in effect in downtown Singapore since 1975. The plan requires motorists who bring automobiles into the downtown area during the morning peak period to purchase special licenses that must be displayed on their vehicles. In addition, downtown parking prices were raised substantially when the licensing plan was put into effect. The plan caused a 77% reduction in the number of automobiles entering the restricted zone during the restricted hours and a 44% reduction in total traffic (including trucks and buses) entering the zone during these hours [74]. The average CO concentration in the restricted zone decreased by approximately 30% during the restricted hours [74].

6.11 *Automobile Restraints*

Automobile restraints are measures whose purpose is to reduce the supplies of parking or roadway facilities that are available for use by automobiles in order to discourage or otherwise control automobile travel.

Vehicle-free zones and traffic cell systems, which were discussed in section 6.4, are examples of automobile restraints. In this section, parking supply restrictions and roadway metering measures will be discussed.

Parking supply restrictions can consist either of reducing the number of existing parking spaces in an area or of controlling the rate of increase in the number of parking spaces in an area where growth is occurring. In many respects, parking supply restrictions are similar to parking pricing measures. Indeed, one of the consequences of restricting the supply of parking spaces in an area may be to cause the price of the remaining spaces to increase, thus making supply restriction an indirect method of raising the price of parking. Like parking pricing measures, parking supply restrictions can be designed to differentiate among geographical areas, times of day, and, in a rough way, trip purposes. For example, if the supply of parking spaces is restricted during the early and midmorning hours of weekdays but not at other times of day or on weekends, then the supply restriction will tend to discourage mainly automobile work trips. Also, parking supply restraints, like parking pricing measures, do not affect trips through an area.

Reducing the supply of existing parking space may present severe implementation difficulties in areas where large proportions of parking spaces are privately owned since the restrictions would entail withdrawing permission for an existing use of private property. These difficulties may be especially severe in cases where productive alternative uses of privately owned parking facilities are not readily available. Moreover, congested areas of cities, where implementation of restraints on automobile travel may seem particularly attractive, often have much unused parking capacity. In such areas, seemingly large reductions in the number of parking spaces may serve mainly to reduce or eliminate excess capacity and have little effect on travel behavior.

Parking supply restrictions that are based on controlling growth in the supply of parking spaces often may be easier to implement than are measures to reduce the supply of existing spaces. Examples of growth restriction measures include denial of zoning permission for using vacant land for parking and zoning restrictions on the numbers of parking spaces that can be provided in newly constructed buildings. The effects of such measures on automobile travel and emissions occur gradually, particularly in areas where there is an excess of existing parking capacity.

There is little quantitative evidence concerning the ability of parking supply restrictions to reduce automobile travel and emissions. Satisfactory modeling techniques for predicting the effects of parking supply restrictions are not yet available, and there have been few direct observations of the effects on travel of changes in supplies of parking spaces. However, the limited evidence that is available tends to support the hypothesis that large reductions in supplies of parking spaces would be effective in reducing automobile travel and emissions, at least in the central areas of cities. In 1972 a strike by parking lot employees in the Pittsburgh area caused the closing of 80% of the parking spaces in downtown Pittsburgh. There was a 24% reduction in the number of automobiles entering the CBD on either CBD-bound or through trips during the morning peak period [75]. In 1971 all parking in downtown Marseilles, France, was prohibited as part of an experiment in traffic restrictions. There was approximately a 40% decrease in the average 8 A.M.–6 P.M. CO concentration in the restricted area during the period of the experiment [39]. In Nagoya, Japan, the volume of traffic entering the CBD during the morning peak hour was reduced by 18% through the implementation of a combination of transit improvement, traffic flow, and parking restraint measures [40]. The parking restrictions are reported to have been the main cause of the reduction in traffic volume [40]. There was a 12% reduction in the number of vehicles entering, leaving, or traveling within the CBD [40].

The roadway metering approach to restraining automobile traffic consists of extending the freeway ramp-metering concept to surface streets. The flow of traffic into an area (e.g., into the CBD of a city) is controlled with traffic signals along the area's boundary and, possibly, at outlying points along the principal streets leading to the area. Bypass lanes for high-occupancy vehicles may be provided at the control points. The traffic signals at the control points meter traffic into the restricted area in accordance with objectives that may include maintaining free flow of traffic in the area or achieving a time advantage for high-occupancy vehicles. An important drawback of roadway metering is that long queues may form at control points. These queues may disrupt traffic that is not traveling to the restricted area and may also cause substantial increases in emissions due to extended idling of the vehicles in the queues.

An experiment in roadway metering was carried out in Nottingham, England, during a 1-yr period starting in mid-1975. In this experiment, traffic signals controlled the rate at which morning peak period traffic leaving two residential zones on the western side of the city entered the main roadway network. In addition, traffic signals on the western boundary of the CBD controlled the rate at which morning peak period traffic entered the CBD from that direction. Bus-only roads and bus lanes enabled buses to avoid delays at the control points. The main objective of the metering plan was to provide a time advantage for buses, relative to automobiles, that would be sufficiently large to attract 10% of morning peak period automobile travelers to transit. The metering plan failed to achieve this objective [76]. It was not possible to create bus-automobile time differentials exceeding roughly 4 min without creating excessively long queues at the control points [76]. (Depending on the control point, 400–800 m of queueing space was available at the entrances to the restricted area.) There was less than a 2% reduction in the volume of morning peak period traffic on controlled radial roadways, and there was a 6% increase in vehicle hours traveled, owing to delays at the control points [76]. Estimates of emissions changes associated with the metering plan have not been reported, but the observation that the plan reduced the average speed of traffic flow by a much larger percentage than it reduced the volume of traffic suggests that the metering plan probably caused CO and HC emissions from morning peak period traffic to increase.

The failure of the Nottingham experiment does not demonstrate that roadway metering necessarily is an ineffective approach to reducing automobile travel and emissions. For example, the long queues that contributed to the failure of the Nottingham metering plan might be avoided by delaying automobiles at several different control points along each major arterial street instead of only at the boundary of the restricted area and at a relatively small number of entrances to the major streets. In addition, metering plans that are implemented under conditions in which there is an opportunity to divert a large number of through automobile trips around the restricted area and plans whose main objective is to reduce congestion in the restricted area, rather than to create a substantial time advantage for transit, may be more successful than the Nottingham plan was. However, the Nottingham experience demonstrates that in designing and implementing roadway metering plans, it is necessary to give considerable attention to the problems of preventing excessively long queues from forming at control points and of preventing speed reductions caused by delays at control points from leading to increases, rather than decreases, in air pollution.

6.12 *Effects of Pricing and Restraint Measures on Business Activity*

Concerns that automobile pricing or restraint measures may have severe adverse effects on business activity often constitute major barriers to the implementation of these measures. Unfortunately, the effects of pricing and restraint measures on business activity are very poorly understood at present. There are no methods available for making useful forecasts of these effects, and the limited evidence that is available from cities where pricing or restraint measures have been implemented is difficult to generalize. The observed effects of these measures have varied greatly, depending on the specific types of measures that were implemented, the circumstances in which they were implemented, and the types of businesses that were involved. Moreover, in many cases it has not been possible to separate the effects on business activity of implementing pricing or restraint measures from the effects of general economic and demographic trends.

The largest body of evidence concerning the effects of pricing or restraint measures on business activity pertains to the effects of vehicle-free zones on retail sales. Establishment of vehicle-free zones has usually increased aggregate retail sales in the restricted areas, although the effects on individual shops can vary greatly, and some shops can be harmed by being included in a vehicle-free zone [39]. Central area traffic cell systems in European cities appear to have had little effect on central area business activity [40]. The Pittsburgh parking strike caused a 6–8% decline in retail business sales in the downtown area and a 60–70% decline in theater patronage [75]. The Singapore licensing plan is believed to have reinforced a previously existing trend toward decentralization of commerical activity in Singapore [74].

For the present, information on the effects on business activity of implementing specific types of pricing or restraint measures in specific settings can be obtained only through experimentation with the measures—that is, by implementing the measures on a trial basis and observing their effects. Considerably more experience with pricing and restraint measures will be needed to make useful forecasting of these measures' effects on business activity possible.

E LAND USE MEASURES

Dispersed, low-density urban land use patterns contribute to dependence of cities on automobile transportation by making the provision of high-quality transit service difficult or impossible (see the discussion in section 6.7) and by making walking an impractical mode for most trips. In addition, dispersed land use patterns tend to generate longer automobile trips than more concentrated patterns do. Accordingly, it is worthwhile to investigate the extent to which automobile-related air pollution in cities might be reduced through the implementation of policies designed to encourage the development of more concentrated urban land use patterns. In particular, it is useful to consider the extent to which automobile-related air pollution could be reduced through the implementation of policies aimed at reversing the long-standing trend of movement of population and employment out of central cities and into suburban areas. Two aspects of this problem need to be addressed. First, it is necessary to determine whether increased centralization of development would, in fact, be beneficial, and further decentralization harmful, to air quality. Second, if it is determined that certain changes in current land use patterns or trends of development would lead to improvements in air quality, it is necessary to identify policy measures whose implementation could be expected to bring about these changes.

The most extensive investigation of the relation between automobile-related air pollution and urban land use that has been conducted to date consists of a computer-based modeling study of a group of hypothetical land use scenarios for the Boston area [77]. The purpose of this study was to estimate the effects that certain land use changes would have on this area's air quality. The study did not attempt to identify or evaluate policies for bringing about changes in land use. The land use scenarios that were modeled included

- centralization employment—40% of suburban industrial employment is is moved to the center of the region, and residential locations remain unchanged;
- decentralization of employment—40% of industrial employment in a large portion of the center of the region is moved to the suburbs, and residential locations remain unchanged;
- centralization of employment and population—40% of suburban industrial employment, retail employment, and population are moved to the center of the region;
- decentralization of employment and population—40% of industrial employment, retail employment, and population in a large portion of the center of the region are moved to the suburbs.

A set of travel demand models was used to estimate the travel that would occur under each of the scenarios. The travel estimates were used to compute CO, HC, and NO_x emissions and annual average CO concentrations for the scenarios. Estimates of emissions and CO concentrations also were developed for the existing pattern of land use. The CO concentrations represent averages over areas of several square miles. Estimates of short-term CO concentrations and of CO concentrations along individual streets were not developed.

Among the scenarios that were investigated, only centralization of both population and employment had significant effects on total regional HC and NO_x emissions from automobiles. Moving 40% of suburban population and employment to the center of the region reduced regional automobile emissions of HC and NO_x by 15 and 18%, respectively [77]. These emissions reductions occurred mainly because centralization of both population and employment moved residential and employment locations closer together than they are under existing conditions, thereby reducing trip lengths. In addition, centralization of population and employment caused a small increase in the proportion of trips using transit. However, centralization of population and employment also increased the traffic density in the center of the Boston region. Consequently, moving 40% of suburban population and employment to the center of the region caused the annual average CO concentration in parts of the center of the region to increase by nearly 100%.

Centralization of employment but not of population and all of the decentralization scenarios that were investigated produced changes of less than 3% in regional automobile emissions of HC and NO_x [77]. Centralization of employment but not of population tends to separate residential and employment locations, thereby increasing trip lengths. Accordingly, this scenario caused a slight increase in total regional automobile emissions of HC and NO_x in comparison to emissions with existing land use patterns. Most of the scenarios that decentralized employment but not population caused HC and NO_x emissions to decrease slightly in comparison to emissions with existing land use patterns, whereas decentralization of both employment and population caused HC and NO_x emissions to increase slightly. Centralization of employment but not of population increased the annual average CO concentration in parts of the center of the region by 20%. The decentralization scenarios reduced the annual average CO concentration in the center of the region by up to 16%, depending on the location in the central area and the scenario, and increased suburban annual average CO concentrations by up to 20%.

The Boston results suggest that land use changes may have considerably smaller effects on O_3 and annual average NO_2 concentrations than on CO concentrations. According to these results, increasing the centralization of population and employment in the Boston region would tend to cause modest reductions in O_3 concentrations and long-term average NO_2 concentrations owing to the reductions in regional motor vehicle emissions of HC and NO_x that such a land use change would produce.[30] However, increased centralization could also cause very large increases in CO concentrations in central areas, where these concentrations are already high. Conversely, decentralization of development, relative to current patterns, could reduce high central area CO concentrations substantially without greatly affecting regional O_3 concentrations. Hence, it cannot be concluded that highly centralized urban land use patterns necessarily are beneficial to air quality or that decentralized patterns necessarily are harmful to air quality.

A study of land use policy for the Washington, DC, area investigated the relation between automobile travel in the region and the geographical distribution within the region of future growth in population and employment [78]. The growth scenarios that were investigated included

- dense center—growth in population and employment take place mainly in the center of the region and in the inner suburbs;
- transit oriented—growth in population and employment take place mainly in corridors that are well served by transit;
- sprawl—population growth takes place mainly in the outer suburbs, whereas employment growth takes place mainly in the center of the region and in the inner suburbs;
- beltway oriented—growth in population and employment occur in the vicinity of the region's circumferential highway (the beltway), which is located approximately 10 mi from the center of Washington;
- current plans—growth takes place in accordance with existing development plans and projections for the region. In this scenario, growth in population and employment take place mainly in the suburbs.

Total regional population and employment were the same in all of the scenarios. The scenarios were assumed to describe the geographical distribution of growth over a 25-yr period of time.

A set of travel demand models was used to compute daily automobile driver trips and VMT for each of the scenarios. Emissions were not computed in this study.

The dense center and transit-oriented scenarios produced considerably greater centralization of population and employment than did the other scenarios. Not surprisingly, the dense center and transit-oriented scenarios also produced fewer daily automobile driver trips and VMT than the other scenarios did. The dense center scenario produced 6% fewer automobile driver trips and 14% fewer VMT than did the current plans scenario. The transit-oriented scenario produced 3% fewer automobile driver trips and 15% fewer VMT than did the current plans scenario. The sprawl scenario, which tended to separate residences and employment by locating the former in the outer suburbs and the latter in the inner suburbs and center of the region, produced 1% more automobile driver trips and 6% more automobile VMT than did the current plans scenario. The beltway-oriented scenario produced the least centralized development patterns and the lowest transit usage of any of the scenarios that were analyzed.

Consequently, it resulted in 8% more automobile driver trips than did the current plans scenario. However, the beltway-oriented scenario also produced a very high concentration of population and employment in the vicinity of the beltway, thereby causing the average trip length to be less in this scenario than in most of the other scenarios. As a result, the beltway-oriented scenario produced 4% fewer automobile VMT per day than did the current plans scenario.

If automobile emissions had been computed in the Washington study, total automobile HC and NO_x emissions almost certainly would have been lower in the dense center and transit-oriented scenarios and higher in the sprawl scenario than they would have been in the current plans scenario. These conclusions, like those of the Boston study, indicate that centralization of both population and employment may cause modest reductions in O_3 concentrations and annual average NO_2 concentrations but that centralization of employment without centralization of population may have a slight adverse effect on these concentrations. It is likely that automobile emissions of HC and NO_x would have been slightly higher in the beltway-oriented scenario than in the current plans scenario, but not substantially so. However, the beltway-oriented scenario also resulted in substantial decentralization of development and an increase in automobile dependence compared to the current plans scenario. Thus, the Washington results, like the Boston results, suggest that further decentralization of urban development in comparison to existing patterns and trends would not necessarily cause appreciable increases in total regional automobile emissions or in O_3 concentrations.

The effects of land use changes on automobile travel in geographical subareas within the Washington region were not investigated in the Washington study. Consequently, it is not possible to use the results of this study to estimate the effects of land use changes on CO concentrations.

The results of the Boston land use study suggest that it may be difficult to identify changes in urban land use patterns that would substantially reduce both central area CO and regional O_3 concentrations. However, the Boston and Washington results, taken together, suggest that certain types of land use changes, such as centralization of employment without centralization of population, may cause increases in both central area CO concentrations and regional O_3 concentrations, although the increases in O_3 concentrations are likely to be very small. The observation that changes in urban land use patterns may not be able to substantially reduce the concentrations of central area CO and regional O_3 simultaneously indicates that in assessing the role of land use changes in achieving reductions in transportation-related air pollution, it may be necessary to make trade-offs between the objective of reducing central area CO concentrations and the objective of reducing regional O_3 concentrations.

Assuming that any needed trade-offs can be made and that desirable changes in the land use patterns in an urban region can be identified, there remains the problem of identifying land use policy measures whose implementation could be expected to achieve the desired changes. This problem constitutes a major obstacle to the achievement of air quality improvements, among other objectives, through planned, regional-scale changes in urban land use patterns. Policy measures that usually are considered to be useful for guiding urban land use on this scale include zoning regulation, taxation policy, and decisions relating to the placement of major transportation facilities and other infrastructure. However, at present it is virtually impossible to predict the long-run, regional-scale effects of most land use policy measures [79, 80].[31] The problems presented by unpredictability are considerably more serious in the case of land use measures than they are in the case of the transportation measures that have been discussed in this chapter. Most of the latter can be reversed or altered relatively easily and quickly if they are found to be having unanticipated and unwanted effects. Even seemingly irreversible transportation measures, such as the construction of new roadways, can be modified to some extent through the application of appropriate operating policies. For example, the traffic volume on a new roadway can be controlled by controlling access to the roadway. In contrast, land use changes tend to be reversible only over periods of several decades, if they are reversible at all.

Since the effects of land use policy measures that are aimed at altering urban land use patterns on a regional scale are, to a large extent, both unpredictable and irreversible, they are also largely uncontrollable. Consequently, such measures cannot be regarded as effective instruments for achieving reductions in concentrations of transportation-related air pollutants. Although it is possible to identify regional-scale changes in urban land use patterns that would tend to reduce concentrations of one or another transportation-related air pollutant, it is not possible to identify policy measures whose implementation would provide a reasonable assurance that these changes would occur.[32]

F SUMMARY AND CONCLUSIONS

Transportation and traffic management measures that may be useful for reducing motor vehicle emissions and improving air quality include traffic engineering measures, transit improvements and carpool incentives, pricing measures, and restraints. Over sufficiently long periods of time, changes in urban regional land use patterns also affect motor vehicle emissions. However, the effects of land use policy measures on urban regional land use patterns are highly uncertain and are difficult or impossible to reverse. Consequently, policy measures aimed at changing urban regional land use patterns cannot be considered to be effective instruments for air quality management at present.

In assessing the ability of transportation and traffic management measures to reduce motor vehicle emissions and improve air quality, it is important to distinguish between the effects of these measures on CO and, possibly, short-term average NO_2 concentrations in small areas and the effects of the measures on regional O_3 and annual average NO_2 concentrations. The results of simulation studies and experience with implemented measures indicate that transportation and traffic management measures can achieve substantial reductions in CO concentrations in small areas, such as the CBDs of cities, and in high-volume corridors. Reductions in small-area CO concentrations of more than 80% have been reported in some cities. It is also likely that transportation and traffic management measures can cause substantial reductions in peak, short-term average NO_2 concentrations near heavily traveled roadways, particularly in areas where O_3 concentrations are high.

The ability of transportation and traffic management measures to reduce regional O_3 and annual average NO_2 concentrations appears to be much more limited. This is due partly to the large proportions of urban regional HC and NO_x emissions (over 50% for each pollutant in the mid-1980s) that are due to nontransportation sources and partly to the difficulty of achieving large reductions in regional automobile emissions through transportation and traffic management. Few transportation and traffic management measures seem likely to be capable of reducing total urban regional automobile emissions of HC and NO_x by more than 15% for each pollutant. Most measures will produce substantially smaller emissions reductions. The corresponding reductions in regional O_3 and annual average NO_2 concentrations are each likely to be less than roughly 10%.

The effects of individual transportation and traffic management measures on emissions can be highly complex. Many measures can cause emissions either to increase or to decrease, depending on the circumstances in which they are implemented. A large number of modeling techniques are available for forecasting the emissions effects of transportation and traffic management measures. However, the state of the art in transportation systems analysis is such that these techniques can yield only rough estimates of the emissions changes that would result from implementing the measures. Moreover, at present, many potentially important side effects of transportation and traffic management measures, such as changes in business activity, cannot be forecast usefully at all. Therefore, in many cases it probably would be wise to implement transportation and traffic management measures initially on an experimental basis and to observe the effects of the measures on traffic flows, emissions, air quality, and other variables of interest. Flexibility should be maintained to modify the measures if their observed effects are inconsistent with the objectives that their implementation was intended to achieve.

References

1

Hill, D. M., Tittemore, L., and Gendell, D., "Analysis of Urban Area Travel by Time of Day," *Highway Research Record*, No. 472, pp. 108–119, 1973.

2

Tittemore, L. H., Birdsall, M. R., Hill, D. M., and Hammond, R. H., *An Analysis of Urban Area Travel by Time of Day*, Report No. FH-11-7519, US Department of Transportation, Washington, DC, January 1972. NTIS Publication No. PB 247776.

3

Horowitz, J. L., and Pernela, L. M., "Comparison of Automobile Emissions Based on Trip Type in Two Metropolitan Areas," *Transportation Research Record*, No. 580, pp. 13–31, 1976.

4

Regional Planning Council, *Transportation Control Plan*, Baltimore, September 1978.

5

Association of Bay Area Governments, Bay Area Air Pollution Control District, and Metropolitan Transportation Commission, *Air Quality Maintenance Plan*, report prepared for public review and comment, San Francisco, December 1977.

6

Metropolitan Washington Council of Governments, *Washington Metropolitan Air Quality Plan*, draft for public review, Washington, DC, September 1978.

7

Horowitz, J. L., and Pernela, L. M., "Analysis of Urban Area Automobile Emissions According to Trip Type," *Transportation Research Record*, No. 492, pp. 1–8, 1974.

8

US Environmental Protection Agency, *Mobile Source Emission Factors*, Report No.

EPA 400/9-78-005, Washington, DC, March 1978. NTIS Publication No. PB 295672.

9
Florian, M., Chapleau, R., Nguyen, S., Achim, C., James-Lefebvre, L., Galarneau, S., Lefebvre, J., and Fisk, C., *Validation and Application of EMME: An Equilibrium Based Two-Mode Urban Transportation Method*, Publication No. 103, Centre de recherche sur les transports, Université de Montréal, Montréal, Québec, July 1978. (An abbreviated version of this report is available in the *Transportation Research Record*, No. 728, pp. 14–23, 1979.)

10
Branston, D., "Link Capacity Functions: A Review," *Transportation Research*, Vol. 10, pp. 223–236, 1976.

11
Florian, M., and Nguyen, S., "A Combined Trip Distribution Modal Split and Assignment Model," *Transportation Science*, Vol. 11, pp. 166–179, 1977.

12
Evans, S., "Derivation and Analysis of Some Models for Combining Trip Distribution and Assignment," *Transportation Research*, Vol. 10, pp. 37–57, 1976.

13
LeBlanc, L. J., Morlok, E. K., and Pierskalla, W. P., "An Efficient Approach to Solving the Road Networks Equilibrium Traffic Assignment Problem," *Transportation Research*, Vol. 9, pp. 309–318, 1975.

14
Nguyen, S., "An Algorithm for the Traffic Assignment Problem," *Transportation Research*, Vol. 4, pp. 391–394, 1974.

15
Gartner, N. H., "Optimal Traffic Assignment with Elastic Demands: A Review. Part I. Analysis Framework," *Transportation Science*, Vol. 14, pp. 174–191, May 1980.

16
Gartner, N. H., "Optimal Traffic Assignment with Elastic Demands: A Review. Part II. Algorithmic Approaches," *Transportation Science*, Vol. 14, pp. 192–208, May 1980.

17
Abdulaal, M., and LeBlanc, L. J., "Methods for Combining Modal Split and Equilibrium Assignment Models," *Transportation Science*, Vol. 13, pp. 292–314, 1979.

18
Domencich, T. A., and McFadden, D., *Urban Travel Demand*, North-Holland/ American Elsevier, New York, 1975.

19
Sheffi, Y., and Daganzo, C. F., "Computation of Equilibrium over Transportation Networks: The Case of Disaggregate Demand Models," *Transportation Science*, Vol. 14, pp. 155–173, May 1980.

20
Florian, M., and Nguyen, S., "An Application and Validation of Equilibrium Trip Assignment Methods," *Transportation Science*, Vol. 10, pp. 374–390, 1976.

21
Eash, R. W., Janson, B. N., and Boyce, D. E., "Equilibrium Trip Assignment: Advantages and Implications for Practice," *Transportation Research Record*, No. 728, pp. 1–8, 1979.

22
Worrall, R. D., Ferlis, R. A., and Lieberman, E., *Network Flow Simulation for Urban Traffic Control System—Phase II, Vol. 1, Technical Report*, Report FHWA-RD-73-83, US Department of Transportation, Washington, DC, March 1974. NTIS Publication No. PB 230760.

23
Worrall, R. D., and Lieberman, E., *Network Flow Simulation for Urban Traffic Control System—Phase II, Vol. 2, Program Documentation for UTCS-1 Network Simulation Model*, Part I, Report FHWA-RD-73-84, US Department of Transportation, Washington, DC, March 1974, NTIS Publication No. PB 230761.

24
Worrall, R. D., and Lieberman, E., *Network Flow Simulation for Urban Traffic Control System—Phase II, Vol. 3, Program Documentation for UTCS-1 Network Simulation Model*, Part II, Report FHWA-RD-73-85, US Department of Transportation, Washington, DC, March 1974. NTIS Publication No. PB 230762.

25
Worrall, R. D., and Lieberman, E., *Network Flow Simulation for Urban Traffic Control System—Phase II, Vol. 4, User's Manual for UTCS-1 Network Simulation Model*, Report FHWA-RD-73-86, US Department of Transportation, Washington, DC, March 1974. NTIS Publication No. PB 230763.

26
Worrall, R. D., *Network Flow Simulation for Urban Traffic Control System—Phase II, Vol. 5, Applications Manual for UTCS-1 Network Simulation Model*, Report FHWA-RD-73-87, US Department of Transportation, Washington, DC, March 1974. NTIS Publication No. PB 230764.

27
Lieberman, E., and Rosenfield, N., *Network Flow Simulation for Urban Traffic Control System—Phase II, Vol. 5, Extension of NETSIM Simulation Model (formerly UTCS-1) to Incorporate Vehicle Fuel Consumption and Emissions*, Report FHWA-RD-77-45, US Department of Transportation, Washington, DC, April 1977. NTIS Publication No. PB 230764.

28
Wicks, D. A., and Lieberman, E. B., *Development and Testing of INTRAS, a Microscopic Freeway Simulation Model, Vol. 1, Program Design, Parameter Calibration and Freeway Dynamics Component Development*, Report FHWA-RD-80-106, US Department of Transportation, Washington, DC, July 1980. NTIS Publication No. PB 258994.

29
Wicks, D. A., and Andrews, B. J., *Development and Testing of INTRAS, a Microscopic Freeway Simulation Model, Vol. 2, User's Manual*, Report FHWA-RD-80-107, US Department of Transportation, Washington, DC, July 1980. NTIS Publication No. PB 258995.

30
Goldblatt, R. B., *Development and Testing of INTRAS, a Microscopic Freeway Simulation Model, Vol. 3, Validation and Application*, Report FHWA-RD-80-108, US Department of Transportation, Washington, DC, July 1980.

31
Wicks, D. A., Andrews, B. J., and Goldblatt, R. B., *Development and Testing of INTRAS, a Microscopic Freeway Simulation Model, Vol. 4, Program Documentation*, Report FHWA-RD-80-109, US Department of Transportation, Washington, DC, July 1980.

32
Jovanis, P. P., May, A. D., and Deikman, A., *Further Analysis and Evaluation of Selected Impacts of Traffic Management Strategies on Surface Streets*, Research Report UCB-ITS-77-9, Institute of Transportation Studies, University of California, Berkeley, October 1977.

33
Cilliers, M. P., Cooper, R., and May, A. D., *FREQ6PL—a Freeway Priority Lane Simulation Model*, Report No. UCB-ITS-RR-78-8, Institute of Transportation Studies, University of California, Berkeley, August 1978.

34
Jovanis, P. P., Yip, W., and May, A. D., *FREQ6PE—a Freeway Priority Entry Control Simulation Model*, Report No. UCB-ITS-RR-78-9, Institute of Transportation Studies, University of California, Berkeley, November 1978.

35
Gershwin, S. B., Ross, P., Gartner, N., and Little, J. D. C., "Optimization of Large Traffic Systems," *Transportation Research Record*, No. 682, pp. 8–15, 1978.

36
Gartner, N. H., Gershwin, S. B., and Little, J. D. C., "Pilot Study of Computer-Based Urban Traffic Management," *Transportation Research*, Vol. 14B, pp. 203–217, 1980.

37
Organization for Economic Cooperation and Development, *Inquiry into the Impact of the Motor Vehicle on the Environment; Automotive Air Pollution and Noise: Implications for Public Policy*, Paris, October 1972.

38
Ingram, G. K., and Fauth, G. R., *TASSIM: A Transportation and Air Shed Simulation Model, Volume 1—Case Study of the Boston Region*, Report No. DOT-OS-30099-5, US Department of Transportation, Washington, DC, May 1974. NTIS Publication No. PB 232933.

39

Orski, C. K., "Car-Free Zones and Traffic Restraints: Tools of Environmental Management," *Highway Research Record*, No. 406, 1972.

40

Organization for Economic Cooperation and Development, *Managing Transport*, Paris, 1979.

41

Koppelman, F. S., "Guidelines for Aggregate Travel Prediction Using Disaggregate Choice Models," *Transportation Research Record*, No. 610, pp. 19–24, 1976.

42

Atherton, T. J., and Suhrbier, J. H., *Urban Transportation Energy Conservation: Analytic Procedures for Establishing Changes in Travel Demand and Fuel Consumption*, Report No. DOE/PE/8628-1, Vol. II, US Department of Energy, Washington, DC, October 1979. NTIS Publication No. DOE/PE/8628-1 (v.2).

43

Atherton, T. J., and Suhrbier, J. H., *Urban Transportation Energy Conservation: Case City Applications of Analysis Methodologies*, Report No. DOE/PE/8628-1, Vol. III, US Department of Energy, Washington, DC, October 1979. NTIS Publication No. DOE/PE/8628-1 (v.3).

44

US Department of Transportation, *Automobile Occupancy*, Report No. 1, Nationwide Personal Transportation Study, April 1972. NTIS Publication No. PB 242985.

45

Horowitz, J. L., "Modifying Urban Transportation Systems to Improve the Urban Environment," *Computers and Operations Research*, Vol. 3, pp. 175–183, 1976.

46

Adler, T. J., and Ben-Akiva, M., "Joint Choice Model for Frequency, Destination, and Travel Mode for Shopping Trips," *Transportation Research Record*, No. 569, pp. 136–150, 1976.

47

Levin, I. P., and Gray, M. J., "Evaluation of Interpersonal Influences in the Formation and Promotion of Carpools," *Transportation Research Record*, No. 724, pp. 35–39, 1979.

48

Margolin, J. B., Misch, M. R., and Stahr, M., "Incentives and Disincentives of Ride Sharing," *Transportation Research Record*, No. 673, pp. 7–15, 1978.

49

McQueen, J. T., Levinsohn, D. M., Waksman, R., and Miller, G. K., *Evaluation of the Shirley Highway Express-Bus-on-Freeway Demonstration Project—Final Report*, Report No. DOT/UMTA 7, US Department of Transportation, Washington, DC, August 1975. NTIS Publication No. PB 247637.

50
Crain, J. L., "Evaluation of a National Experiment in Bus Rapid Transit," *Transportation Research Record*, No. 546, pp. 22–29, 1975.

51
Crain and Associates, *San Bernardino Freeway Express Busway—Evaluation of Mixed Mode Operations*, report prepared for the Southern California Association of Governments, Los Angeles, July 1978.

52
Wagner, F. A., and Gilbert, K., *Transportation System Management: An Assessment of Impacts*, report prepared for the US Department of Transportation, Urban Mass Transportation Administration, November 1978. NTIS Publication No. PB 294986.

53
Cambridge Systematics, Inc., *Carpool Incentives: Analysis of Transportation and Energy Impacts*, Report No. FEA/D-76/391, Federal Energy Administration, Washington, DC, June 1976. US Government Printing Office Stock No. 041-018-00124-7.

54
Horowitz, J., unpublished modeling studies of travel behavior in the Washington, DC, area. (In these studies, the Washington area work trip mode choice model that is described in reference [53] was used to estimate the travel behavior of a sample of Washington area workers under different sets of travel conditions. The data used in forming the estimates were obtained from the Washington area transportation study.)

55
Peat, Marwick, Mitchell & Co., *Socioeconomic Impacts of Alternative Transportation Control Plans for the San Diego Air Quality Control Region*, US Environmental Protection Agency, Washington, DC, November 1974.

56
Adler, T., and Ben-Akiva, M., "A Theoretical and Empirical Model of Trip Chaining Behavior," *Transportation Research*, Vol. 13B, pp. 243–257, 1979.

57
Horowitz, J., "An Aggregate Supply Model for Urban Bus Transit," *Transportation Research Record*, No. 626, pp. 12–15, 1977.

58
Horowitz, J., *Transit Requirements for Achieving Large Reductions in Los Angeles Area Automobile Travel*, Report No. EPA/400-11/76-001, US Environmental Protection Agency, Washington, DC, November 1976. NTIS Publication No. PB80-199375.

59
American Institute of Planners and Motor Vehicle Manufacturers Association of the US, Inc., *Urban Transportation Factbook*, Part 1, Washington, DC, and Detroit, March 1974. (The population data presented in this document have been assembled from the 1970 US census.)

60

Charles River Associates, Inc., *Regional Management of Automotive Emissions: The Effectiveness of Alternative Policies for Los Angeles*, Report No. EPA-600/5-77-014, US Environmental Protection Agency, Washington, DC, December 1977. NTIS Publication No. PB 281213.

61

Bhatt, K., "Comparative Analysis of Urban Transportation Costs," *Transportation Research Record*, No. 559, pp. 101–116, 1976.

62

Englisher, L. S., and Sobel, K. L., "Methodology for the Analysis of Local Paratransit Options," *Transportation Research Record*, No. 650, pp. 18–24, 1977.

63

Altshuler, A., *The Urban Transportation System: Politics and Policy Innovation*, MIT Press, Cambridge, MA, 1979.

64

US Department of Transportation and US Environmental Protection Agency, *How to Prepare the Transportation Portion of Your State Air Quality Implementation Plan*, Washington, DC, November 1978.

65

Levinson, H. S., Hoey, W. F., Sanders, D. B., and Houston, F., *Bus Use of Highways: State of the Art*, Report No. 143, National Cooperative Highway Research Program, 1973.

66

US Department of Transportation, *Home-to-Work Trips and Travel*, Report No. 8, Nationwide Personal Transportation Study, April 1973. NTIS Publication No. PB 242892.

67

Dewees, D. N., "Travel Cost, Transit and Control of Urban Motoring," *Public Policy*, Vol. 24, pp. 59–79, Winter 1976.

68

Elliot, W., "The Los Angeles Affliction: Suggestions for a Cure," *The Public Interest*, No. 38, pp. 119–128, Winter 1975.

69

Horowitz, J., "Pricing the Use of the Automobile to Achieve Environmental and Energy Goals," in *Urban Transportation Economics*, Special Report 181, Transportation Research Board, Washington, DC, 1978.

70

Horowitz, J., unpublished modeling studies of travel behavior in the Washington, DC, area. (In these studies, the nonwork travel demand model described in reference [71] was used to estimate the nonwork travel behavior of a sample of Washington area households under different sets of travel conditions. The data used in forming the estimates were obtained from the Washington area transportation study.)

71
Horowitz, J., "A Utility-Maximizing Model of the Demand for Multi-Destination Non-Work Travel," *Transportation Research*, Vol. 14B, pp. 369–386, December 1980.

72
Fauth, G. R., Gomez-Ibanez, J. A., Howitt, A. M., Kain, J. F., and Wilkins, H. C., *Central Area Auto Restraint: A Boston Case Study*, Research Report R78-2, Department of City and Regional Planning, Harvard University, Cambridge, MA, December 1978. NTIS Publication No. PB 290913.

73
Horowitz, J., "Random Utility Models of Urban Non-Work Travel Demand: A Review," *Papers of the Regional Science Association*, Vol. 45, pp. 125–137, 1980.

74
Watson, P. L., and Holland, E. P., *Relieving Traffic Congestion: The Singapore Area License Scheme*, World Bank Staff Working Paper No. 281, The World Bank, Washington, DC, June 1978.

75
Hoel, L. A., and Roszner, E. S., *The Pittsburgh Parking Strike*, report prepared for the US Department of Transportation, December 1972. NTIS Publication No. PB 213798.

76
Vincent, R. A., and Layfield, R. E., *Nottingham Zones and Collar Study—Overall Assessment*, TRRL Laboratory Report 805, Transport and Road Research Laboratory, Crowthorne, England, 1977.

77
Ingram, G. K., and Pellechio, A., *Air Quality Impacts of Changes in Land Use Patterns: Some Simulation Results for Mobile Source Pollutants*, Discussion Paper D76-2, Department of City and Regional Planning, Harvard University, Cambridge, MA, January 1976.

78
Roberts, J. S., *Energy, Land Use and Growth Policy: Implications for Metropolitan Washington*, report prepared for the Metropolitan Washington Council of Governments, Washington, DC, June 1975.

79
Gomez-Ibanez, J. A., *Transportation Policy and Urban Land Use Control*, Discussion Paper D75-10, Department of City and Regional Planning, Harvard University, Cambridge, MA, November 1975.

80
Putman, S. H., *The Interrelationships of Transportation Development and Land Development*, Vol. 1, report prepared for the US Department of Transportation, Washington, DC, June 1973.

7 Atmospheric Dispersion[1]

7.1 Introduction

This chapter deals with the problem of developing quantitative models of the relations between emissions and atmospheric concentrations of air pollutants. The models are called atmospheric dispersion or diffusion models. Their main purpose is to enable predictions to be made of the changes in pollutant concentrations that would be caused by specified changes in emissions.

Suppose that pollutants are being emitted at various locations and at various rates in an area. Once emitted, the pollutants are dispersed by winds, and they may react together chemically to form new compounds or to deplete the emitted ones. The basic problem that will be addressed in this chapter is to express the concentration of each pollutant species at any given location and time in terms of

- the rates and locations of emissions,
- meteorological factors, such as wind speeds and directions,
- the rates of any chemical reactions that may occur.

Three main approaches to dispersion modeling will be discussed. The first approach consists of using the continuity equation of mathematical physics to develop a description of the physical and chemical processes that govern the relations between emissions and concentrations. This is often called the "eulerian approach" to dispersion modeling. The second approach consists of using a probabilistic description of the motions of pollutant particles in the atmosphere to derive expressions for pollutant concentrations. This is called the "lagrangian approach" to dispersion modeling. The third approach is statistically based. It consists of attempting to infer relations between pollutant emissions and concentrations from observations of changes in concentrations that occur when emissions or meteorological conditions change. The first approach provides a detailed representation of atmospheric physical and chemical processes. The most important application of this approach is in modeling the formation, destruction, and dispersion of chemically reactive pollutants, such as O_3 and its precursors. The second approach often yields models that are simpler computationally than models based on the first approach are, and it avoids certain approximations that restrict the ability of models based on the first approach to treat concentrations near strong emissions sources. However, the second approach does not readily accommodate complex chemical reactions, such as those involved in O_3 formation. Models based on the second approach are used frequently for estimating pollutant concentrations near roadways and for computing long-term (e.g., annual) average concentrations. The third approach is less highly developed than the other two are and is used less frequently in practice. However, under suitable conditions it offers the possibility of enabling accurate emissions-concentration relations to be developed with a minimum of mathematical complexity.

7.2 *The Continuity Equation and Its Application to Atmospheric Dispersion*[2]

The continuity equation is a mathematical statement of the law of conservation of mass. The equation states that the rate of increase of the mass of a pollutant in a small volume element of the atmosphere is equal to the sum of

- the net rate at which pollutant mass enters the volume element from surrounding volume elements due to transport by winds,
- the net rate at which pollutant mass diffuses into the volume element due to collisions with randomly moving air molecules (molecular diffusion),
- the net rate at which pollutant mass is created within the volume element by chemical reactions,
- the rate of pollutant emissions into the volume element from emissions sources.

The term "net rate" refers to the difference between the rate at which a process causes the mass of pollutant in a volume element to increase and the rate at which the same process causes the mass of pollutant in the volume element to decrease. For example, the wind moves pollutant mass into a volume element from its surroundings, thereby tending to cause the mass of pollutant in the volume element to increase. However, the wind also moves pollutant mass out of the volume element and into the surroundings, thereby tending to cause the mass of pollutant in the volume element to decrease.

Let t denote time and let $\mathbf{x} = (x_1, x_2, x_3)$ be a vector that specifies the three-dimensional, rectangular coordinates of a point in space. In addition, define the following symbols:

$C_i(\mathbf{x}, t)$ = the concentration of pollutant species i at the point $\mathbf{x}$ and time t.

$u_j(\mathbf{x}, t)$ = the component of the wind velocity in direction j ($j = 1, 2, 3$) at location $\mathbf{x}$ and time t. The directions j correspond to the coordinate directions x_1, x_2, and x_3.

D_i = the molecular diffusivity of pollutant species i. D_i is a constant that is related to the rate at which the concentration of species i changes due to molecular diffusion.

$R_i(C_1, C_2, \ldots, C_n)$ = the rate at which the concentration of pollutant species i changes due to chemical reactions when the concentrations of all reacting species are given by $C_1, C_2, \ldots, C_n$.

$S_i(\mathbf{x}, t)$ = the rate at which the concentration of species i at point $\mathbf{x}$ and time t is changing due to emissions from sources at point $\mathbf{x}$ and time t.

Then the continuity equation for pollutant species i is

$$\frac{\partial C_i}{\partial t} + \sum_j \frac{\partial}{\partial x_j}(u_j C_i) = D_i \sum_j \frac{\partial^2 C_i}{\partial x_j^2} + R_i + S_i. \tag{7.1}$$

The first term on the left is the net rate of change of the concentration of pollutant i at point $\mathbf{x}$ and time t. The second term on the left is minus the rate of change of the concentration of i due to transport by winds. The first term on the right is the rate of change of the concentration of i due to molecular diffusion, and the second and third terms on the right are the rates of change of the concentration of i due to chemical reactions and emissions from sources, respectively. If the point $\mathbf{x}$ is considered to be surrounded by a small volume element of size δV, then the terms of equation (7.1) multiplied by δV correspond to the rates of change of pollutant mass described in the previous paragraph.

When several pollutant species are present, the dispersion of each is described by an equation of the form (7.1). If the pollutants react together chemically, then the equations describing the dispersion of each species are coupled together through the chemical reaction terms R_i. Equation (7.1) then becomes a system of coupled, simultaneous, partial differential equations for the concentrations of the various pollutants.

In principle, equation (7.1) completely describes the dispersion of pollutants in the atmosphere. If the wind velocity components $u_j(\mathbf{x}, t)$, chemical reaction rates R_i, and emissions rates S_i are known, then this equation can be solved to yield the pollutant concentrations as functions of time at each point in space.[3] However, in practice the necessary knowledge of the wind velocity and emissions rates cannot be obtained. There are at least two reasons for this. First, the atmosphere is turbulent, meaning that the wind velocity fluctuates rapidly and unpredictably between spatial locations and instants of time. It is not possible to measure all of these velocity fluctuations. Second, emissions rates fluctuate in ways that are not completely predictable or measurable owing to factors such as unpredictable and (for practical purposes) unobservable fluctuations in traffic flow conditions, vehicle operating conditions, and operating conditions of stationary sources. Because it is not possible to obtain complete information on the wind velocity and emissions rates, equation (7.1) cannot be solved for pollutant concentrations in the real atmosphere. Moreover, since fluctuations in the wind velocity and emissions rates are transmitted to pollutant concentrations through equation (7.1), pollutant concentrations themselves are not completely predictable or measurable.

The problem posed by the unavailability of complete information on wind velocity and emissions rates can be dealt with in the following way. The wind velocity and emissions rates can be considered to be sums of measurable or predictable components and unmeasurable, unpredictable ones. The former components are called "deterministic," and the latter are called "random." For example, suppose that a measurement of the time-averaged wind velocity is available for a certain time period and location but that measurements of the instantaneous velocity are not available. Then, for that time period and location, the deterministic component of the wind velocity is the time-averaged velocity. The random component is the difference between the instantaneous (but unknown) and time-averaged velocities. Similar concepts apply to emissions rates. Since the deterministic components of the wind velocity and emissions rates are, by definition, either measurable or predictable, an equation that related suitably defined deterministic components of pollutant concentrations to the deterministic components of the wind velocity and emissions rates could be solved. It is convenient for this purpose to define the deterministic components of pollutant concentrations as ensemble average concentrations. (Ensemble averages consist of averages over repeated observations that are made under identical measurable conditions.) The random components of pollutant concentrations are then the differences between the instantaneous and ensemble average concentrations. As will be discussed in the following paragraphs, an equation relating ensemble average pollutant concentrations to the deterministic components of the wind velocity and emissions rates can be derived from equation (7.1) if suitable approximations are made.

To begin the derivation of the equation, define the following notation:

$\langle C_i \rangle$ = the deterministic component of C_i.
C_i' = the random component of C_i.
$\bar{u}_j$ = the deterministic component of u_j.
u_j' = the random component of u_j.
$\bar{S}_i$ = the deterministic component of S_i.
S_i' = the random component of S_i.

The notation $\langle \; \rangle$ is used to denote ensemble averaging. A bar over a symbol denotes averaging over time and, possibly, space, as occurs when measurements are made. The instantaneous pollutant concentrations, wind velocity, and emissions rates are related to their respective deterministic and random components by

$$C_i = \langle C_i \rangle + C_i', \tag{7.2a}$$

$$u_j = \bar{u}_j + u_j', \tag{7.2b}$$

$$S_i = \bar{S}_i + S_i'. \tag{7.2c}$$

The random components of pollutant concentrations satisfy

$$\langle C_i' \rangle = 0 \tag{7.3}$$

by definition. In addition, it is assumed that

$$\langle u_j' \rangle = 0, \tag{7.4a}$$

$$\langle S_i' \rangle = 0. \tag{7.4b}$$

If equations (7.2) are substituted into equation (7.1), the following equation is obtained:

$$\begin{aligned} &\frac{\partial C_i}{\partial t} + \frac{\partial C_i'}{\partial t} + \sum_j \frac{\partial}{\partial x_j} [(\bar{u}_j + u_j')(\langle C_i \rangle + C_i')] \\ &= D_i \sum_j \frac{\partial^2 \langle C_i \rangle}{\partial x_j^2} + D_i \sum_j \frac{\partial^2 C_i'}{\partial x_j^2} \\ &\quad + R_i(\langle C_1 \rangle + C_1', \ldots, \langle C_n \rangle + C_n') + \bar{S}_i + S_i'. \end{aligned} \tag{7.5}$$

Taking ensemble averages on both sides of this equation and making use of equations (7.3) and (7.4) yields

$$\begin{aligned} &\frac{\partial \langle C_i \rangle}{\partial t} + \sum_j \frac{\partial}{\partial x_j} (\bar{u}_j \langle C_i \rangle) + \sum_j \frac{\partial}{\partial x_j} \langle u_j' C_i' \rangle \\ &= D_i \sum_j \frac{\partial^2 \langle C_i \rangle}{\partial x_j^2} + \langle R_i(\langle C_1 \rangle + C_1', \ldots, \langle C_n \rangle + C_n') \rangle + \bar{S}_i. \end{aligned} \tag{7.6}$$

This equation contains no random terms. However, it contains a large number of unknown quantities. If the pollutants in question do not react together chemically, then for each pollutant species i, there are four unknown quantities: $\langle C_i \rangle$ and $\langle u_j' C_i' \rangle$ $(j = 1, 2, 3)$. If chemical reactions occur, then R_i typically is a nonlinear function, so that $\langle R_i \rangle$ depends not only on the $\langle C_i \rangle$ but also on terms such as $\langle C_i' C_j' \rangle$ $(i, j = 1, \ldots, n)$. The latter terms thus become unknown quantities in equation (7.6), in addition to the four previously identified unknown quantities. Hence, for each pollutant species i, equation (7.6) contains at least four unknown quantities, depending on whether chemical reactions occur. Since there are more unknown quantities than there are equations, the system of equations represented by (7.6) cannot be solved.

The problem of having more unknown quantities than equations in the system of equations represented by (7.6) is known as the closure problem of turbulent dispersion. This problem usually is dealt with by making two approximations. The first approximation, which is called the mixing length approximation, consists of assuming that the terms $\langle u_j' C_i' \rangle$ can be written in the form

$$\langle u_j' C_i' \rangle = -\sum_k K_{jk} \frac{\partial \langle C_i \rangle}{\partial x_k}, \tag{7.7}$$

where the K_{jk} are a set of functions of space and, possibly, time called eddy diffusivities. The eddy diffusivities can be evaluated using meteorological theory and measurements of atmospheric turbulence. Equation (7.7) amounts to representing the effects of turbulence on pollutant transport by winds as a diffusion process. The second approximation consists of assuming that the effects of the random fluctuation terms in $\langle R_i \rangle$ are negligible. In other words, it is assumed that $\langle R_i(\langle C_1 \rangle + C_1', \ldots, \langle R_n \rangle + C_n') \rangle$ can be replaced by $R_i(\langle C_1 \rangle, \ldots, \langle C_n \rangle)$. Given these approximations, equation (7.6) becomes

$$\begin{aligned} \frac{\partial \langle C_i \rangle}{\partial t} + \sum_j \frac{\partial}{\partial x_j} (\bar{u}_j \langle C_i \rangle) = {} & \sum_j \sum_k \frac{\partial}{\partial x_j} \left(K_{jk} \frac{\partial \langle C_i \rangle}{\partial x_k} \right) \\ & + D_i \sum_j \frac{\partial^2 \langle C_i \rangle}{\partial x_j^2} + R_i(\langle C_1 \rangle, \ldots, \langle C_n \rangle) \\ & + \bar{S}_i. \end{aligned} \tag{7.8}$$

The molecular diffusion term in this equation typically is assumed to be negligible in comparison to the other terms and is dropped. The eddy diffusivities K_{jk} usually are assumed to be nonzero only if $j = k$. In addition, it is usually assumed that the atmosphere can be treated as an incompressible fluid, so that $\sum_j \partial \bar{u}_j / \partial x_j = 0$. Given these assumptions and approximations, equation (7.8) becomes

$$\frac{\partial \langle C_i \rangle}{\partial t} + \sum_j \overline{u}_j \frac{\partial \langle C_i \rangle}{\partial x_j} = \sum_j \frac{\partial}{\partial x_j}\left(K_{jj}\frac{\partial \langle C_i \rangle}{\partial x_j}\right) + R_i(\langle C_1 \rangle, \ldots, \langle C_n \rangle) + \overline{S}_i(\mathbf{x}, t). \tag{7.9}$$

Equation (7.9) is called the semiempirical equation of atmospheric diffusion. It contains only the single unknown quantity $\langle C_i \rangle$ and, hence, can be solved if the chemical reaction rates and the deterministic components of the wind velocity and emissions rates are known. All of these quantities can be measured or approximated. Although equation (7.9) has the same general structure as the continuity equation (7.1), equation (7.9) is not a rigorous statement of the law of conservation of mass as equation (7.1) is. Rather, equation (7.9) has been derived from equation (7.1) by making a series of approximations whose accuracy must now be investigated.

The most important approximations that have been made in deriving equation (7.9) are

- the effects of random concentration fluctuations on chemical reaction rates are ignored;
- atmospheric turbulence is treated as a diffusion process by means of the mixing length approximation.

In addition, it is important to note that equation (7.9) expresses pollutant concentrations in terms of ensemble averages. Since these correspond to averages of repeated concentration observations that are made under identical measured conditions, it is not possible to measure them in practice. Rather, concentration measurements made in practice correspond to averages over time and, possibly, space. Therefore, in order to use equation (7.9) to compute measured concentrations, it is necessary to assume that spatial and temporal averages of concentrations are closely approximated by ensemble averages.

The first approximation listed above is valid if the chemical reactions among air pollutants are sufficiently slow not to be affected substantially by concentration fluctuations caused by atmospheric turbulence. In other words, the time scales of the chemical reactions must be long compared to the time scale of the turbulent fluctuations. Using a somewhat less intuitive set of arguments, it can be shown that the mixing length approximation is valid if the time and distance scales associated with significant changes in ensemble average concentrations are large compared to the time and distance scales associated with atmospheric turbulence [1, 5]. Finally, the approximation that ensemble average concentrations are equivalent to spatially and temporally averaged concentrations is valid if the distances and times used in computing spatially and temporally averaged concentrations are sufficiently large and if the ensemble average concentrations vary little over these times and distances. Thus, both the mixing length approximation and the assumption that ensemble averages are good approximations of spatial-temporal averages require that ensemble average concentrations (and, therefore, spatially and temporally averaged concentrations) vary gradually over space and time.

The condition that ensemble and spatial-temporal average concentrations vary gradually over space is likely to be violated frequently near large emissions sources. Accordingly, equation (7.9) is most likely to be valid away from such sources. Simulation studies of the turbulent diffusion of air pollutants have tended to confirm this conclusion [6].

Air pollutant dispersion models based on equation (7.9) are often called "eulerian" or "grid" models. The term "eulerian" means that pollutant concentrations at a given location and time are derived from consideration of the wind velocity, emissions rates, and pollutant concentrations at or near that location and time. The term "grid model" arises from the method used to solve equation (7.9). It is not possible in practice to solve this equation analytically in terms of continuous spatial and temporal variables $\mathbf{x}$ and t. To achieve an approximate solution to the equation, the geographical region being studied is divided into discrete grid cells, and time is divided into discrete, nonoverlapping intervals. Concentrations are treated as being constant within grid cells and time intervals, and numerical methods are used to approximate the continuous space and time solution to equation (7.9) with concentrations within grid cells and time intervals.

7.3 Implementation of Models Based on the Semiempirical Diffusion Equation

Air pollutant dispersion models based on equation (7.9) are used mainly for studying the formation, destruction, and dispersion of O_3 and other constituents of photochemical smog, including estimating the changes in O_3 and NO_2 concentrations that would result from changes in HC and NO_x emissions. Several computerized models of photochemical smog formation, based on equation (7.9), have been developed and implemented [7–12]. Models based on equation (7.9) also have been proposed for computing CO concentrations near highways [13–16].

However, these models are rarely used in practice. Rather, computations of CO concentrations near highways usually are carried out with lagrangian models that are computationally simpler than models based on equation (7.9) are. Lagrangian models are discussed in sections 7.5–7.7. (Lagrangian models cannot be used for modeling photochemical smog since they cannot treat the complex chemical reactions involved in smog formation.) Apart from computational complexity, CO models based on equation (7.9) have the disadvantage that large changes in CO concentrations can take place over very short distances near high-volume roadways. When this occurs, the validity of the mixing length approximation used in deriving equation (7.9) cannot be guaranteed.

Dispersion models based on equation (7.9), or eulerian grid models, provide more realistic descriptions of atmospheric physics and chemistry and are sensitive to a wider range of emissions control policies than are other currently available dispersion models for photochemical smog. (Other models are discussed in sections 7.4 and 7.8.) In addition to providing means for estimating the effects on concentrations of changes in aggregate regional emissions levels, eulerian grid models can be used for estimating the effects of changes in the spatial distribution, distribution according to time of day, and chemical composition of emissions.

Eulerian grid models of photochemical smog require substantial quantities of input data. These data include the following:

- A spatially and temporally resolved emissions inventory for each pollutant species represented in the model. This usually means that the emissions inventory must specify emissions according to the grid cell and time interval in which they occur. The size of a grid cell typically is in the range 2×2–10×10 km^2, depending on the model, and each time interval typically is 1 hr long. In addition to specifying the distribution of emissions according to grid cell and time interval, it is necessary to disaggregate HC emissions into several different chemical categories of hydrocarbons and to disaggregate NO_x emissions into NO and NO_2.
- Meteorological data. The wind velocity must be estimated for each grid cell and time interval. Since it is not feasible to measure the wind velocity with this degree of spatial and temporal detail, wind velocity data usually are estimated by interpolating velocities among measurement sites. Nonetheless, it usually is necessary to carry out special wind velocity measurements in order to develop a sufficiently dense set of measured values to enable the interpolation to be carried out. Measurements of wind velocity are needed aloft as well as at ground level. In addition to the wind velocity, it is necessary to measure the mixing height (the maximum vertical distance through which pollutants freely disperse) and the intensity of solar radiation. The latter information is

needed because of the importance of solar radiation in driving the chemical reactions that lead to formation of photochemical smog. Other meteorological data that can be required include the air temperature as a function of height above ground, air pressure, and relative humidity.

- Initial and boundary conditions. The formation and dispersion of air pollutants can be modeled only for relatively short time periods (typically less than one full day) and within bounded geographical areas (e.g., a 100 × 100-km^2 area centered on a large city). However, emissions that occur prior to the time period being modeled (e.g., during the hours immediately preceding the modeled period and, under stagnant meteorological conditions, during the previous day) can cause substantial pollutant concentrations to be present at the beginning of this time period. These initial concentrations can significantly influence pollutant concentrations during the modeled time period. Therefore, the input data to the model must include the pollutant concentration in each grid cell at the beginning of the time period being modeled (initial conditions). Similarly, pollutants that are transported into the modeled geographical area by winds can have significant effects on pollutant concentrations in this area. To account for these effects, it is necessary to specify the pollutant concentrations on the upwind boundaries of the modeled area as functions of location and time. The specified concentrations are called boundary conditions.[4] When pollutant concentrations in current or past years are being computed, it is possible to estimate initial and boundary concentrations by interpolating concentration values between measurement sites. Frequently, however, special measurement programs are needed to obtain sufficiently dense sets of measurements and to disaggregate concentrations into hydrocarbon categories, NO, and NO_2. When concentrations are being predicted for a future year, specification of initial and boundary concentrations can present serious difficulties. This is because the initial and boundary concentrations depend on emissions in future years, but this dependence is not treated by the dispersion model. Therefore, it is often necessary to use plausible but unverified assumptions to develop initial and boundary concentrations for future years. For example, initial concentrations might be assumed to change in proportion to emissions levels between the current year and the future year for which predictions are being made. Boundary concentrations might be set at background levels on the assumption that any upwind communities will reduce emissions sufficiently to prevent transported pollutants from significantly affecting concentrations in the modeled area.

Concentration estimates obtained from grid dispersion models can be highly sensitive to errors in input data. Sensitivity studies carried out with one such model have indicated that variations in wind velocities and mixing heights over plausible ranges of uncertainty can cause changes in computed regional maximum O_3 concentrations of more than 30% [7, 12]. A sensitivity study carried out with another model compared the O_3

concentrations that were computed when emissions were distributed among grid cells in proportion to the residential populations of the cells (a plausible shortcut approach to distributing emissions) with the concentrations that were computed when emissions were distributed in accordance with detailed estimates of emissions levels within each cell. Depending on aggregate emissions levels, the computed regional maximum O_3 concentration was 15–25% lower when the population distribution of emissions was used than when the detailed distribution was used [17]. These results show that it is necessary to use considerable care in preparing input data for grid dispersion models. In typical applications, data preparation can require several months of effort and cost several hundred thousands of dollars [18]. Detailed discussions of data preparation for eulerian grid dispersion models of photochemical smog are available in references [3, 4, 7, 18–21].

The accuracy of eulerian grid dispersion models of photochemical smog can be investigated by comparing pollutant concentrations computed by the models with concentrations measured in urban air. For most practical air quality planning purposes, the most important outputs of these investigations are comparisons of the computed and measured values of the maximum 1-hr average O_3 and NO_2 concentrations that occur during a day at any of the locations at which measurements are available.[5] Table 7.1 shows computed and measured maximum O_3 concentrations for several cities and computed and measured maximum NO_2 concentrations for one city. In all of the tabulated cases, the differences between the computed and measured concentrations are less than 40%, and in most cases they are less than 20%. These differences reflect errors both in the dispersion models themselves and in input data to the models. It is worth noting that the errors in concentrations computed at individual measurement sites are considerably larger than are the errors in computed maximum concentrations over all measurement sites [7]. Hence, from a practical point of view, the current generation of eulerian grid dispersion models of photochemical smog is more useful for studying the effects of emissions changes on regional maximum pollutant concentrations than for studying the effects of emissions changes on concentrations at specific locations.

7.4 *Simplified Models Based on the Semiempirical Diffusion Equation*

Eulerian grid dispersion models of photochemical smog are complex and difficult to use, owing primarily to the large quantities of detailed input data that they require and, secondarily, to the need to solve the multi-

Table 7.1 Comparison of maximum 1-hr average O_3 and NO_2 concentrations observed at urban monitoring sites with maximum concentrations computed with an eulerian grid dispersion model[a]

City	Day	Maximum 1-hr average O_3 concentration (pphm)			Maximum 1-hr average NO_2 concentration (pphm)		
		Observed	Computed	Percent difference[b]	Observed	Computed	Percent difference[b]
Los Angeles	26 June 1974	34	47	+38	26	18	−31
	27 June 1974	49	58	+18	36	32	−11
	4 August 1975	32	40	+25	19	16	−16
Sacramento	28 June 1976	16	17	+6			
	24 August 1976	13	11	−15			
Denver	29 July 1975	11	13	+18			
	28 July 1976	16	15	−6			

a. Source: reference [7]. Blanks signify information not available.
b. Percent difference computed as 100(computed maximum − observed maximum)/(observed maximum).

dimensional system of partial differential equations represented by equation (7.9). Dispersion models less complex than eulerian grid models can be obtained by making further simplifying approximations and assumptions beyond those already made in deriving equation (7.9). The resulting simplified models cannot be expected to have the accuracy, flexibility, or generality of eulerian grid models, but the simplified models are, nonetheless, useful for many purposes.

One type of simplification consists of assuming that the air above a city can be divided into nonoverlapping, vertical columns that move intact and without distortion along trajectories that are determined by the wind speed and direction (see figure 7.1). It is then possible to model the dispersion of pollutants in individual air columns, rather than having to deal with dispersion in the entire mass of air over a city. To insure that air columns remain intact, the following assumptions must be made:

- There is no horizontal diffusion of pollutants across column boundaries. This assumption is needed because horizontal diffusion causes the air

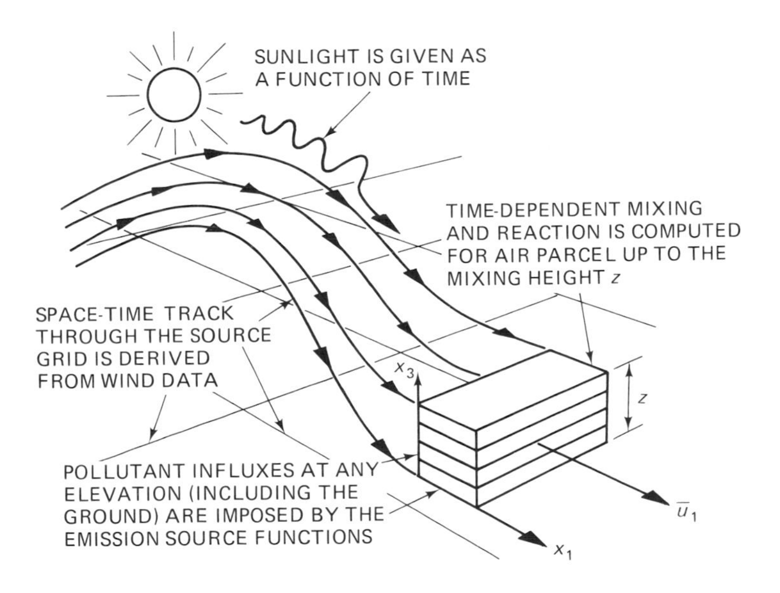

Figure 7.1 Schematic diagram of a dispersion model for a moving air column. (Adapted with permission from reference [22]: Eschenroeder, A. Q., and Martinez, J. R., in *Photochemical Smog and Ozone Reactions,* R. F. Gould, ed., *Advances in Chemistry Series,* No. 113, American Chemical Society, Washington, DC, 1972, p. 108.)

in a vertical column to mix with the surrounding air so that the column loses its identity. Mathematically, the assumption is implemented by setting the horizontal eddy diffusivity components equal to zero. It is assumed that within air columns the horizontal distribution of pollutants is uniform.

- The wind velocity does not vary with height. This assumption is needed because if the wind velocity varies with height, the speed and direction of movement of the air in a vertical column will also vary with height. Consequently, layers of the column that are at different heights will move with different speeds and in different directions, and the column will not remain intact.

In addition, it is usually assumed that the vertical component of the wind velocity is negligibly small.[6]

Given these assumptions, a differential equation governing the formation, destruction, and dispersion of pollutants within a vertical air column can be derived from equation (7.9) as follows: Let x_1 and x_2 in equation (7.9) correspond to the horizontal coordinate axes, and let x_3 correspond to the vertical axis. Then the foregoing assumptions imply that $K_{11} = K_{22} = \bar{u}_3 = 0$. Thus, equation (7.9) becomes

$$\frac{\partial\langle C_i\rangle}{\partial t} + \bar{u}_1 \frac{\partial\langle C_i\rangle}{\partial x_1} + \bar{u}_2 \frac{\partial\langle C_i\rangle}{\partial x_2} = \frac{\partial}{\partial x_3}\left(K_{33}\frac{\partial\langle C_i\rangle}{\partial x_3}\right) + R_i(\langle C_1\rangle, \ldots, \langle C_n\rangle) + \bar{S}_i(\mathbf{x}, t). \tag{7.10}$$

Equation (7.10) applies to any air mass for which the simplifying assumptions on which it is based are valid. Let a vertical air column be such an air mass. In addition, at any instant of time, let the coordinate axes be oriented so that the x_1 axis is parallel to the wind direction at the location of the air column. This causes the component of the wind velocity in the x_2 direction to be zero. Hence, equation (7.10) can be written as

$$\frac{\partial\langle C_i\rangle}{\partial t} + \bar{u}_1 \frac{\partial\langle C_i\rangle}{\partial x_1} = \frac{\partial}{\partial x_3}\left(K_{33}\frac{\partial\langle C_i\rangle}{\partial x_3}\right) + R_i + \bar{S}_i. \tag{7.11}$$

Finally, transform equation (7.11) to a coordinate system that moves in the x_1 direction with speed $\bar{u}_1$. In other words, the equation is transformed to a coordinate system that moves with the air column that is being modeled. Then, equation (7.11) becomes

$$\frac{\partial\langle C_i\rangle}{\partial t} = \frac{\partial}{\partial x_3}\left(K_{33}\frac{\partial\langle C_i\rangle}{\partial x_3}\right) + R_i(\langle C_1\rangle, \ldots, \langle C_n\rangle) + \bar{S}_i(x_3, t). \tag{7.12}$$

Equation (7.12) describes the formation, destruction, and dispersion of pollutants within a single vertical column of air, subject to the simplifying assumptions that were made in deriving this equation. The equation is called a "trajectory model" since it describes pollutant concentrations along the trajectory that a vertical air column follows as it is moved by winds. Operational dispersion models based upon equation (7.12) are described in references [7, 22–24].

The simplifying assumptions used in deriving equation (7.12) from equation (7.9) are most likely to be valid over smooth terrain with moderate to high wind speeds and relatively smooth distributions of emissions sources [23]. Investigations of the effects of violations of the simplifying assumptions have suggested that neglect of horizontal diffusion is not likely to cause errors in computed concentrations of more than roughly 10% [25]. However, violation of the assumption that the wind velocity is independent of height can cause errors in computed concentrations of more than 50%, and neglect of vertical winds can cause errors of several hundred percent [25]. Thus, the results obtained from trajectory models based upon equation (7.12) can be highly erroneous under conditions in which the simplifying assumptions regarding wind velocities are violated.

Apart from the effects of the simplifying assumptions upon which equation (7.12) is based, trajectory models are less general than eulerian grid models are because the former class of models computes pollutant concentrations only along the trajectory followed by an air column, whereas the latter class of models computes concentrations throughout an urban area. Although the geographical coverage of trajectory models can be expanded by computing concentrations along many trajectories, performing these computations for sufficiently many trajectories to develop estimates of pollutant concentrations throughout entire urban areas is not likely to be feasible in practice. Thus, trajectory models are likely to be useful mainly for estimating concentrations downwind of major emissions sources (e.g., to enable the effects of changes in the sources' emissions rates to be estimated), rather than for regional air quality studies. In contrast, eulerian grid models can be used both for estimating concentrations downwind of major sources and for regional studies. However, trajectory models require considerably less input data than eulerian grid models do since the inputs to trajectory models need describe emissions and initial conditions only for the trajectories of the air columns being modeled, rather than throughout an entire urban area. Moreover, boundary conditions for trajectory models need to be specified only at the tops of the air columns that are modeled since horizontal movement of pollutants into the columns is precluded by the simplifying assumptions used in deriving the models.

A limited test of the accuracy of maximum O_3 concentrations computed with a trajectory model has been carried out using three trajectories in St. Louis, Missouri [23]. The maximum O_3 concentration along each trajectory as computed with the model was compared to an estimate of the true maximum O_3 concentration that was obtained by interpolating measured concentrations between measurement sites. The model underestimated the maximum O_3 concentrations along the three trajectories by 20–30% [23]. As in the case of the table 7.1 estimates of the accuracy of an eulerian grid model, the results obtained with the trajectory model reflect errors in the input data for the model, as well as errors in the structure of the model itself.

Trajectory models simpler than those based on equation (7.12) can be obtained by making the approximation that vertical dispersion is instantaneous. This causes pollutant concentrations to be uniform throughout the air column being modeled. When this simplification is made, the concentration in the column can change due to changes in the mixing height, emissions into the column, and chemical reactions that take place in the column. A differential equation describing the concentrations of pollutants in the column under these conditions can be derived as follows.

Let $C_i(t)$ denote the concentration of pollutant i in the column at time t; A, the cross-sectional area of the column; and $z(t)$, the mixing height (which is the same as the height of the column) at time t. In addition, let $C_{i,\mathrm{top}}(t)$ denote the concentration of pollutant i just above the top of the column at time t; $S_i(t)$, the flux of pollutant i into the column from emissions sources (expressed in units of pollutant mass per unit area per unit time); and $R_i(C_1, \ldots, C_n)$, the time rate of change of the concentration of pollutant i due to chemical reactions.[7] At time t, the volume of the air column is $Az(t)$, and the mass of pollutant i in the column is $Az(t)C_i(t)$. Let Δt denote a very short interval of time. Suppose that by time $t + \Delta t$ the height of the column has increased by an amount Δz to $z + \Delta z$. Then the volume of the column is now $A[z(t) + \Delta z]$. The increase in the height of the column increases the mass of pollutant i in the column by an amount $A\,\Delta z\,C_{i,\mathrm{top}}(t)$.[8] In addition, emissions during the time interval t to $t + \Delta t$ increase the mass of pollutant i in the column by an amount $AS_i(t)\,\Delta t$, and chemical reactions during this time interval change the mass of pollutant i in the column by an amount $R_i Az(t)\,\Delta t$. The change in the total mass of pollutant i in the column is the sum of the mass changes due to the change in the mixing height, emissions, and chemical reactions. Hence, at time $t + \Delta t$, the total mass of pollutant i in the column is

$$\text{Mass}_i(t + \Delta t) = Az(t)C_i(t) + A\,\Delta z\,C_{i,\text{top}}(t) + AS_i(t)\,\Delta t + R_iAz(t)\,\Delta t. \tag{7.13}$$

The concentration of pollutant i in the column at time $t + \Delta t$ is the mass of pollutant i in the column at time $t + \Delta t$ divided by the volume of the column at that time, or

$$C_i(t + \Delta t) = \frac{Az(t)C_i(t) + A\,\Delta z\,C_{i,\text{top}}(t) + AS_i(t)\,\Delta t + R_iAz(t)\,\Delta t}{A[z(t) + \Delta z]}. \tag{7.14}$$

The change in the concentration of pollutant i during the time interval t to $t + \Delta t$ is

$$\Delta C_i = C_i(t + \Delta t) - C_i(t), \tag{7.15a}$$

or

$$\Delta C_i = \frac{[C_{i,\text{top}}(t) - C_i(t)]\,\Delta z + S_i(t)\,\Delta t + R_iz(t)\,\Delta t}{z(t) + \Delta z}. \tag{7.15b}$$

Dividing equation (7.15b) by Δt and taking the limit as $\Delta t \rightarrow 0$, the following differential equation for $C_i(t)$ is obtained:

$$\frac{dC_i}{dt} = \frac{1}{z}\frac{dz}{dt}(C_{i,\text{top}} - C_i) + \frac{S_i}{z} + R_i. \tag{7.16}$$

Equation (7.16) constitutes a dispersion model for pollutant i in the air column, subject to the simplifying assumption that pollutant concentrations are uniform throughout the column and provided that the mixing height increases with time or is constant. If the mixing height decreases over time, then changes in the mixing height do not cause changes in the concentrations of pollutants in the air column. Hence, equation (7.16) must be modified by dropping the first term on its right-hand side. In practice, mixing heights usually increase from morning to afternoon, which is the time period of greatest interest in studies of photochemical smog. Hence, equation (7.16) as given is applicable to most situations occurring in practice. Operational dispersion models based on equation (7.16) are described in references [7, 26, 27].

Another way to simplify equation (7.9) is to treat the air between the ground and the top of the mixing layer as a large box in which pollutants disperse instantaneously. Pollutant concentrations are then uniform throughout the box. Concentrations in the box can change due to transport of pollutants into the box from its surroundings or out of the box to the surroundings, increases in the mixing height, emissions into the box, and chemical reactions that take place within the box (see figure 7.2). This approach leads to a simplified dispersion model that describes

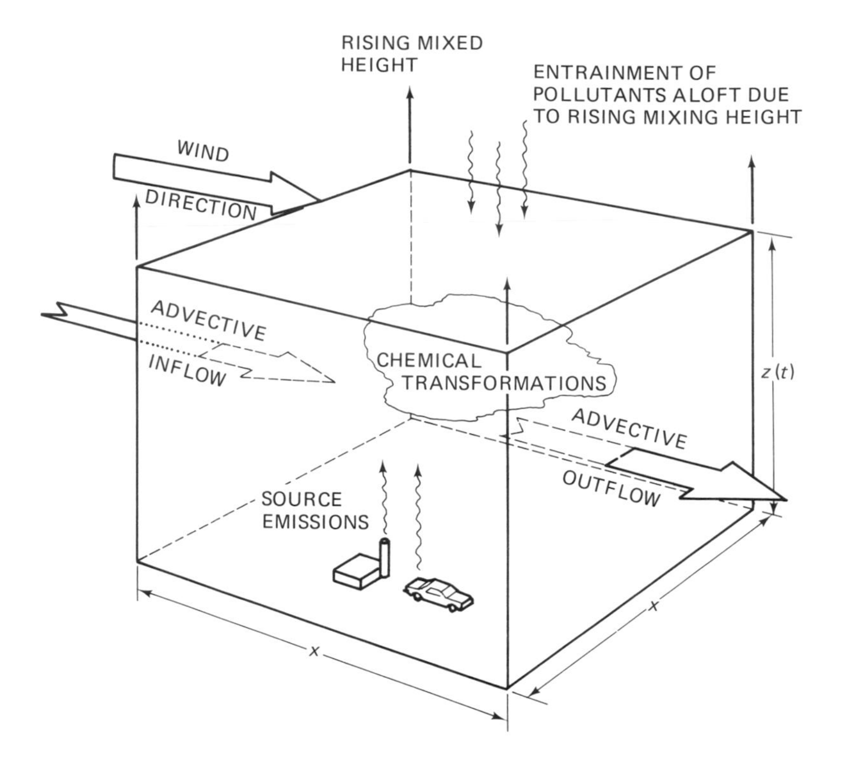

Figure 7.2 Schematic diagram of processes affecting pollutant concentrations in a box. (Adapted with permission from reference [30])

pollutant concentrations over an entire urban area rather than only within a single air column.

A differential equation that describes pollutant concentrations in the box can be derived as follows. The time rate of change of C_i, the concentration of pollutant i in the box, is equal to the sum of the time rates of change of C_i due to the four processes that can change pollutant concentrations in the box. The time rates of change of C_i due to increases in the mixing height, emissions, and chemical reactions are given by the terms on the right-hand side of equation (7.16). Hence, an equation for the time rate of change of C_i can be obtained by adding to the right-hand side of equation (7.16) a term that gives the time rate of change of C_i due to transport of pollutant i into and out of the box. This term can be derived in the following way.

Assume as in figure 7.2 that the box has a square horizontal cross section and that the wind direction is perpendicular to one side of the box. Also assume that the wind velocity is the same everywhere in the box. Let u be the average wind speed in the box; $x = \sqrt{A}$, the width of the box; and $C_{i,b}(t)$, the concentration of pollutant i just outside of the upwind boundary of the box. Then during the time interval t to $t + \Delta t$, a volume of air equal to $xz(t)u\,\Delta t$ is transported into the box, and an equal volume of air is transported out. Therefore, a mass of pollutant i equal to $C_{i,b}(t)\,xz(t)\,u\,\Delta t$ is transported into the box, and a mass of pollutant i equal to $C_i(t)\,xz(t)\,u\,\Delta t$ is transported out. The change in the mass of pollutant i in the box due to this transport is $[C_{i,b}(t) - C_i(t)]\,xz(t)\,u\,\Delta t$. Since the four processes that change pollutant concentrations in the box operate independently, the time rates of change of the concentration of pollutant i due to each process can be computed as if the other processes were not present. Therefore, assume that transport is the only process operating to change pollutant concentrations in the box. Then the mass of pollutant i in the box at time $t + \Delta t$ is

$$\text{Mass}_i(t + \Delta t) = Az(t)C_i(t) + [C_{i,b}(t) - C_i(t)]\,xz(t)\,u\,\Delta t. \tag{7.17}$$

The concentration of pollutant i at time $t + \Delta t$ is

$$C_i(t + \Delta t) = \frac{Az(t)C_i(t) + [C_{i,b}(t) - C_i(t)]\,xz(t)\,u\,\Delta t}{Az(t)}. \tag{7.18}$$

Therefore, the change in the concentration of pollutant i due to transport during the time interval t to $t + \Delta t$ is

$$(\Delta C_i)_{\text{transport}} = [C_{i,b}(t) - C_i(t)]\,u\,\Delta t/x. \tag{7.19}$$

Dividing equation (7.19) by Δt and taking the limit as $\Delta t \to 0$, the following equation is obtained for the time rate of change of the concentration of pollutant i due to transport:

$$\left(\frac{dC_i}{dt}\right)_{\text{transport}} = (C_{i,b} - C_i)\,u/x. \tag{7.20}$$

A differential equation for the concentration of pollutant i in the box can now be obtained by adding the right-hand side of equation (7.20) to the right-hand side of equation (7.16). The resulting equation is

$$\frac{dC_i}{dt} = \frac{1}{x}(C_{i,b} - C_i)\,u + \frac{1}{z}\frac{dz}{dt}(C_{i,\text{top}} - C_i) + \frac{S_i}{z} + R_i. \tag{7.21}$$

This equation constitutes a dispersion model for pollutant i in the box of air over a city, subject to the simplifying assumptions that the average wind velocity and the concentrations of pollutants are uniform throughout the box, and provided that the mixing height is constant or increases over time. If the the mixing height decreases over time, then equation (7.21) must be modified by dropping the second term on its right-hand side since decreases in the mixing height do not cause pollutant concentrations in the box to change. Equation (7.21) is frequently called a "box model" of dispersion.[9]

Box models require the same general classes of input data as eulerian grid dispersion models require. However, box models do not require the data to be spatially disaggregated. They operate from regional emissions levels (disaggregated according to time of day and chemical species), regional average wind speeds (according to time of day), and regional average initial and boundary concentrations. Thus, box models achieve a considerable simplification of input data, relative to eulerian grid models. In addition, box models are computationally simpler than eulerian grid models are since box models require solving systems of ordinary differential equations, rather than the partial differential equations that must be solved for eulerian grid models. Examples of operational box models are given in references [28–31].

Since box models are not sensitive to spatial variations in pollutant concentrations, these models cannot reasonably be expected to provide accurate estimates of concentrations at specific locations. Rather, box models should be expected to provide estimates of spatially averaged pollutant concentrations in urban areas. The ability of box models to estimate spatially averaged, ground-level concentrations of O_3 and NO_2 in St. Louis and Houston is illustrated in figures 7.3 and 7.4. In these figures, concentrations computed with box models are compared with concentrations measured at monitoring stations in the St. Louis and Houston metropolitan areas. It is reasonable to compare the concentrations obtained from the box models with spatially averaged values of the concentrations that were measured at the monitoring stations in the two cities. In St. Louis, the maximum computed O_3 concentration and the maximum value of the spatially averaged measured concentrations are virtually equal, although the computed and measured maxima occur at different times of day. In Houston, the maximum computed O_3 concentration and the maximum value of the average measured concentration differ by roughly 25%. The maximum computed and maximum average measured NO_2 concentrations differ by roughly 25% in St. Louis and 10% in Houston.

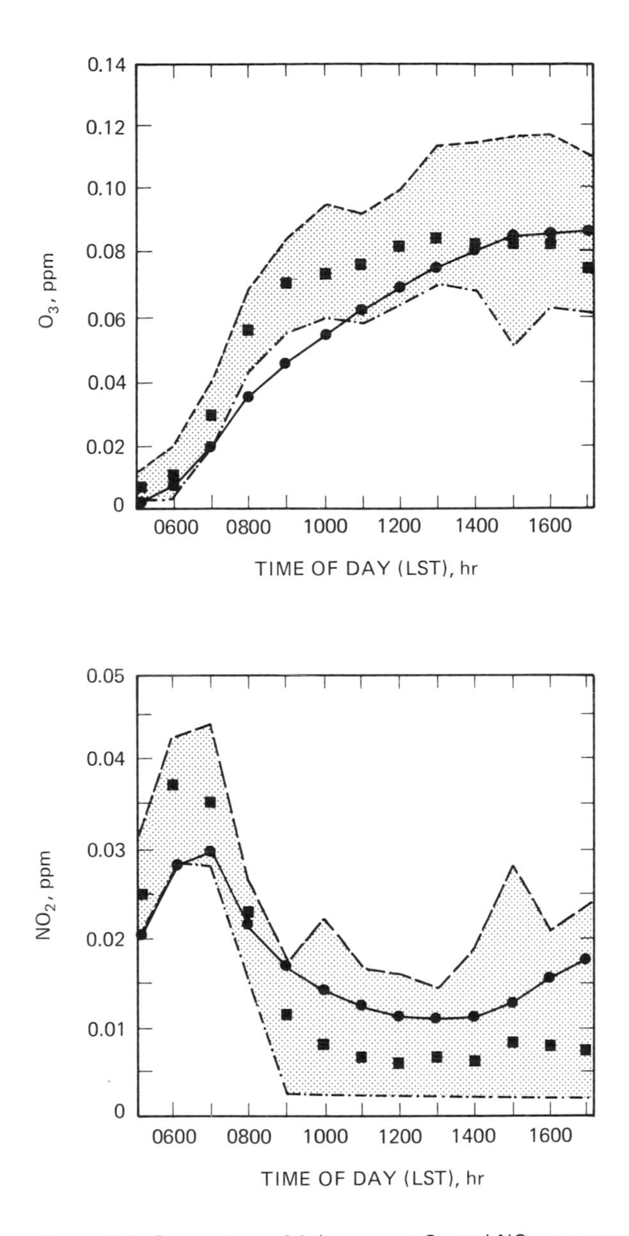

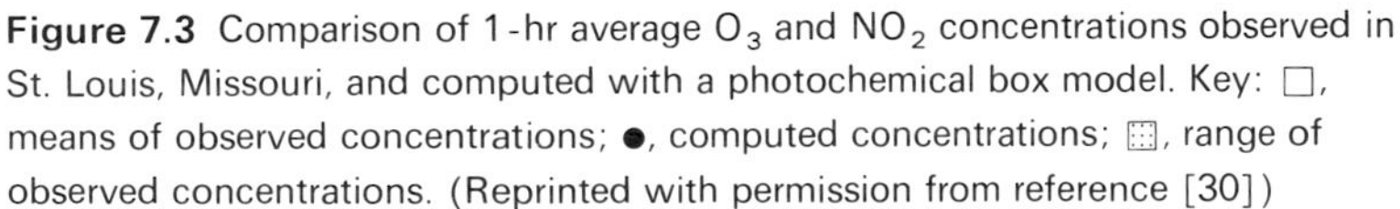

Figure 7.3 Comparison of 1-hr average O_3 and NO_2 concentrations observed in St. Louis, Missouri, and computed with a photochemical box model. Key: □, means of observed concentrations; ●, computed concentrations; ⬚, range of observed concentrations. (Reprinted with permission from reference [30])

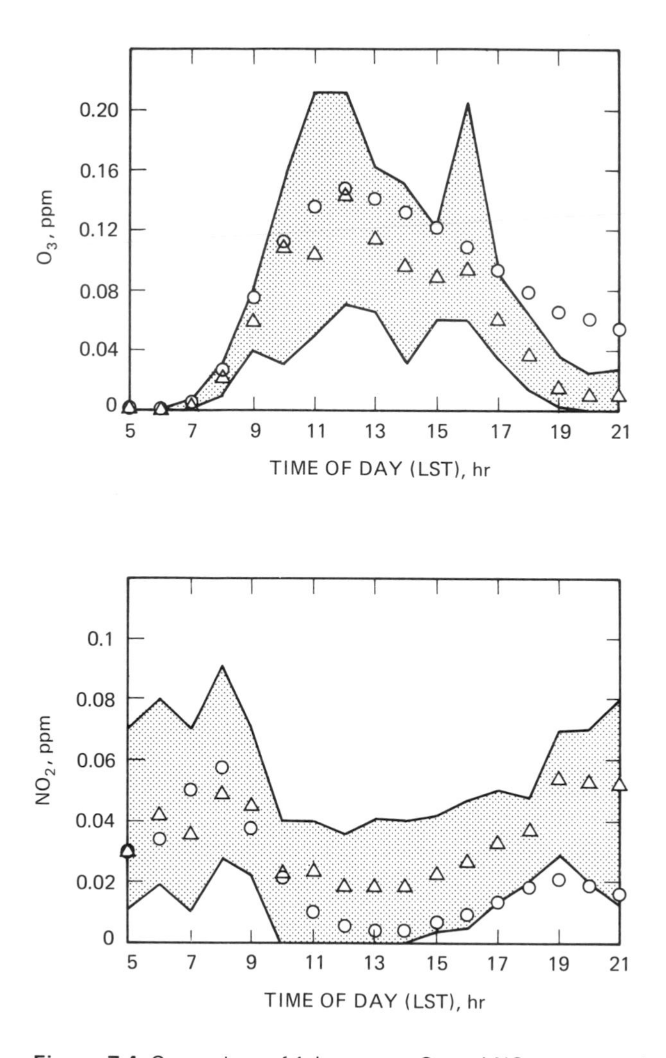

Figure 7.4 Comparison of 1-hr average O_3 and NO_2 concentrations observed in Houston, Texas, and computed with a photochemical box model. Key: △, means of observed concentrations; ○, computed concentrations; ▦, range of observed concentrations. (Reprinted with permission from reference [31])

7.5 The Lagrangian Approach to Dispersion Modeling[10]

Recall that the mass concentration of a pollutant in the air is defined as the mass of pollutant per unit volume of air. Equivalently, it is the number of pollutant particles per unit volume, multiplied by the mass of a particle.[11] Keeping this definition in mind, suppose that the exact trajectories followed by the individual particles of a pollutant as they move through the air were known. Then the mass concentration of the pollutant in a small volume element of air at a given time could be computed in the following way. Let δV denote the volume of the small volume element. Count the number of particles of the pollutant that are in the volume element at the given time (a particle is in the volume element at the given time if its trajectory passes through the element at that time), multiply this number by the mass of a particle, and divide the product by δV. Of course, in practice it is not possible to determine the exact trajectories of the individual particles of a pollutant since the number of particles is very large and their motions are strongly influenced by random events. However, in many cases of practical interest, it is possible to develop probabilistic descriptions of particles' trajectories. These descriptions can be used to estimate pollutant concentrations by a method that amounts to a probabilistic generalization of the procedure that has just been described.

The use of probabilistic descriptions of particles' trajectories to estimate pollutant concentrations is called the "lagrangian approach" to dispersion modeling. The lagrangian approach can be used in connection with pollutants that are chemically inert in the air or that undergo first-order chemical reactions. Only inert pollutants are treated here in order to minimize the complexity of the discussion. The lagrangian approach does not treat pollutants that undergo complex chemical reactions such as the reactions that are involved in the formation of photochemical smog.

To develop the lagrangian approach, it is necessary to define what is meant by a probabilistic description of a pollutant particle's trajectory and to relate this description to pollutant concentrations. An exact description of the ith particle's trajectory consists of a mathematical function $x_i(t)$ that specifies the particle's position $\mathbf{x}$ at each time t. By analogy, the trajectory of particle i can be described probabilistically by the probability density function $p_i(\mathbf{x}, t)$ of the particle's location at time t. Let δV denote the volume of a small volume element of air. Then, the probability that particle i is located in this volume element at time t, $P_i(\delta V, t)$, is

$$P_i(\delta V, t) = \int_{\delta V} p_i(\mathbf{x}, t)\, d\mathbf{x}, \tag{7.22}$$

where the symbol $d\mathbf{x}$ denotes three-dimensional integration and the integral extends over the volume element under consideration.

The ensemble average concentration of a pollutant can be expressed in terms of the probability densities of the locations of individual particles as follows. Let $Z_i(\delta V, t)$ be a random variable that is defined by

$$Z_i(\delta V, t) = \begin{cases} 1 & \text{if particle } i \text{ is in the volume element } \delta V \text{ at time } t \\ 0 & \text{otherwise.} \end{cases} \tag{7.23}$$

Note that $Z_i(\delta V, t)$ has the binomial probability distribution and that the probability that $Z_i(\delta V, t)$ equals 1 is $P_i(\delta V, t)$. Therefore, the ensemble average value of $Z_i(\delta V, t)$ is also $P_i(\delta V, t)$. The total number of pollutant particles in volume element δV at time t, $N(\delta V, t)$, is given by

$$N(\delta V, t) = \sum_i Z_i(\delta V, t), \tag{7.24}$$

where the sum extends over all pollutant particles that are in existence at time t. Since the mean of the sum of random variables is equal to the sum of the means of the variables, the ensemble average number of pollutant particles in δV at time t is given by

$$\langle N(\delta V, t) \rangle = \sum_i \langle Z_i(\delta V, t) \rangle \tag{7.25a}$$

or

$$\langle N(\delta V, t) \rangle = \sum_i P_i(\delta V, t). \tag{7.25b}$$

Let m denote the mass of a pollutant particle. Then, the ensemble average value of the pollutant mass in volume element δV is $m\langle N(\delta V, t) \rangle$. The ensemble average pollutant concentration in the volume element δV, $\langle C(\delta V, t) \rangle$, is the ensemble average mass divided by the size of the volume element. Thus,

$$\langle C(\delta V, t) \rangle = \frac{m}{\delta V} \sum_i P_i(\delta V, t). \tag{7.26}$$

As the volume element δV shrinks in size, the quantity $P_i(\delta V, t)/\delta V$ approaches the probability density $p_i(\mathbf{x}, t)$, where $\mathbf{x}$ is the location of the center of the element. Therefore, the ensemble average pollutant concentration at location $\mathbf{x}$ and time t can be written as

$$\langle C(\mathbf{x}, t) \rangle = m \sum_i p_i(\mathbf{x}, t), \tag{7.27}$$

where the sum is carried out over all pollutant particles that are in existence at time t.

Equation (7.27) provides a means for relating pollutant concentrations to probabilistic descriptions of individual particles' trajectories. However, this equation cannot be used directly to compute pollutant concentrations since it does not relate concentrations to the locations and strengths of emissions sources and no means has been provided for evaluating the probability density functions $p_i(\mathbf{x}, t)$. The problem of relating pollutant concentrations to source locations and strengths can be dealt with in the following way. Let $q(\mathbf{x}, t | \mathbf{x}', t')$ denote the probability density that a particle is at location $\mathbf{x}$ at time t, given that it was at location $\mathbf{x}'$ at time t'. This function provides a probabilistic description of the trajectory of a particle whose location $\mathbf{x}'$ at a given time t' is known. Since all particles of a pollutant are identical, they would all follow the same trajectory if they were all at the same location at some initial time. Therefore, the function q can be assumed to be the same for all particles. [Note that the probability densities $p_i(\mathbf{x}, t)$ cannot be assumed to be the same for all particles unless all particles were at the same location at some past time.] Using the laws of conditional probability, p_i and q can be related by

$$p_i(\mathbf{x}, t) = \int q(\mathbf{x}, t | \mathbf{x}', t') p_i(\mathbf{x}', t')\, d\mathbf{x}', \tag{7.28}$$

where the integral extends over all possible $\mathbf{x}'$ values.

Let $t_0 \leq t$ be an initial time, and divide all particles present at time t into two classes: those that were in the air at time t_0 and those that had not yet been emitted at time t_0. Let $\langle C(\mathbf{x}, t) \rangle_0$ denote the ensemble average pollutant concentration at location $\mathbf{x}$ and time t due to particles that were in the air at time t_0. Then by applying equations (7.27) and (7.28) to these particles, $\langle C(\mathbf{x}, t) \rangle_0$ can be written in the form

$$\langle C(\mathbf{x}, t) \rangle_0 = \int q(\mathbf{x}, t | \mathbf{x}', t_0) \langle C(\mathbf{x}', t_0) \rangle d\mathbf{x}'. \tag{7.29}$$

Now consider particles that were emitted during the time period t_0 to t. Let $S(\mathbf{x}', t')$ denote the time rate of change of the pollutant concentration at location $\mathbf{x}'$ and time t' due to emissions at $\mathbf{x}'$ and t'. Let $\langle C(\mathbf{x}, t) \rangle_e$ denote the ensemble average pollutant concentration at location $\mathbf{x}$ and time t due to particles that were emitted at times between t_0 and t. Then $\langle C(\mathbf{x}, t) \rangle_e$ is related to the function S by

$$\langle C(\mathbf{x}, t) \rangle_e = \int \int_{t_0}^{t} q(\mathbf{x}, t | \mathbf{x}', t') S(\mathbf{x}', t')\, dt'\, d\mathbf{x}', \tag{7.30}$$

where the outer integral extends over all possible $\mathbf{x}'$ values. The total ensemble average pollutant concentration at location $\mathbf{x}$ and time t,

$\langle C(\mathbf{x}, t) \rangle$, is the sum of the partial concentrations $\langle C(\mathbf{x}, t) \rangle_0$ and $\langle C(\mathbf{x}, t) \rangle_e$. Therefore, the total concentration is given by

$$\langle C(\mathbf{x}, t) \rangle = \int q(\mathbf{x}, t|\mathbf{x}', t_0) \langle C(\mathbf{x}', t_0) \rangle d\mathbf{x}' + \int\int_{t_0}^{t} q(\mathbf{x}, t|\mathbf{x}', t') S(\mathbf{x}', t') \, dt' \, d\mathbf{x}'. \tag{7.31}$$

Equation (7.31) forms the basis of lagrangian dispersion models for air pollutants. This equation relates the pollutant concentration at a given location and time to the concentrations at an initial time, the locations and strengths of emissions sources, and the function q that describes the trajectories of pollutant particles, given the particles' locations at the initial time. Compared to the eulerian dispersion equation (7.9), equation (7.31) has the disadvantage of not being able to treat complex chemical reactions. However, in contrast to equation (7.9), equation (7.31) is not based on the mixing length approximation and, therefore, is valid near strong emissions sources.[12] Thus, if the transition probability density function $q(\mathbf{x}, t|\mathbf{x}', t')$ were known, equation (7.31) would provide a means for computing the concentrations of chemically unreactive pollutants in the vicinities of strong emissions sources (e.g., for computing CO concentrations near high volume roadways). The problem of determining $q(\mathbf{x}, t|\mathbf{x}', t')$ is discussed in the two sections.

7.6 *Gaussian Plume Models*

The function $q(\mathbf{x}, t|\mathbf{x}', t')$ can be determined most easily if it is assumed that the atmosphere is stationary and homogeneous. The atmosphere is stationary if the probability distribution of the wind velocity is independent of time, and it is homogeneous if the probability distribution of the wind velocity is independent of location. Let $i = 1, 2, 3$ denote the three axes of a rectangular coordinate system, and let $\bar{u}_i$ denote the component of the mean wind velocity in direction i. Then, if the atmosphere is stationary and homogeneous, $q(\mathbf{x}, t|\mathbf{x}', t')$ is given by

$$q(\mathbf{x}, t|\mathbf{x}', t') = \frac{1}{(2\pi)^{3/2}\sigma_1\sigma_2\sigma_3} \cdot \exp\left\{-\frac{1}{2}\sum_{i=1}^{3}\left[\frac{x_i - x_i' - \bar{u}_i(t - t')}{\sigma_i}\right]^2\right\} \tag{7.32}$$

where for each i, $\sigma_i = \sigma_i(t - t')$ is a function of the time difference $t - t'$ [1, 2, 32, 33]. The functions σ_i must be evaluated through measurements or simulation of turbulent dispersion. These functions give the standard deviations of the coordinates of a pollutant particle's location, relative to the coordinates that the particle would have if it moved with the mean wind velocity. Thus, the σ functions characterize the extent to which a particle's location at a given time is affected by atmospheric turbulence.

The assumption that the atmosphere is stationary and homogeneous implies that it is unbounded in all directions. Hence, equation (7.32) does not take account of the presence of the surface of the earth. The equation can be modified to incorporate the presence of the earth's surface if the following additional assumptions are made:

1. The earth's surface is a plane. For specificity, assume that it is the plane defined by the equation $x_3 = 0$.
2. The vertical component of the mean wind velocity, u_3, is zero.
3. The atmosphere is stationary and homogeneous above the earth's surface. In other words, the earth's surface does not affect the wind velocity above the surface.
4. Pollutant particles that hit the earth's surface are perfectly reflected. This means that a collision with the surface causes no change in the horizontal components of a particle's velocity and causes the vertical component of the velocity to change sign but not magnitude.

If these assumptions are made, then the equation for q with the presence of the earth's surface taken into account is

$$q(\mathbf{x}, t|\mathbf{x}', t') = \frac{1}{(2\pi)^{3/2}\sigma_1\sigma_2\sigma_3} \cdot \exp\left\{-\frac{1}{2}\sum_{i=1}^{2}\left[\frac{x_i - x_i' - \bar{u}_i(t-t')}{\sigma_i}\right]^2\right\} \cdot \left\{\exp\left[-\frac{1}{2}\left(\frac{x_3 - x_3'}{\sigma_3}\right)^2\right] + \exp\left[-\frac{1}{2}\left(\frac{x_3 + x_3'}{\sigma_3}\right)^2\right]\right\}, \quad x_3 \geq 0. \tag{7.33}$$

In this equation, x_3 represents the distance above the surface of the earth and x_1 and x_2 are horizontal coordinates. It is customary to assume that the coordinate axes are oriented so that the x_1 direction coincides with the direction of the mean wind. Therefore, if the mean wind speed is $\bar{u}$ then $\bar{u}_1 = \bar{u}$ and $\bar{u}_2 = 0$. When the coordinate axes are oriented in this manner, the axis along the direction of the wind is labeled x, the axis that is perpendicular to the wind direction but parallel to the ground (the crosswind horizontal axis) is labeled y, and the vertical axis is labeled z (see figure 7.5). Given these conventions, equation (7.33) becomes

$$q(\mathbf{x}, t|\mathbf{x}', t') = \frac{1}{(2\pi)^{3/2}\sigma_x\sigma_y\sigma_z}\exp\left\{-\frac{1}{2}\left[\frac{x - x' - \bar{u}(t-t')}{\sigma_x}\right]^2 - \frac{1}{2}\left[\frac{y - y'}{\sigma_y}\right]^2\right\} \cdot \left\{\exp\left[-\frac{1}{2}\left(\frac{z - z'}{\sigma_z}\right)^2\right]\right.$$

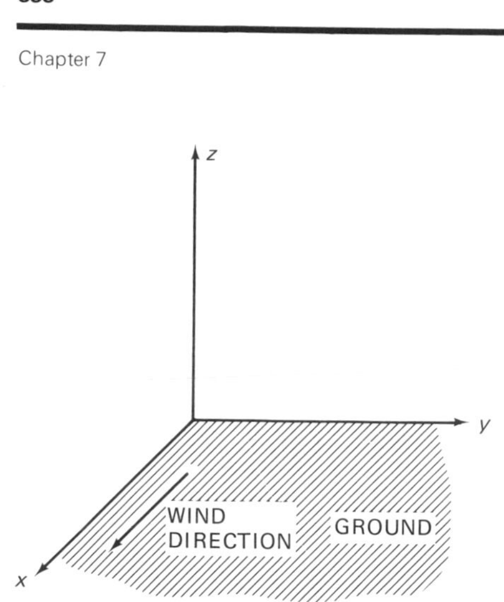

Figure 7.5 Coordinate axes for equation (7.34).

$$+ \exp\left[-\frac{1}{2}\left(\frac{z+z'}{\sigma_z}\right)^2\right]\Bigg\}, \qquad z \geq 0. \tag{7.34}$$

Most practical applications of equation (7.34) begin by using it, together with equation (7.31), to compute the pollutant concentration due to a continuously emitting point source. This is done in the following way. Assume that the source is located on the z axis at a distance H above the ground. Then the source strength function $S(x, y, z, t)$ is given by

$$S(x, y, z, t) = Q\,\delta(x)\,\delta(y)\,\delta(z - H), \tag{7.35}$$

where δ denotes the Dirac delta function and Q is the mass of pollutant emitted by the source per unit time. Assume that the initial time t_0 is in the infinite past so that initial pollutant concentrations have completely dissipated by time t and have no effects on current concentrations. Then, by substituting equations (7.34) and (7.35) into equation (7.31) and performing the integration over spatial coordinates, the following equation for $\langle C(x, y, z, t)\rangle$ is obtained:

$$\begin{aligned}\langle C(x, y, z, t)\rangle = \int_{-\infty}^{t} \frac{Q}{(2\pi)^{3/2}} \frac{1}{\sigma_x \sigma_y \sigma_z} \exp\Bigg\{-\frac{1}{2}\left[\frac{x - \bar{u}(t-t')}{\sigma_x}\right]^2 \\ -\frac{1}{2}\left(\frac{y}{\sigma_y}\right)^2\Bigg\} \cdot \Bigg\{\exp\left[-\frac{1}{2}\left(\frac{z-H}{\sigma_z}\right)^2\right] \\ + \exp\left[-\frac{1}{2}\left(\frac{z+H}{\sigma_z}\right)^2\right]\Bigg\}\, dt', \qquad z \geq 0.\end{aligned} \tag{7.36}$$

Since σ_x, σ_y, and σ_z are functions of $t - t'$, this equation cannot be simplified further unless an additional assumption is made. The usual assumption is that the mean wind speed $\bar{u}$ is sufficiently large to make the effects of turbulent diffusion in the direction of the wind (the x direction) negligible in comparison to the effects of transport by the mean wind. This is equivalent to assuming that $\sigma_x = 0$. When this assumption is made, the integrand in equation (7.36) becomes

$$\frac{Q}{2\pi}\frac{1}{\sigma_y\sigma_z}\delta[x - \bar{u}(t - t')]\exp\left[-\frac{1}{2}\left(\frac{y}{\sigma_y}\right)^2\right] \cdot \left\{\exp\left[-\frac{1}{2}\left(\frac{z-H}{\sigma_z}\right)^2\right] + \exp\left[-\frac{1}{2}\left(\frac{z+H}{\sigma_z}\right)^2\right]\right\}, \tag{7.37}$$

where δ is, again, the Dirac delta function. With this simplification, the integral in equation (7.36) can be performed to obtain

$$\langle C(x, y, z, t)\rangle = \frac{Q}{2\pi\bar{u}\sigma_y\sigma_z}\exp\left[-\frac{1}{2}\left(\frac{y}{\sigma_y}\right)^2\right] \cdot \left\{\exp\left[-\frac{1}{2}\left(\frac{z-H}{\sigma_z}\right)^2\right] + \exp\left[-\frac{1}{2}\left(\frac{z+H}{\sigma_z}\right)^2\right]\right\}, \quad x, z \geq 0, \tag{7.38}$$

where σ_y and σ_z are evaluated at $t - t' = x/\bar{u}$. Upwind of the source ($x < 0$) the concentration is zero.

Equation (7.38) is called the "gaussian plume dispersion equation" for a continuous point source of emissions. It forms the basis of a large number of dispersion models that have received widespread use in practical air quality analyses, including models for estimating CO concentrations near roadways [34–41] and models for estimating annual average NO_2 concentrations [42].[13]

A simple gaussian model for estimating CO concentrations near a roadway can be developed by treating each lane of the roadway as a straight, infinitely long, line source of emissions. Each line source can be considered to be composed of a continuum of infinitesimal point sources. The CO concentration due to each of the point sources can be computed using equation (7.38), and the concentration due to the entire line source can be computed by integrating over the concentrations due to the point sources. Finally, the concentration due to the entire roadway can be computed by summing the concentrations due to the line sources representing each lane.

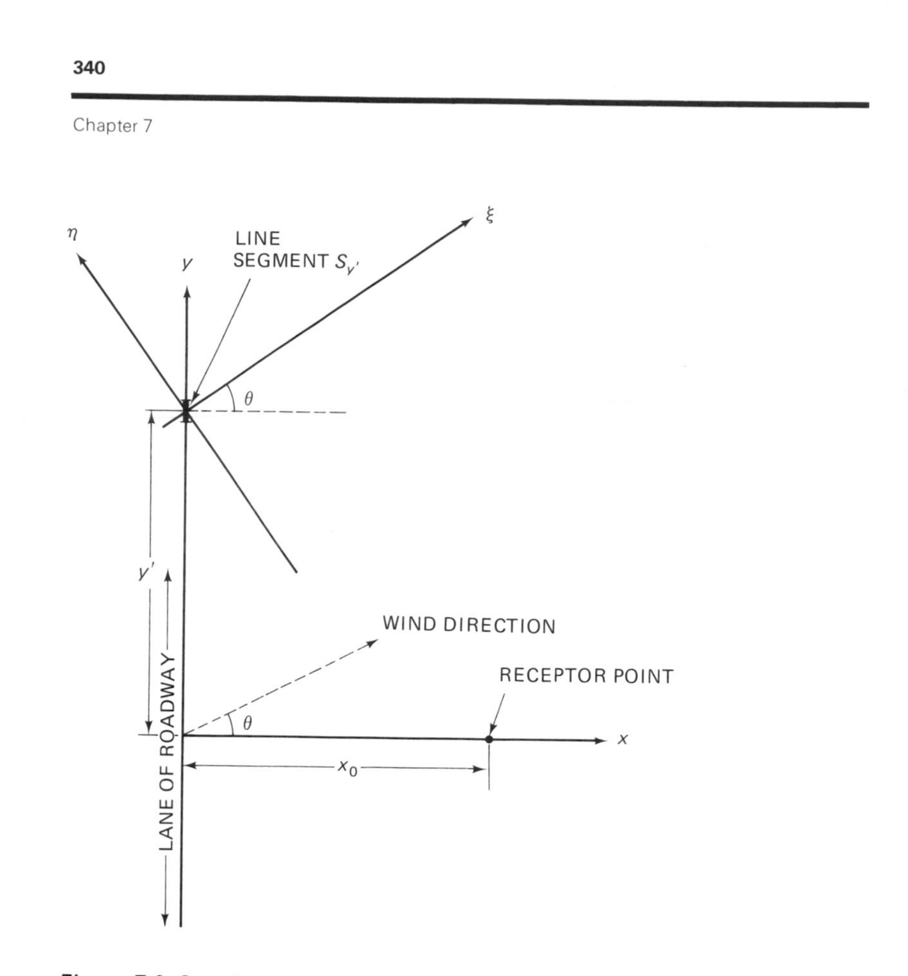

Figure 7.6 Coordinate systems for a simple roadway dispersion model.

As an illustration of this method, consider its use for computing the ground-level CO concentration due to a single lane of a roadway. Let the lane coincide with the y axis of a coordinate system, and let the x axis be parallel to the ground and perpendicular to the lane (see figure 7.6). Let the lane be located at ground level [$H = 0$ in the notation of equation (7.38)]. The point for which the concentration of CO is being computed (the receptor point) also is at ground level ($z = 0$) and can be assumed to be on the x axis at a positive distance x_0 downwind from the center of the lane. Let the wind direction make an angle θ with respect to the x axis. Consider the lane to be a line source that is concentrated on its centerline. Let Q_L denote the CO emissions rate per unit length along this line. Divide the line source into a series of very short, contiguous, nonoverlapping, line segments. Let $\delta y'$ be the length of a line segment that is centered on the point $(x, y, z) = (0, y', 0)$. The emissions rate of this segment (in units of mass per unit time) is $Q_L \, \delta y'$. Referring to Figure 7.6, let $(\xi, \eta, 0)$ denote the coordinates of the receptor point relative to a coordinate system that has its origin at the center of the line segment $\delta y'$ and that has one horizontal axis (the ξ axis) oriented parallel to the wind direction and one horizontal axis (the η axis) oriented perpendicular to the wind direction. The vertical axis of this coordinate system is perpendicular to the ground. Note that equation (7.38) is expressed relative to such a coordinate system. [In equation (7.38), the x axis corresponds to the ξ axis in figure 7.6 and the y axis corresponds to the η axis. The vertical or z axes in equation (7.38) and figure 7.6 are the same.]

From equation (7.38), the CO concentration $\delta\langle C(y')\rangle$ at the receptor point due to emissions from the line segment $\delta y'$ is

$$\delta\langle C(y')\rangle = \frac{2Q_L}{2\pi\bar{u}} \frac{1}{\sigma_V(\xi)\sigma_H(\xi)} \exp\left[-\frac{1}{2}\left(\frac{\eta}{\sigma_H(\xi)}\right)^2\right] \delta y', \tag{7.39}$$

where the functions σ for the crosswind (η) and vertical (z) directions are denoted by σ_H and σ_V, respectively. In addition, σ_H and σ_V are expressed as functions of the downwind distance from the source, ξ, rather than of the downwind travel time $\xi/\bar{u}$. Note that ξ and η are related to x_0 and y' by

$$\xi = x_0 \cos\theta - y' \sin\theta, \tag{7.40a}$$

$$\eta = -x_0 \sin\theta - y' \cos\theta. \tag{7.40b}$$

Therefore, $\delta\langle C(y')\rangle$ can be written as

$$\delta\langle C(y)\rangle = \frac{2Q_L}{2\pi\bar{u}} \frac{1}{\sigma_V(x_0\cos\theta - y'\sin\theta)\sigma_H(x_0\cos\theta - y'\sin\theta)} \cdot \exp\left\{-\frac{1}{2}\left[\frac{x_0\sin\theta + y'\cos\theta}{\sigma_H(x_0\cos\theta - y'\sin\theta)}\right]^2\right\} \delta y'. \tag{7.41}$$

The CO concentration due to the entire lane is the sum of the concentrations due to all of the small line segments composing the lane. If the lengths of these segments are made to approach zero, then the sum becomes an integral. The CO concentration at the receptor point due to the entire lane can then be written as

$$\langle C(x_0)\rangle = \frac{2Q_L}{2\pi\bar{u}} \int_{-\infty}^{x_0\cot\theta} \frac{1}{\sigma_V(x_0\cos\theta - y'\sin\theta)\sigma_H(x_0\cos\theta - y'\sin\theta)} \cdot \exp\left\{-\frac{1}{2}\left[\frac{x_0\sin\theta + y'\cos\theta}{\sigma_H(x_0\cos\theta - y'\sin\theta)}\right]^2\right\} dy'. \tag{7.42}$$

The upper limit of this integral is $x_0 \cot \theta$, rather than infinity, because line segments for which $y' > x_0 \cot \theta$ are downwind of the receptor point.

If the wind direction is perpendicular to the roadway ($\theta = 0$), then the integral in equation (7.42) can be expressed in a closed analytic form. The CO concentration at the receptor point is

$$\langle C(x_0, \theta = 0) \rangle = \frac{2Q_L}{\sqrt{2\pi}\bar{u}\sigma_V(x_0)}. \tag{7.43}$$

If $\theta \neq 0$, then the integral does not have a closed analytic form. However, if the wind angle is less than roughly 75°, the integral can be approximated closely with an analytic function to obtain [43]

$$\langle C(x_0, \theta) \rangle = \frac{2Q_L}{\sqrt{2\pi}\bar{u} \cos \theta \, \sigma_V(x_0/\cos \theta)} \tag{7.44}$$

Equations (7.42)–(7.44) constitute a very simple roadway dispersion model. Most gaussian roadway dispersion models that are used in practical applications are based on elaborations of these equations that enable the models to treat, in an approximate way, such factors as:

- roadways that have finite lengths or are not straight, and intersections;
- elevated and depressed roadways;
- the effects of turbulence generated by the movement of vehicles on roadways.

Examples of methods for treating these factors are given in references [37, 39, 40].

The basic gaussian dispersion equation (7.38) assumes that the wind velocity is stationary and that the emissions rates of all relevant sources are constant. These assumptions imply that ensemble average concentrations are constant over time. Therefore, pollutant concentrations estimated from equation (7.38) and from models based on equation (7.38) do not refer to specific averaging times. However, in practice the assumptions of a stationary wind velocity and constant emissions rates are not tenable for extended time periods. Therefore, when computations of 8-hr average CO concentrations and annual average NO_2 concentrations are performed with gaussian models, the averaging period of interest is divided into shorter time intervals (typically 1-hr intervals). The gaussian dispersion model is applied to each interval separately, and the resulting estimates of the average concentrations within the short intervals are averaged together to obtain estimates of concentrations for longer averaging times.

Before investigating the accuracy of dispersion models based on the gaussian plume equation, it is useful to assess the validity of the assumptions that were used in deriving this equation. These assumptions include the following:

1. *The atmosphere is stationary.* As has already been discussed, this assumption is not valid over extended time periods. It is acceptable as an approximation over brief time periods (probably less than 1 hr).
2. *The initial time t_0 in equation (7.31) can be taken to be in the infinite past.* This assumption insures that initial concentrations have dissipated by the current time and do not affect current concentrations. The assumption is valid if the time period required for initial concentrations to dissipate is short compared to the time period over which the atmosphere can be assumed to be stationary and compared to the averaging time for which pollutant concentrations are being estimated. (If this averaging time has been divided into smaller time intervals in order to improve the quality of the stationarity approximation, then the dissipation time must be short compared to the durations of these intervals.) The dissipation time is roughly equal to the maximum travel time at the mean wind velocity from the relevant sources to the receptor point. This time increases as source-to-receptor distances increase. Therefore, the assumption of an initial time in the infinite past is most likely to be valid for receptors that are close to the relevant sources.
3. *The surface of the earth is a plane.* This assumption implies that models based on the gaussian plume equation cannot treat dispersion over complex terrain, in the vicinities of tall buildings, or in street canyons.
4. *The atmosphere is homogeneous above the surface of the earth, and there are no vertical winds.* The homogeneity assumption is violated near the ground owing to frictional effects that cause the wind speed to decrease with decreasing height above the ground. The assumption of no vertical winds is violated in the vicinity of complex terrain, tall buildings, or other obstructions to smooth air flow. In addition, roadway sections that are elevated or depressed can cause inhomogeneities in air flow and vertical air movements, even if the surrounding terrain is flat and unobstructed. It may be possible to compensate to some extent for violations of the assumptions of a homogeneous atmosphere with no vertical winds by making appropriate empirical choices of the σ functions in the gaussian plume equation and by replacing the true source height with a suitably chosen effective source height [32, 40]. Nonetheless, violations of these assumptions represent potentially serious sources of error in gaussian models.
5. *The effects of turbulent diffusion in the direction of the wind are negligible in comparison to the effects of transport by the wind.* This assumption implies that models based on the gaussian plume equation are not valid at low wind speeds. This is a serious limitation

of these models since high pollutant concentrations occur at low wind speeds.

It can be seen that one of the assumptions upon which the gaussian plume equation is based (homogeneity of the atmosphere) is not valid near the earth's surface, and that the other assumptions are valid only under limited conditions. Consequently, models derived from the gaussian plume equation can be expected to provide only rough estimates of pollutant concentrations.

Studies of the accuracy of gaussian, roadway dispersion models have confirmed this expectation. During the middle and late 1970s, several studies were carried out in which measurements of pollutant concentrations near roadways were compared with estimates obtained from gaussian, roadway dispersion models that were then in widespread use [44–47]. These studies consistently found that the dispersion models tended to give poor estimates of measured concentrations. The estimated and observed concentrations frequently differed by more than a factor of two. The models tended to perform particularly poorly when the wind speed was low, the wind direction was parallel or nearly parallel to the roadway, and the atmosphere was stable. As a result of these studies, efforts have been made recently to improve the accuracy of gaussian, roadway dispersion models. These efforts have been directed mainly toward obtaining improved estimates of the functions σ_H and σ_V, improving the representation of the effects of vehicle-induced turbulence, and, in at least one case, adjusting the effective source height to account for the tendency of warm exhaust gases to rise [40, 41, 48, 49]. In most cases these improvements have been achieved by fitting the parameters of gaussian models (e.g., the σ functions) to observations of pollutant concentrations and meteorological variables. The restrictive assumptions on which gaussian models are based have not been altered. Table 7.2 summarizes the accuracy of several gaussian models of both the old and the new generations. The table is based on a set of dispersion experiments in which concentrations of a tracer material (sulfur hexafluoride) were measured in the vicinity of a ground-level roadway in flat, open terrain [50–52]. The tabulated results suggest that the new models are considerably more accurate than the old ones are and that the concentrations estimated by the new models usually are within a factor of two of the measured concentrations.

7.7 A Lagrangian Alternative to Gaussian Models

If the restrictive assumptions used in deriving the gaussian plume disper-

Table 7.2 Comparisons of measured pollutant concentrations near a roadway with concentrations computed with gaussian dispersion models[a]

			Percentages of observations with			
Model	Reference	Number of observations	0.5 obs ≤ computed ≤ 2.0 obs	0.95 obs ≤ computed ≤ 1.05 obs	1.05 obs ≤ computed ≤ 2.0 obs	0.5 obs ≤ computed ≤ 0.95 obs
Old generation models						
HIWAY	[37]	594	56	5	37	14
AIRPOL-4	[38]	594	61	7	27	27
CALINE-2	[39]	509[b]	52	5	28	19
New generation models						
HIWAY-2	[41]	594	85	11	44	30
GM	[48]	594	87	10	29	48

a. Adapted with permission from references [47, 49].
b. Only observations with wind speed exceeding 1 m/sec were used.

sion equation are not made, then it is no longer possible to obtain a closed form, analytic expression for the transition probability function $q(\mathbf{x}, t|\mathbf{x}', t')$. However, it is possible to estimate this function numerically by means of Monte Carlo simulation.

To develop the simulation approach, let $\delta V(\mathbf{x}^*)$ denote the volume of a small volume element of air that is centered on the point $\mathbf{x}^*$. Suppose that a large number of pollutant particles are released from the point $\mathbf{x}'$ at time t'. Define $Z_n(\mathbf{x}^*, t^*|\mathbf{x}', t')$ by

$$Z_n(\mathbf{x}^*, t^*|\mathbf{x}', t') = \begin{cases} 1 & \text{if the } n\text{th particle that was released from point } \mathbf{x}' \\ & \text{at time } t' \text{ is in volume element } \delta V(\mathbf{x}^*) \text{ at time } t^* \\ 0 & \text{otherwise.} \end{cases} \tag{7.45}$$

The ensemble average value of $Z_n(\mathbf{x}^*, t^*|\mathbf{x}', t')$ is $q(\mathbf{x}^*, t^*|\mathbf{x}', t')\,\delta V(\mathbf{x}^*)$, the probability that a particle released from $\mathbf{x}'$ at time t' is in $\delta V(\mathbf{x}^*)$ at time t^*. Therefore, $q(\mathbf{x}^*, t^*|\mathbf{x}', t')\,\delta V(\mathbf{x}^*)$ can be written in the form

$$q(\mathbf{x}^*, t^*|\mathbf{x}', t')\,\delta V(\mathbf{x}^*) = \lim_{N \to \infty} (1/N) \sum_{n=1}^{N} Z_n(\mathbf{x}^*, t^*|\mathbf{x}', t') \tag{7.46}$$

or

$$q(\mathbf{x}^*, t^*|\mathbf{x}', t') = [1/\delta V(\mathbf{x}^*)] \lim_{N \to \infty} (1/N) \sum_{n=1}^{N} Z_n(\mathbf{x}^*, t^*|\mathbf{x}', t'). \tag{7.47}$$

Equation (7.47) implies that the value of the function $q(\mathbf{x}^*, t^*|\mathbf{x}', t')$ can be evaluated with arbitrary accuracy if the values of the functions $Z_n(\mathbf{x}^*, t^*|\mathbf{x}', t')$ can be evaluated for a large number of pollutant particles.

The functions $Z_n(\mathbf{x}^*, t^*|\mathbf{x}', t')$ can be evaluated if the trajectories of individual pollutant particles released from $\mathbf{x}'$ at time t' are known. As has already been discussed, deterministic knowledge of these trajectories cannot be obtained owing to the random nature of particles' movements. However, to implement equation (7.47), it is not necessary to have deterministic information concerning the functions $Z_n(\mathbf{x}^*, t^*|\mathbf{x}', t')$. Information concerning the statistical properties of these functions suffices. This information can be obtained by simulating the effects of atmospheric turbulence on particles' movements.

To set up the simulation, let $\mathbf{u}(\mathbf{x}, t)$ denote the wind velocity at the point $\mathbf{x}$ and time t. $\mathbf{u}(\mathbf{x}, t)$ includes both the deterministic and the random components of the wind velocity. The equation of motion of a particle is given by

$$d\mathbf{x}/dt = \mathbf{u}(\mathbf{x}, t), \tag{7.48}$$

where $\mathbf{x}$ in this equation denotes the particle's location at time t. Let time be divided into a series of very short, discrete intervals of duration Δt. Let $\mathbf{x}_0$ denote the location of a particle at the initial time $t = t_0$, and let $\mathbf{x}_k$ denote its position at time $t = t_k = k\,\Delta t$ (i.e, after k time intervals have elapsed). Then $\mathbf{x}_k$ is given approximately by the recursion formula

$$\mathbf{x}_k = \sum_{i=0}^{k-1} \mathbf{u}(\mathbf{x}_i, t_i)\,\Delta t. \tag{7.49}$$

The approximation can be made arbitrarily accurate by reducing the duration of the time interval Δt. Now separate the wind velocity $\mathbf{u}$ into its deterministic component $\bar{\mathbf{u}}$ and its random component $\mathbf{u}'$. Then equation (7.49) can be written as

$$\mathbf{x}_k = \sum_{i=0}^{k-1} [\bar{\mathbf{u}}(\mathbf{x}_i, t_i) + \mathbf{u}'(\mathbf{x}_i, t_i)]\,\Delta t. \tag{7.50}$$

Given equation (7.50), the trajectory of a pollutant particle can be simulated by generating a sequence of appropriately distributed random numbers to represent the random component of the wind velocity. To be appropriately distributed, the random numbers must have statistical properties that are consistent with observable statistical properties of atmospheric turbulence. Procedures for generating random numbers with the desired properties are described in references [53, 54].

To estimate the value of the function $q(\mathbf{x}^*, t^* | \mathbf{x}', t')$ by simulation, the initial locations of a large number of particles are set equal to $\mathbf{x}'$, and the initial time t_0 in equations (7.48) and (7.49) is set equal to t'. Equation (7.50) then is used to simulate the trajectory of each particle over sufficiently many time intervals Δt to enable t_k, the total duration of the simulated trajectory, to exceed t^*, the time for which the function q is being evaluated. At time t^*, the simulated trajectory of each particle is examined to determine whether the particle is in the volume element $\delta V(\mathbf{x}^*)$, and the value of the function $Z_n(\mathbf{x}^*, t^* | \mathbf{x}', t')$ for that particle is set equal to zero or one accordingly. After simulating the trajectories of a large number N of particles, the value of $q(\mathbf{x}^*, t^* | \mathbf{x}', t')$ is estimated using equation (7.47). By repeating the estimation of $q(\mathbf{x}^*, t^* | \mathbf{x}', t')$ for a large number of initial points $\mathbf{x}'$ and initial times t', sufficient information about the function q can be obtained to enable the ensemble average concentration $\langle C(\mathbf{x}^*, t^*) \rangle$ to be estimated using equation (7.31).[14]

The simulation approach to estimating pollutant concentrations is considerably more general than gaussian plume models are. Unlike gaussian plume models, the simulation approach can treat inhomogeneous and nonstationary atmospheric conditions, low wind speeds, and complex terrain. In addition, the simulation approach does not require the use of the mixing length approximation that is used in deriving the semiempirical diffusion equation (7.9). Although it is not possible to treat complex chemical reactions with the simulation approach (owing to the inability of lagrangian models, generally, to treat such reactions), it is possible to develop an approximate treatment of the reaction of NO with O_3 [reaction (3.4)] [54, 55]. This approximate treatment enables the simulation approach to be used to estimate short-term average NO_2 concentrations near roadways [54].

The input data required to implement the simulation approach include the mean wind velocity as a function of spatial location and time and certain statistical characteristics of atmospheric turbulence. Obtaining these data can require extensive meteorological measurements, particularly in situations where topographical features or man-made structures interrupt the smooth flow of air. Methods for obtaining the required input data are discussed in references [54, 56].

A model for estimating pollutant concentrations near roadways that is based on the simulation approach is described in reference [54]. The model has been used to estimate concentrations of sulfur hexafluoride tracer material in the vicinity of a ground-level roadway in flat, open terrain. Comparisons of the estimated concentrations with measured ones suggest that the model's accuracy under these conditions is similar to that of the new-generation gaussian models listed in table 7.2 [54].[15] The accuracy of the model under more complex topographical conditions has not been investigated.

In summary, the theoretical foundations of the simulation approach to dispersion modeling are considerably firmer than are the foundations of gaussian plume models. However, there has not been sufficient experience with the simulation approach to assess its value as a practical modeling technique.

7.8 *Statistical Models*

The dispersion models that are discussed in the preceding sections were developed mainly from detailed theoretical consideration of the physical processes that affect pollutant dispersion under turbulent conditions. Relatively little use was made of direct observations of relations between pollutant concentrations, meteorological variables, and emissions rates. This order of reliance on physical theory and on observations is reversed in the development of statistical models. These models are developed by using statistical techniques, such as regression analysis, and observations of pollutant concentrations, meteorological conditions, and emissions rates to infer functional relations between pollutant concentrations and the other variables. The theoretical physical considerations that are involved in the development of these models usually are elementary and often are qualitative. For example, physical theory may be used to identify potentially useful explanatory variables for a statistical model, but not to identify the mathematical form of the functional relation between pollutant concentrations and the explanatory variables.

Statistical models offer the possibility of describing relatively complex physical situations with simple mathematical relations and easily measured explanatory variables. The Tiao and Hillmer model [equation (3.1)] for CO concentrations at a monitoring site near the San Diego Freeway is an example of such a model. This model was developed by applying nonlinear regression analysis to observations of CO concentrations, wind speeds and directions, and traffic densities at the monitoring site [57, 58]. The model has a simple functional form and contains only three, easily measured explanatory variables (wind speed, wind direction, and—as a surrogate for the emissions rate along the freeway—traffic density). Nonetheless, the model yields good estimates of the observed CO concentrations (see figures 3.10–3.12).

The simple Tiao and Hillmer statistical model provides a description of CO concentrations at the San Diego Freeway site that can be duplicated by models based on fundamental physical considerations only at the cost of substantial complexity. Unlike a gaussian model, the Tiao and Hillmer model is valid at low wind speeds. Moreover, it captures the tendency of CO concentrations at the monitoring site to increase as the component of the wind velocity that is perpendicular to the freeway increases from 0 to roughly 3 mi/hr. Since the Tiao and Hillmer model is not gaussian, a model derived from physical theory that is equivalent to the Tiao and Hillmer model almost certainly would have to be a Monte Carlo simulation model and, therefore, considerably more complex than the Tiao and Hillmer model. Moreover, highly detailed measurements of the wind velocity near the freeway would be needed to enable a simulation model to duplicate the Tiao and Hillmer model's description of the dependence of CO concentrations on the perpendicular component of the wind velocity.

Statistical models also have been developed for estimating O_3 concentrations.[16] However, these models are considerably less satisfactory than is the Tiao and Hillmer CO model. Most statistical models for O_3 (see references [59, 60, 62], for example) assume the relation between O_3 concentrations and HC and NO_x emissions to be of the form

$$C_{O_3} = f(X_{HC}, X_{NO_x}), \tag{7.51}$$

where C_{O_3} is an indicator of the O_3 concentration in the geographical area of interest, such as the daily maximum 1-hr average concentration or the probability that this concentration exceeds a specified value. X_{HC} and X_{NO_x}, respectively, are indicators of HC and NO_x emissions. The function f is identified by fitting curves to observations of C_{O_3} and the X variables. Typical X variables are 6–9 A.M. average HC and NO_x concentrations in geographical areas that are upwind of the area for which O_3 concentrations are being estimated, or 6–9 A.M. or total daily HC and NO_x emissions in the upwind areas. When HC and NO_x concentrations are used as the X variables, their purpose is to function as surrogates for HC and NO_x emissions in the areas where the concentrations are measured.

Models based on equation (7.51) are unsatisfactory because they seriously oversimplify the relation between O_3 concentrations and precursor emissions. One way that they oversimplify this relation is by not including meteorological factors as explanatory variables. This is a particularly serious oversimplification when the X variables represent HC and NO_x concentrations. As has already been noted, these concentrations are used as surrogates for HC and NO_x emissions. It is assumed that a change in HC or NO_x emissions would produce a proportional change in the corresponding X variable and that this change in the X variable would change C_{O_3} in the manner that is specified by the function f. However, virtually all weekday-to-weekday variations in O_3, HC, and NO_x concentrations in cities are caused by variations in meteorological conditions. Hence, in the process of statistically identifying the function f, the X variables are likely to act as surrogates for meteorological variables as well as for emissions. If this occurs, the effects of changes in meteorological conditions will be reflected in the values of the parameters of the function f and, possibly, in the mathematical form of this function. The resulting statistical model will not provide a correct description of the relation between O_3 concentrations and HC and NO_x emissions, even if it is true that changes in HC and NO_x emissions produce proportional changes in HC and NO_x concentrations.

The omission of meteorological variables from equation (7.51) also can produce systematic errors in a statistical model when the X variables represent emissions upwind of the O_3 monitoring site. For example, suppose that O_3 at a particular monitoring site is an increasing function of upwind emissions when meteorological conditions are held constant and a decreasing function of the wind speed when upwind emissions are held constant. Suppose, also, that the wind speed and direction in the vicinity of the site are correlated so that low-emissions areas tend to be upwind when the wind speed is low, and high-emissions areas tend to be upwind when the wind speed is high. Then, in fitting equation (7.51) to observations of C_{O_3} and upwind emissions the emissions variables (X) will function, in part, as surrogates for the omitted wind speed variable. The resulting statistical model will tend to underestimate the effects of emissions on O_3 concentrations.

Even if the problems associated with the absence of meteorological explanatory variables in equation (7.51) were not present, this equation would not adequately describe the relation between O_3 concentrations and HC and NO_x emissions. If the X variables represent 6–9 A.M. emissions or concentrations, then the equation does not describe the effects of emissions that occur at other times of day. If the X variables represent total daily emissions in a source area, then the equation does not adequately account for the temporal relations between O_3 concentrations and HC and NO_x emissions. Moreover, expanding the set of explanatory variables in equation (7.51)—for example, to include meteorological variables and variables characterizing HC and NO_x emissions at several different times of day—would not necessarily produce a satisfactory statistical model for O_3. To produce a satisfactory model, it also would be necessary to identify a physically plausible and computationally tractable functional specification for relating O_3 concentrations to the expanded set of explanatory variables. Because of the complexity of the processes involved in O_3 formation, it is highly unlikely that such a function could be identified through the use of curve-fitting techniques alone. These difficulties make it useful to seek an approach other than pure curve fitting for developing statistical models for O_3.

Many of the problems of existing statistical models for O_3 are due to the fact that these models usually are not related in any direct way to quantitative physical and chemical theories of O_3 formation. Rather, the models are developed in the hope that O_3 concentrations can be explained using somewhat arbitrarily selected explanatory variables and relatively simple functional forms that can be identified easily through the use of graphical techniques or regression analysis. Although this type of approach can work well for CO (as is illustrated by the Tiao and Hillmer model), the complexities introduced by the photochemistry of O_3 formation make it much less useful for developing models for O_3. An alternative approach consists of using physical and chemical theory to identify the explanatory variables of a model for O_3 and the mathematical form of the functional relation between these variables and O_3 concentrations. Statistical methods then are used only to estimate the values of unknown constant parameters in this relation. Although this approach is not yet well developed, the results that were obtained in an initial application of the approach suggest that it may ultimately prove to be useful.

The initial application consisted of using statistical techniques to estimate the values of the parameters of a box model for the concentrations of CO, O_3, NO, and NO_2 in downtown Los Angeles [63]. In this model, pollutant concentrations above the mixing layer are assumed to be negligible, pollutant concentrations on the upwind boundary of the box are assumed to be proportional to concentrations inside the box, and the number of vehicles operating in the vicinity of downtown Los Angeles is used as a surrogate for CO and NO emissions. NO_2 and O_3 are treated as pollutants that have no emissions sources. The equations of the model, which can be obtained easily from equation (7.21), are

$$\frac{d[CO]}{dt} = -k_u u[CO] - \frac{I_z}{z}\frac{dz}{dt}[CO] + \frac{k_T T}{z}, \tag{7.52a}$$

$$\frac{d[NO]}{dt} = -k_u u[NO] - \frac{I_z}{z}\frac{dz}{dt}[NO] + \frac{k_{TNO} k_T T}{z} + R_{NO}, \tag{7.52b}$$

$$\frac{d[NO_2]}{dt} = -k_u u[NO_2] - \frac{I_z}{z}\frac{dz}{dt}[NO_2] + R_{NO_2}, \tag{7.52c}$$

$$\frac{d[O_3]}{dt} = -k_u u[O_3] - \frac{I_z}{z}\frac{dz}{dt}[O_3] + R_{O_3}, \tag{7.52d}$$

where t denotes the time of day, [CO] denotes the concentration of CO and the other quantities in brackets are defined similarly, u denotes the wind speed, z denotes the mixing height, T is the traffic variable, and the R's denote the rates of change of the various pollutant concentrations due to chemical reactions. The quantity I_z is defined by

$$I_z = \begin{cases} 1 & \text{if } dz/dt > 0 \\ 0 & \text{otherwise.} \end{cases} \tag{7.53}$$

The functional forms of the chemical reaction terms R are determined from a simple model of the chemical processes involved in O_3 formation [63]. The quantities k_u, k_T, and k_{TNO} are constants whose values, along with the values of several constants that appear in the chemical model, are estimated by fitting the model to observations of pollutant concentrations. By comparing equations (7.52) with equation (7.21), it can be seen that k_T and k_{TNO} are determined by emissions rates, and that the value of k_u is determined by the physical dimensions of the box and by pollutant concentrations on the upwind boundary of the box.

The explanatory variables of the model of equations (7.52) are t, u, z, T, the initial concentrations of CO, NO, NO_2, and O_3, the concentration of HC, and the intensity of solar radiation. The latter two variables enter the model through the chemical reaction rates. All of the explanatory variables, except t and the initial pollutant concentrations, and all of the dependent variables are considered to be functions of the time of day t. The model does not include a procedure for relating HC concentrations to HC emissions. It would be incorrect to assume that HC concentrations at a given time of day necessarily are directly proportional or even linearly related to HC emissions at earlier times.

The model of equations (7.52) is unusual for a statistical model in that it is specified in terms of differential equations, rather than in terms of algebraic functions. Nonetheless, it is possible to estimate the values of the model's constant parameters through the use of nonlinear least squares techniques. Of course, the computations are considerably more cumbersome than they would be if the model were specified in terms of simple algebraic functions. Figure 7.7 illustrates the ability of the fitted model (7.52) to reproduce the average values of the hourly pollutant concentrations that were observed in downtown Los Angeles during 1972. The modeled concentrations are in close agreement with the observed ones.

It is not difficult to identify improvements that could be made in the model of equations (7.52). For example, an equation should be included in the model to relate HC concentrations to HC emissions. Moreover, the agreement between observed and modeled concentrations that is shown in figure 7.7 does not, by itself, imply that the model produces accurate forecasts of the effects on O_3 concentrations of changes in the values of the explanatory variables. However, the results shown in figure 7.7 do suggest that it may be possible to develop useful models for photo-chemically reactive pollutants by using what is essentially a two-phase process. In the first phase, the general mathematical form of the functional relation between pollutant concentrations and a set of explanatory variables is developed through consideration of the physical and chemical processes involved in pollutant formation and dispersion. This relation is likely to contain a variety of constant parameters whose values are not known a priori. In the second phase, the values of the parameters are estimated by fitting the model to observations of pollutant concentrations and the explanatory variables.

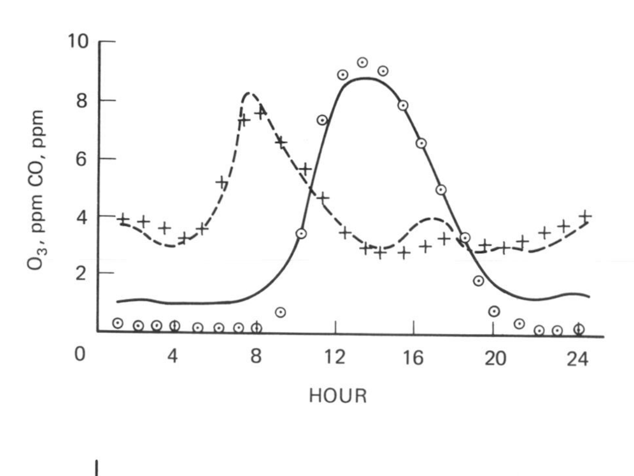

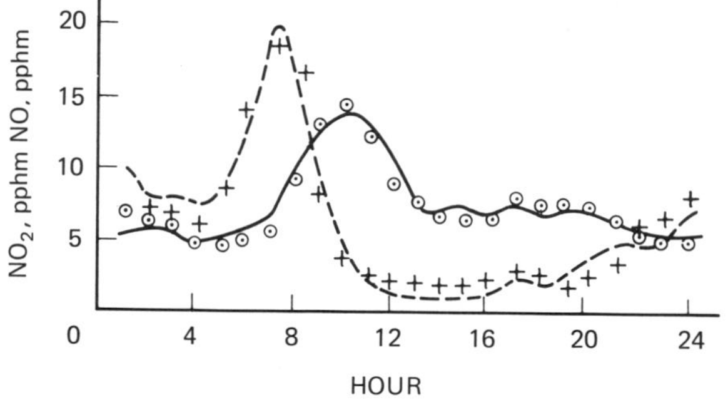

Figure 7.7 Comparison of pollutant concentrations observed in downtown Los Angeles, California, with concentrations computed by a box model with statistically estimated parameters. Key (top): — — —, observed CO; + + +, computed CO; ———, observed O_3; ⊙ ⊙ ⊙, computed O_3. Key (bottom): — — —, observed NO; + + +, computed NO; ———, observed NO_2; ⊙ ⊙ ⊙, computed NO_2. (Reprinted with permission from reference [63])

Statistical models do not include among their explanatory variables all of the factors that influence pollutant concentrations at a location. As a result, it is not possible to use a statistical model that was developed for one location to estimate pollutant concentrations at another location. For example, the Tiao and Hillmer model of equation (3.1) contains no variables describing site geometry or atmospheric stability. Rather, the values of such variables are represented implicitly through the values of the model's parameters and through the relation of atmospheric stability to wind speed. At another location, the implicit variables describing site geometry would not necessarily have the same values that they have at the San Diego Freeway site to which the model pertains, nor would the relation between wind speed and atmospheric stability necessarily be the same. Hence a statistical model for CO concentrations at another location would be likely to have different parameter values and, possibly, a different functional form from the Tiao and Hillmer model.

The nontransferability among locations of statistical models that lack locational explanatory variables can be illustrated by comparing the Tiao and Hillmer model with a model for CO concentrations at a site on the median strip of the San Diego Freeway, 75 ft away from the site to which the Tiao and Hillmer model applies. The model for the median site is given by [64]:

$$[\mathrm{CO}] = (\alpha + kT)\exp(-\gamma\sqrt{\mathrm{WS}_{\perp}}), \tag{7.54}$$

where [CO] is the 1-hr average CO concentration at the site in units of ppm, T is the traffic density on the freeway in the vicinity of the site in units of veh/mi, and $\mathrm{WS}_{\perp}$ is the component of the wind velocity that is perpendicular to the freeway in units of mi/hr. The quantities α, k, and γ are constant parameters whose values were estimated by nonlinear regression to be 2.17, 0.047, and 0.16, respectively, for summer weekdays in 1977. The model for CO concentrations on the median strip of the San Diego Freeway does not have the same functional form as the Tiao and Hillmer model for CO concentrations beside the freeway, thus illustrating the nontransferability of the two models among measurement sites.

If CO concentration measurements are available at several locations near a roadway, then the coordinates of these locations can be used as explanatory variables in a statistical model for CO concentrations near the roadway. The model then can be used to estimate CO concentrations over a range of locations in the vicinities of the locations at which measurements are available. A statistical model for CO concentrations along a line that is perpendicular to the San Diego Freeway and that passes through the monitoring sites to which the Tiao and Hillmer model and equation (7.54) apply is given in references [64, 65].

In addition to being nontransferable among geographical locations, statistical models usually cannot be used reliably to estimate pollutant concentrations when the values of the explanatory variables or other relevant factors that are not included among the explanatory variables are substantially beyond the ranges of values represented in the data sets used to develop the models.

Because of their nontransferability, statistical models cannot be used to estimate pollutant concentrations at locations or under conditions for which the observations of pollutant concentrations, meteorological conditions, and other variables needed for model development are unavailable. For example, a statistical model could not be used to forecast the CO concentration that would be likely to occur near a not yet constructed roadway after that roadway has been built and opened to traffic. A model that was developed directly from consideration of the physical processes involved in dispersion of pollutants could be used to make such forecasts if sufficient information regarding wind velocities in the vicinity of the proposed roadway and likely traffic flow conditions on it were available.

The foregoing discussion implies that statistical models for estimating pollutant concentrations are considerably more restricted in their applicability than are models that are developed directly from dispersion theory. The value of a statistical model lies not in its generality, but in its potential ability to provide a relatively simple but usefully accurate description of dispersion in the particular situation for which it has been developed.

References

1

Seinfeld, J. H., *Air Pollution: Physical and Chemical Fundamentals*, McGraw-Hill, New York, 1975.

2

Lamb, R. G., and Seinfeld, J. H., "Mathematical Modeling of Urban Air Pollution: General Theory," *Environmental Science and Technology*, Vol. 7, pp. 253–261, March 1973.

3

Reynolds, S. D., Roth, P. M., and Seinfeld, J. H., "Mathematical Modeling of Photochemical Air Pollution—I. Formulation of the Model," *Atmospheric Environment*, Vol. 7, pp. 1033–1061, 1973.

4
Seinfeld, J. H., Reynolds, S. D., and Roth, P. M., "Simulation of Urban Air Pollution," in *Advances in Chemistry Series*, Vol. 113, pp. 58–100, 1972.

5
Lamb, R. G., "Note on Application of *K*-Theory to Turbulent Diffusion Problems Involving Nonlinear Chemical Reactions," *Atmospheric Environment*, Vol. 7, pp. 257–263, 1973.

6
Lamb, R. G., Chen, W. H., and Seinfeld, J. H., "Numerico-Empirical Analyses of Atmospheric Diffusion Theories," *Journal of Atmospheric Sciences*, Vol. 32, pp. 1794–1807, September 1975.

7
Reynolds, S. D., Reid, L., Hillyer, M., Killus, J. P., Tesche, T. W., Pollack, R. I., Anderson, G. E., and Ames, J., *Photochemical Modeling of Transportation Control Strategies—Volume I. Model Development, Performance Evaluation and Strategy Assessment*, report prepared by Systems Applications, Inc., for The Federal Highway Administration, Washington, DC, March 1979.

8
Ames, J., Myers, T. C., Reid, L. E., Whitney, D. C., Golding, S. H., Hayes, S. R., and Reynolds, S. D., *The User's Manual for the S.A.I. Airshed Model*, US Environmental Protection Agency, Research Triangle Park, NC, August 1978.

9
MacCracken, M. C., Wuebbles, D. J., Walton, J. J., Duewer, W. H., and Grant, K. E., "The Livermore Regional Air Quality Model: I. Concept and Development," *Journal of Applied Meteorology*, Vol. 17, pp. 254–272, March 1978.

10
Duewer, W. H., MacCracken, M. C., and Walton, J. J., "The Livermore Regional Air Quality Model: II. Verification and Sample Application in the San Francisco Bay Area," *Journal of Applied Meteorology*, Vol. 17, pp. 273–311, March 1978.

11
Association of Bay Area Governments, *Application of Photochemical Models, Vol. II, Applicability of Selected Models for Addressing Ozone Control Strategy Issues*, Report No. EPA 450/4-79-026, US Environmental Protection Agency, Research Triangle Park, NC, December 1979. NTIS Publication No. PB80-227623.

12
Anderson, G. E., Hayes, S. R., Hillyer, M. J., Killus, J. P., and Mundbur, P. V., *Air Quality in the Denver Metropolitan Region 1974–2000*, Report No. EPA 908/1-77-002, US Environmental Protection Agency, Denver, May 1977. NTIS Publication No. PB 271894.

13
Danard, M. B., "Numerical Modeling of Carbon Monoxide Concentrations near Highways," *Journal of Applied Meteorology*, Vol. 11, pp. 947–957, 1972.

14
Ragland, K. W., and Pierce, J. J., "Boundary Layer Model for Air Pollutant Concen-

trations Due to Highway Traffic," *Journal of the Air Pollution Control Association*, Vol. 25, pp. 48–51, January 1975.

15

Eskridge, R. E., and Hunt, J. C. R., "Highway Modeling. Part I: Prediction of Velocity and Turbulence Fields in the Wake of Vehicles," *Journal of Applied Meteorology*, Vol. 18, pp. 387–400, April 1979.

16

Eskridge, R. E., Binkowski, F. S., Hunt, J. C. R., Clark, T. L., and Demerjian, K. L., "Highway Modeling. Part II: Advection and Diffusion of SF_6 Tracer Gas," *Journal of Applied Meteorology*, Vol. 18, pp. 401–412, April 1979.

17

Association of Bay Area Governments, *Application of Photochemical Models, Vol. III, Recent Sensitivity Tests and Other Applications of the LIRAQ Model*, Report No. EPA 450/4-79-027, US Environmental Protection Agency, Research Triangle Park, NC, December 1979. NTIS Publication No. PB81-112898.

18

Association of Bay Area Governments, *Application of Photochemical Models, Vol. I, The Use of Photochemical Models in Urban Ozone Studies*, Report No. EPA-450/4-79-025, US Environmental Protection Agency, Research Triangle Park, NC, December 1979. NTIS Publication No. PB80-192495.

19

Roth, P. M., Roberts, P. J. W., Liu, M. K., Reynolds, S. D., and Seinfeld, J. H., "Mathematical Modeling of Photochemical Air Pollution—II. A Model and Inventory of Pollutant Emissions," *Atmospheric Environment*, Vol. 8, pp. 97–130, 1974.

20

Reynolds, S. D., and Seinfeld, J. H., "Interim Evaluation of Strategies for Meeting Ambient Air Quality Standard for Photochemical Oxidant," *Environmental Science and Technology*, Vol. 9, pp. 433–447, May 1975.

21

Ranzieri, A.J., and Shirley, E. C., "Examination of Regional Photochemical Models by a User," in *Assessing Transportation-Related Air Quality Impacts*, Special Report No. 167, Transportation Research Board, Washington, DC, 1976, pp. 75–89.

22

Eschenroeder, A. Q., and Martinez, J. R., in *Photochemical Smog and Ozone Reactions*, R. F. Gould, ed., *Advances in Chemistry Series* No. 113, pp. 101–168, American Chemical Society, Washington, DC, 1972.

23

Lurman, F., Golden, D., Lloyd, A. C., and Nordsieck, R. A., *A Lagrangian Photochemical Air Quality Simulation Model—Adaptation to the St. Louis-RAPS Data Base, Vol. I, Model Formulation*, Report No. EPA-600/8-79-015a, US Environmental Protection Agency, Research Triangle Park, NC, June 1979. NTIS Publication No. PB 300470.

24

Lurman, F., Golden, D., Lloyd, A. C., and Nordsiek, R. A., *A Lagrangian Photo-*

chemical Air Quality Simulation Model—Adaptation to the St. Louis-RAPS Data Base, Vol. II, User's Manual, Report No. EPA-600/8-79-015b, US Environmental Protection Agency, Research Triangle Park, NC, June 1979. NTIS Publication No. PB 300471.

25

Liu, M. K., and Seinfeld, J. H., "On The Validity of Grid and Trajectory Models of Urban Air Pollution," *Atmospheric Environment*, Vol. 9, pp. 555–574, 1975.

26

US Environmental Protection Agency, *Uses, Limitations and Technical Bases of Procedures for Quantifying Relationships Between Photochemical Oxidants and Precursors*, Report No. EPA-450/2-77-021a, Research Triangle Park, NC, November 1977. NTIS Publication No. PB 278412.

27

US Environmental Protection Agency, *Procedures for Quantifying Relationships between Photochemical Oxidants and Precorsors: Supporting Documentation*, Report No. EPA-450/2-77-021b, Research Triangle Park, NC, February 1978. NTIS Publication No. PB 280058.

28

Hanna, S. R., "A Simple Dispersion Model for The Analysis of Chemically Reactive Pollutants," *Atmospheric Environment*, Vol. 7, pp. 803–817, 1973.

29

Hanna, S. R., "Application of a Simple Model of Photochemical Smog," *Proceedings of the Third International Clean Air Congress*, VDI Verlag, Dusseldorf, 1973, pp. B72–B74.

30

Schere, K. L., and Demerjian, K. L., "A Photochemical Box Model for Urban Air Quality Simulation," *Proceedings of The 4th Joint Conference on Sensing Environmental Pollutants*, American Chemical Society, 1978, pp. 427–433.

31

Demerjian, K. L., and Schere, K. L., "Applications of a Photochemical Box Model for O_3 Air Quality in Houston, Texas," in *Ozone/Oxidants Interactions with the Total Environment, II*, Air Pollution Control Association, Pittsburgh, 1979.

32

Lamb, R. G., "Mathematical Principles of Turbulent Diffusion Modeling," in A. Longhetto, ed., *Atmospheric Planetary Boundary Layer Physics. Developments in Atmospheric Science*, 11, Elsevier, Amsterdam, pp. 173–210, 1980.

33

Batchelor, G. K., "Diffusion in a Field of Homogeneous Turbulence, I. Eulerian Analysis," *Australian Journal of Scientific Research*, Vol. 2, pp. 437–450, 1949.

34

Johnson, W. B., Ludwig, F. L., Dabberdt, W. F., and Allen, R. J., "An Urban Diffusion Simulation Model for Carbon Monoxide," *Journal of the Air Pollution Control Association*, Vol. 23, pp. 490–498, June 1973.

35
Dabberdt, W. F., Ludwig, F. L., and Johnson, W. B., "Validation and Applications of An Urban Diffusion Model for Vehicular Pollutants," *Atmospheric Environment*, Vol. 7, pp. 603–618, 1973.

36
Ludwig, F. L., Simmon, P. B., Sandys, R. C., Bobick, J. C., Deiders, L. R., and Mancuso, R. L., *User's Manual for the APRAC-2 Emissions and Diffusion Model*, US Environmental Protection Agency, San Francisco, June 1977. NTIS Publication No. PB 275459.

37
Zimmerman, J. R., and Thompson, R. S., *User's Guide for HIWAY: A Highway Air Pollution Model*, Report No. EPA-650/4-74-008, US Environmental Protection Agency, Research Triangle Park, NC, February 1975. NTIS Publication No. PB 239944.

38
Carpenter, W. A., and Clemana, G. G., *Analysis and Comparative Evaluation of AIRPOL-4*, Report No. VHTRC 75-R55, Virginia Highway and Transportation Research Council, Charlottesville, VA, 1975.

39
Ward, C. E., Ranzieri, A. J., and Shirley, E. C., *CALINE 2—an Improved Microscale Model for the Diffusion of Air Pollutants from a Line Source*, Report No. CA-DOT-TL-7218-1-76-23, California Department of Transportation, Sacramento, November 1976.

40
Benson, P. E., *CALINE 3—a Versatile Dispersion Model for Predicting Air Pollutant Levels near Highways and Arterial Streets*, Report No. FHWA/CA/TL-79/23, California Department of Transportation, Sacramento, November 1979.

41
Petersen, W. B., *User's Guide for HIWAY-2: A Highway Air Pollution Model*, Report No. EPA-600/8-80-018, US Environmental Protection Agency, Research Triangle Park, NC, May 1980. NTIS Publication No. PB80-227556.

42
Busse, A. D., and Zimmerman, J. R., *User's Guide for the Climatalogical Dispersion Model*, Report No. EPA-R4-73-024, US Environmental Protection Agency, Research Triangle Park, NC, 1973. NTIS Publication No. PB 227346.

43
Calder, K. L., "On Estimating Air Pollution Concentrations from a Highway in an Oblique Wind," *Atmospheric Environment*, Vol. 7, pp. 863–868, 1973.

44
Chock, D. P., "General Motors Sulfate Dispersion Experiment: Assessment of the EPA HIWAY Model," *Journal of the Air Pollution Control Association*, Vol. 27, pp. 39–45, January 1977.

45
Noll, K. E., Miller, T. L., and Claggett, M., "A Comparison of Three Highway Line

Source Dispersion Models," *Atmospheric Environment*, Vol. 12, pp. 1323–1329, 1978.

46

Rao, S. T., Keenan, M., Sistla, G., and Samson, P., *Dispersion of Pollutants near Highways. Data Analysis and Model Evaluation*, Report No. EPA-600/4-79-011, US Environmental Protection Agency, Research Triangle Park, NC, February 1979. NTIS Publication No. PB 296625.

47

Rao, S. T., Sistla, G., Keenan, M. T., and Wilson, J. S., "An Evaluation of Some Commonly Used Highway Dispersion Models," *Journal of the Air Pollution Control Association*, Vol. 30, pp. 239–242, March 1980.

48

Chock, D. P., "A Simple Line-Source Model for Dispersion near Roadways," *Atmospheric Environment*, Vol. 12, pp. 823–829, 1978.

49

Rao, S. T., and Keenan, M. T., "Suggestions for Improvement of the EPA-HIWAY Model," *Journal of the Air Pollution Control Association*, Vol. 30, pp. 247–256, March 1980.

50.

Cadle, S. H., Chock, D. P., Monson, P. R., and Heuss, J. M., "General Motors Sulfate Dispersion Experiment: Experimental Procedures and Results," *Journal of the Air Pollution Control Association*, Vol. 27, pp. 33–38, 1977.

51

Chock, D. P., "General Motors Sulfate Dispersion Experiment—an Overview of the Wind, Temperature and Concentration Fields," *Atmospheric Environment*, Vol. 11, pp. 553–559, 1977.

52

Chock, D. P., "An Overview of the General Motors Sulfate Dispersion Experiment," *Transportation Research Record*, No. 670, pp. 36–43, 1978.

53

Lamb, R. G., "A Scheme for Simulating Particle Pair Motions in Turbulent Fluid," *Journal of Computational Physics*, Vol. 39, pp. 329–346, 1981.

54

Lamb, R. G., Hogo, H., and Reid, L. E., *A Lagrangian Approach to Modeling Air Pollutant Dispersion*, Report No. EPA-600/4-79-023, US Environmental Protection Agency, Research Triangle Park, NC, April 1979. NTIS Publication No. PB 296095.

55

Lamb, R. G., and Shu, W. R., "A Model of Second Order Chemical Reactions in Turbulent Fluid—Part I. Formulation and Validation," *Atmospheric Environment*, Vol. 12, pp. 1685–1694, 1978.

56

Binkowski, F. S., "A Simple Semi-Empirical Theory for Turbulence in the Atmospheric Surface Layer," *Atmospheric Environment*, Vol. 13, pp. 247–253, 1979.

57
Tiao, G. C., and Hillmer, S. C., "Statistical Models for Ambient Concentrations of Carbon Monoxide, Lead and Sulfate Based on the LACS Data," *Environmental Science and Technology*, Vol. 12, pp. 820–828, July 1978.

58
Tiao, G. C., and Hillmer, S. C., "Statistical Analysis of the Los Angeles Catalyst Study Data—Rationale and Findings," in *The Los Angeles Catalyst Study Symposium*, Report No. EPA-600/4-77-036, US Environmental Protection Agency, Research Triangle Park, NC, June 1977, pp. 415–460. NTIS Publication No. PB 278305.

59
Myrabo, L. N., Wilson, K. R., and Trijonis, J. C., "Survey of Statistical Models for Oxidant Air Quality Prediction," in *Assessing Transportation-Related Air Quality Impacts*, Special Report 167, Transportation Research Board, Washington, DC, 1976, pp. 46–62.

60
Trijonis, J. C., "Economic Air Pollution Control Model for Los Angeles County in 1975," *Environmental Science and Technology*, Vol. 8, pp. 811–826, September 1974.

61
Trijonis, J. C., *Empirical Relationships between Atmospheric Nitrogen Dioxide and Its Precursors*, Report No. EPA-600/3-78-018, US Environmental Protection Agency, Research Triangle Park, NC, February 1978. NTIS Publication No. PB 278547.

62
Merz, P. H., Painter, L. J., and Ryason, P. R., "Aerometric Data Analysis—Time Series Analysis and Forecast and an Atmospheric Smog Diagram," *Atmospheric Environment*, Vol. 6, pp. 319–342, 1972.

63
Tiao, G. C., Phadke, M. S., and Box, G. E. P., "Some Empirical Models for the Los Angeles Photochemical Smog Data," *Journal of the Air Pollution Control Association*, Vol. 26, pp. 485–490, May 1976.

64
Ledolter, J., Tiao, G. C., Graves, S. B., Jian-tu, H., and Hudak, G. B., *Statistical Analysis of the Los Angeles Catalyst Study Data*, Report No. EPA-600/4-79-070, US Environmental Protection Agency, Research Triangle Park, NC, October 1979. NTIS Publication No. PB80-144439.

65
Ledolter, J., and Tiao, G. C., "Statistical Models for Ambient Air Pollutants, with Special Reference to the Los Angeles Catalyst Study (LACS) Data," *Environmental Science and Technology*, Vol. 10, pp. 1233–1240, October 1979.

Notes

Chapter 1

1

The description of the US air quality standards given here represents the situation in 1980. Standards are established by the US Environmental Protection Agency in accordance with the Clean Air Act (42 USC 7401, et seq., as amended by Public Law 95–95, 91 Stat 685). Legally, the standards are required to be set at concentrations below which no harmful effects to health (in the case of the primary standards) or plants, animals, materials, visibility, soils, water, climate, and things of economic value (in the case of the secondary standards) occur. The primary standards must include margins of safety to guard against the possibility that harmful health effects occur below the currently established concentration thresholds for such effects. However, it is virtually certain that there are no concentration thresholds for some harmful effects of air pollution. Hence, strict observance of the legal requirements would necessitate setting some air quality standards at concentrations of zero, thereby depriving the standards of any operational significance. This problem has not been addressed in US law either through legislation or as a matter of formal administrative policy. Therefore, the relation between the air quality standards and the harm caused by various concentrations of air pollutants cannot be specified precisely. The interpretation of the standards that is given in the text is a practical one and should not be treated as being legally or scientifically precise.

Chapter 2

1

These pollutants sometimes are referred to as "criteria pollutants" because the US Environmental Protection Agency has published documents called "air quality criteria" that describe the pollutants' properties and harmful effects.

2

"Volatile organic compounds" (VOC) is a more precise term that sometimes is used in place of the term "hydrocarbons."

3

Two possible exceptions to this statement are worthy of note. First, the hydrocarbon ethylene is harmful to plants at concentrations that are occasionally reached in the atmosphere. Second, some of the organic compounds that are emitted by motor vehicles, among other sources, have been shown in laboratory studies to be mutagenic or carcinogenic. At present, it is not known whether motor vehicle emissions of such compounds constitute a significant risk to public health.

4
Nitrous oxide (N_2O) is produced in large quantities by certain soil bacteria but is not usually considered to be an air pollutant.

5
The reason for the adjective "photochemical" is that sunlight plays an important role in initiating and maintaining these chemical reactions. See section 3.3.

6
Photochemical oxidants consist of photochemical reaction products, other than NO_2, that are capable of oxidizing potassium iodide in solution. Nitric acid and PAN are examples of oxidants. Photochemical smog is a less precise term that refers to the entire mixture of compounds formed by the reactions that produce O_3, including such nonoxidant compounds as formaldehyde.

7
See reference [23] for a description of methods for estimating the emissions of individual sources. The estimation of vehicular emissions is discussed in detail in section 5.7.

8
As will be discussed in chapter 3, HC in the atmosphere participates in a large number of chemical reactions. The effects of these reactions are not represented in the acetylene tracer method, so this method gives only an approximate indication of the proportion of atmospheric HC that is due to motor vehicle exhaust.

9
These statistics are based on air quality data obtained from the US Environmental Protection Agency. Cases of exceeding the air quality standard for NMHC are not reported here for reasons that are explained in footnote c of table 1.1. For the purposes of this discussion, metropolitan areas are defined to be standard metropolitan statistical areas (SMSAs), and nonurban counties are defined to be counties that are not in SMSAs. Definitions of metropolitan areas can differ greatly among air quality analyses. Readers who wish to compare the statistics presented here with those found in other sources should first make certain that the various statistics pertain to the same geographical areas.

10
It is beyond the scope of this book to reach conclusions as to whether the social benefits of achieving the US air quality standards exceed the social costs. The air quality standards are used here as practical guides for distinguishing between concentrations that are sufficiently low to be ignored for most purposes and concentrations that are sufficiently high to justify considering corrective action. It is in this sense that pollutant concentrations exceeding the standards are identified as problems.

11
The emissions standards for new motor vehicles and techniques for controlling emissions from these vehicles are discussed in chapter 5.

12
Specifically, the federal studies have estimated that in 1990 the air quality standards may be exceeded in roughly 5 to 39 or more geographical areas known as "air quality control regions" (AQCRs). Many of these AQCRs include several SMSAs and urbanized areas.

13
Note that emissions control requirements that existed in 1977 were not necessarily implemented in 1977. For example, emissions control standards for post-1978 model year automobiles existed in 1977, but obviously were not implemented in that year. The projections in table 2.7 include the effects of delayed implementation of 1977 emissions control requirements.

Chapter 3

1
However, exceptions to this generalization can occur. See equation (2.1) and the subsequent discussion for an example.

2
In addition, traffic flow conditions, such as the speed of traffic flow, affect the relation between CO concentrations and traffic volumes by affecting emissions rates. See sections 5.6 and 5.7.

3
The relation between motor vehicles' CO emissions and the air temperature is discussed in section 5.6.

4
Statistical models of CO concentrations also have been developed for downtown Los Angeles [5], several locations in New Jersey [4], and Toronto [7].

5
I thank Walter F. Dabberdt for suggesting the following explanation to me.

6
The relations between vehicles' emissions rates and traffic flow conditions are discussed in more detail in sections 5.6 and 5.7.

7
A free radical is a highly reactive molecular fragment.

8
This sequence is based on reference [23]. Reactions (3.5) and (3.7) are lumped reactions that encompass several elementary reaction steps each. Thus, the sequence (3.5) to (3.7) is a schematic, rather than literal, representation of a chemical mechanism for NO oxidation.

9
In addition to assuming that all emissions occur before the reaction period begins, rather than gradually throughout the reaction period, the system of figure 3.20 includes the following important simplifications. First, the only HC species emitted into the system are propylene and butane along with a mixture of aldehydes, whereas urban atmospheres contain vast numbers of different hydrocarbon species.

The reaction period in figure 3.20 is limited to 10 hr, whereas in the real atmosphere the reactions may continue for several days. The only meteorological effect included in figure 3.20 is a gradual lifting of an inversion layer, whereas the real atmosphere also has diffusion and transport processes caused by atmospheric turbulence and winds.

10

The examples in table 3.2 do not prove, and it is not true, that percentage reductions in O_3 concentrations always are less than the corresponding percentage reductions in precursor emissions.

11

Aldehydes do not oxidize potassium iodide and, therefore, are not oxidants. However, they are important constituents of photochemical smog. In addition to being produced in photochemical reactions, aldehydes are present as primary pollutants in motor vehicle exhaust. NO_2, which also is an important constituent of photochemical smog, is one of the principal transportation pollutants and is treated separately from other smog constituents throughout this book.

12

Because O_3 and other smog constituents have different molecular weights, the ratios of their concentrations do not have the same values when the concentrations are expressed in volumetric units (e.g., ppm) as when the concentrations are expressed in mass units (e.g., $\mu g/m^3$). Therefore, it is necessary to specify the concentration units when specifying the concentration ratios.

13

Because of the rapid consumption of O_3 by reaction (3.4), O_3 monitors that are used to characterize the general levels of O_3 concentrations in a city should not be located next to heavily traveled roadways or other large NO sources.

14

Although high O_3 concentrations in the eastern United States normally are associated with high atmospheric pressure, not all high-pressure systems cause high O_3. High O_3 concentrations comparable to those found in the rural eastern United States have not been found along the paths of high-pressure systems in the northern great plains. This is presumably due to the lack of large cities or other major sources of precursors in the northern plains area.

15

However, as explained in section 3.2, these measures can be effective in reducing CO concentrations.

16

The laboratory environment consists of a device called a "smog chamber." This is basically a large box in which photochemically reactive materials can be contained, irradiated, and measured. The smog chamber permits atmospheric photochemical processes to be simulated and studied under controlled conditions and without the complicating effects of variations in meteorological conditions.

Chapter 4

1

At concentrations far above those occurring in polluted air, CO, O_3, and NO_2 are fatal. The discussion in this chapter is restricted to the effects of pollutants at or near atmospheric concentrations. Examples of CO, O_3, and NO_2 concentrations that are near the high ends of their respective atmospheric ranges are given in table 2.2.

2

The units given with these variables are illustrative. Obviously, other units are possible.

3

The blood is saturated with COHb when all of the oxygen-binding sites of the hemoglobin are occupied by CO.

4

Angina pectoris is a cardiovascular disease in which the circulation of blood to the heart is impaired. Mild exercise or excitement causes pressure or pain in the chest because the supply of oxygen to the heart muscle is inadequate. In iliofemoral occlusive arterial disease, the flow of blood to the legs is impaired, and exercise causes intermittent claudication, or leg pain and weakness.

5

Examples of possibly significant factors that were not measured in one or more of the Los Angeles studies are occupational and socioeconomic status, variations in weather conditions, smoking histories, and the incidence of acute respiratory diseases such as influenza.

6

The effects of smoking on COHb levels are described later in this section.

7

Examples of indicators of pulmonary function are respiratory frequency, forced vital capacity (the volume of air that can be expelled after full inspiration), 1-sec forced expiratory volume, peak forced expiratory flow rate, airway resistance, and residual volume (the volume of air remaining in the lungs after full expiration). Examples of blood chemical indicators include erythrocyte (red blood cell) fragility and the concentrations and activity levels of various blood enzymes.

8

It has been established clinically that exposure to O_3 at concentrations near the upper end of the atmospheric range increases the sensitivity of nonasthmatics to certain respiratory irritants. A clinical study suggesting that exposure to low concentrations of NO_2, which also is an oxidizing agent, increases asthmatics' sensitivity to other irritants is described in section 4.5.

9

O_3 is not an eye irritant at atmospheric concentrations, but other constituents of photochemical smog, such as aldehydes and PAN, are. A dose-response relation is a functional relation that specifies the incidence or severity of an effect as a function of the levels of one or more causative factors.

10
However, the eye discomfort found in some studies certainly was caused by substances other than O_3, since O_3 at atmospheric concentrations is not an eye irritant.

11
The effects of O_3 exposure that have been observed in clinical studies almost certainly also occur outside of the laboratory, but the effects may occur at different concentrations under laboratory and nonlaboratory conditions.

12
Exposure of human subjects to 0.4 ppm of O_3 for 4 hr has been found to impair the bactericidal capabilities of leucocytes (white cells) taken from venous blood [84]. The relation of this effect to impairment of alveolar macrophages and its significance, if any, for human susceptibility to infections are unknown.

13
An epidemiological study that found a statistical association between atmospheric oxidant concentrations and the occurrence of mild respiratory diseases such as colds is not convincing in this respect, owing to its failure to distinguish adequately between the effects of O_3 and those of potentially confounding factors [61].

14
One-hour average NO_2 concentrations in cities usually are below 1,000 $\mu g/m^3$ (0.53 ppm) and rarely exceed 1,500 $\mu g/m^3$ (0.80 ppm).

15
The concentrations reported in the original publications from the Chattanooga study [101–103] were determined subsequently to be incorrect [104]. The concentrations cited here are from a later publication.

16
Examples of data imperfections include some studies' use of retrospective information on the occurrence of respiratory diseases in households and the use of range type or NO_2 concentrations measured in rooms of houses or at fixed outdoor sites as surrogates for the concentrations to which individuals were exposed.

17
A much more difficult form of the market approach involves comparing wage rates in cities with different levels of air quality and attempting to infer the wage premium that must be paid to induce people to live and work in relatively highly polluted cities. See references [123, 124] for discussions of the problems of this form of the market approach.

18
See reference [125] for a discussion of applications of the enumeration-and-evaluation method and the market approach to stationary source pollution.

19
The reduction in oxidant concentrations was not specified in reference [126] but has been estimated from data presented there.

20
The survey instrument stated the health benefits to be a reduction in eye irritation and a 2–3-yr increase in average life span.

21
The details of the estimation procedures as well as certain technical assumptions and issues associated with the housing price form of the market approach are discussed in references [123, 127].

Chapter 5

1
Much of the discussion in this section is based on references [1, 2]. Readers are referred to these sources for more detailed presentations.

2
Emissions from motorcycles with less than 50 cc displacement are not regulated.

3
These results and the exhaust emissions reduction estimates given in the following paragraphs apply to emissions measured in low-altitude cities, such as Los Angeles, St. Louis, and Washington, DC. Emissions rates and the effects of emissions control devices are not the same in high-altitude cities, such as Denver, as they are in low-altitude cities, owing to the reduced oxygen content of the air at high altitudes.

4
Exceptions to this statement include diesels, because of their low HC and CO emissions, and motorcycles, for which there are not evaporative emissions standards. In addition, the emissions reductions achieved in California, which has different emissions standards from the rest of the nation, and in high-altitude areas will be somewhat different from those cited here.

5
The 3%/yr VMT growth rate is illustrative. Actual VMT growth rates may be more or less than 3%/yr, depending on such factors as the price of fuel, geographical location, and the time period.

6
The EPA uses two driving cycles to measure automobile fuel economy. The LA-4 cycle is used for the urban-driving fuel economy test. The highway-driving fuel economy test is based on a cycle that contains more high-speed driving and less idling than the LA-4 cycle does. The EPA fuel economy measurements tend to differ from actual over-the-road fuel economy in ways that depend on the fuel economy and model year of the vehicle in question. In addition, weight and mileage adjustments do not necessarily compensate for the effects of changes in all factors other than emissions controls that affect fuel economy. Hence, the EPA method fuel economy measurements provide only rough indications of the effects of emissions controls on fuel economy and of trends in over-the-road fuel economy. See reference [14] for a detailed discussion of the causes and extent of differences between fuel economy as measured by the EPA method and over-the-road fuel economy.

7
However, as will be illustrated shortly, idle emissions tests are not sensitive indicators of excessive NO_x emissions. In practice, many I/M programs based on idle tests do not include testing of NO_x emissions.

8
In reality, a vehicle's emissions in the absence of I/M are likely to fluctuate over time in response to voluntary maintenance that the vehicle receives. However, in these examples it is assumed that the emissions changes caused by voluntary maintenance are small in comparison to the changes caused by I/M, so that the effects of voluntary maintenance can be ignored.

9
The premaintenance emissions of the Portland vehicles are less than the emissions of the Eugene vehicles because I/M had been in effect in Portland for 2 yr before the evaluation of the program began.

10
The term "emissions rate" refers both to mass emitted per unit time and mass emitted per unit distance traveled. In this discussion, statements concerning emissions rates are accompanied by units (mass per unit time or mass per unit distance traveled) whenever it is necessary to specify a particular meaning of the term. Statements about emissions rates that do not include units apply to both meanings of the term.

11
Persons interested in obtaining the latest estimates of the coefficients of the EPA's Modal Analysis Model should contact the Office of Mobile Source Air Pollution Control, US Environmental Protection Agency, 2565 Plymouth Road, Ann Arbor, Michigan 48105.

12
The traffic density is defined as the number of vehicles per unit length of roadway.

13
The traffic volume on a short section of roadway is defined as the number of vehicles that pass over the section per unit time. The traffic density on the section is the number of vehicles per unit length on the section. The traffic density multiplied by the speed of traffic on the section equals the traffic volume on the section. The density of emissions of a particular pollutant on the section (in units of mass of pollutant per unit length of roadway per unit time) equals the average emissions rate of the individual vehicles passing over the section (in units of mass of pollutant per unit time) multiplied by the traffic density on the section. Alternatively, the emissions density is the sum of the individual emissions of all of the vehicles that are in the section during a unit of time, divided by the length of the section.

14
The form given here is the one being used currently for the estimation of motor vehicle emissions. In reference [5], the form $e = G(F + H)$ is used. This form permits the same simplifications that equations (5.13) do.

Chapter 6

1

Motor vehicles that are used for personal transportation include motorcycles and some vans and light trucks in addition to automobiles. Most of the discussion in this chapter is phrased in terms of measures for reducing emissions from automobiles. However, the measures and the conclusions that are reached apply to the entire class of vehicles that are used for personal transportation.

2

For the purposes of this discussion, peak periods are defined as 2-hr periods in the morning and afternoon during which peak traffic flows occur. Eight-hour periods that consist largely of early morning hours (before 6 A.M.) usually do not have high 8-hr average CO concentrations and are excluded from this discussion.

3

It is likely that large changes in the spatial or temporal distributions of HC and NO_x emissions can bring about large changes (both increases and decreases) in O_3 concentrations, even if total emissions do not change. For example, delaying early morning emissions until later in the day would tend to reduce O_3 concentrations by reducing the number of daylight hours in which these emissions can react chemically to form O_3. There has been little research into the quantitative effects on O_3 concentrations of spatially or temporally redistributing precursor emissions. However, the limited evidence that is available suggests that transportation and traffic management measures whose effects on total emissions are small are unlikely to alter the distributions of emissions sufficiently to cause substantial changes in O_3 concentrations.

4

There is preliminary evidence suggesting that g/veh mi NO_x emissions rates from diesel vehicles may decrease with increasing average speed of travel, at least up to roughly 35 mi/hr. Thus, the NO_x speed-emissions relations for traffic streams containing sufficiently large proportions of diesel vehicles may be decreasing functions of speed at low speeds. In addition, it should be noted that although NO_x emissions from gasoline-powered vehicles tend to increase with increasing average speed, there is considerable scatter about this trend. Hence, it is possible to find pairs of high- and low-speed driving conditions in which higher NO_x emissions occur at the low speed than at the high speed.

5

It is assumed in these examples that average speeds are equal in both directions of traffic flow. If the speeds are different in different directions, then the effects of speed changes must be computed separately for each direction.

6

The physical capacity of a link is the largest traffic volume that the link can carry. The practical capacity is a parameter with units of traffic volume that appears in many formulations of link performance functions. The practical capacity of a link does not constitute an upper bound on the volume of traffic that the link can carry.

However, the speed of traffic flow on the link decreases rapidly as the traffic volume increases above the link's practical capacity.

7
The most important simplification that has been made in developing the model of equations (6.4)–(6.6) consists of using a very small set of variables (i.e., traffic volumes, speeds, and link capacities) to describe travel demand and transportation system performance. Inclusion of more variables would make the model more realistic, but would also make it more complicated without substantially altering the conclusions that will be reached.

8
This performance function has been adapted from the results of traffic measurements that were made in Winnipeg, Manitoba [9, 10].

9
When route choice is the only travel demand dimension that is included in the equilibration framework, network equilibration often is called equilibrium traffic assignment.

10
Although vehicle-free zones and traffic cells are discussed here as traffic engineering measures, they are also simple forms of traffic restraints. See part D of this chapter for further discussion of traffic restraints.

11
Of course, the emissions reductions that can be achieved in principle by transit improvements and carpool incentives are very large. For example, a carpooling program that succeeded in doubling average automobile occupancy would reduce automobile emissions by roughly 50%. However, for reasons that will be explained in this part, it is highly unlikely that such large emissions reductions can be achieved in practice.

12
Utility maximization is not the only conceivable behavioral paradigm, but it is one that has sufficient flexibility to encompass a large variety of decision processes while still giving useful guidance for the development of travel demand models. Utility maximization is the only paradigm that has thus far proved useful for developing travel demand models.

13
An analysis based on a single "typical" traveler cannot be expected to yield precisely the same results that a more elaborate analysis based on a large sample of travelers would yield. However, the single-traveler results are similar to the results of more elaborate, multitraveler analyses in terms of their general magnitudes and main qualitative features, and obtaining them requires considerably less computational effort than do multitraveler analyses. These characteristics of the single-traveler approach make it very useful for exploring some of the key characteristics of the relation between service quality and mode choice. Later in this section, the results of several computer-based, multitraveler analyses will be presented.

14
However, exceptions to this generalization can arise. See example 6.6 in section 6.8.

15
Due to rounding of the mode choice probabilities in table 6.8, the automobile driver trip reductions cited in the text may differ slightly from those that would be obtained from the tabulated probabilities.

16
The discussion in this section does not address the possibility that transit improvements combined with sufficiently large increases in the costs of automobile travel or with other types of restraints on automobile travel may be capable of achieving emissions reductions that are significant on an urban regional scale. The effects of pricing and restraint measures are discussed in part D of this chapter.

17
Although the mode choice probabilities in table 6.8 have been computed using an assumed carpool size of two persons, the tabulated probabilities would not be significantly changed if three-person carpools were assumed. In addition, the reductions in automobile driver trips achieved by carpool incentives are not highly sensitive to the level of transit service quality and, for example, would not be significantly affected by using the case 3 and 8 transit service levels in the examples in place of the transit service levels that have been used.

18
Although the Shirley Highway and San Bernardino Freeway bus services are still operating, the demonstration and evaluation phases of these projects have ended. Therefore, the projects are discussed here in the past tense.

19
No attempt is made here to assign monetary values to resources such as travel time that are not conventionally valued in monetary terms. In an analysis of the net social benefits of providing transit service, it would be necessary to assign monetary values to such resources. However, the present discussion is not intended to provide such an analysis. Rather, its objective is to describe certain aspects of the relations among transit service quality, ridership levels, and monetary costs that have important bearings on the emissions reductions that can be achieved through transit service improvements.

20
These studies were carried out in the late 1960s. More recent data that could be used to operate the transit supply model for these cities are not available. However, it is unlikely that developments since the late 1960s have substantially altered the conclusions that are reached in this section.

21
The New York City area, which has an unusually high population density by US standards, and where automobile ownership is difficult and expensive, is a possible exception to this generalization.

22
Strictly speaking, V_A depends on transit service quality since improvements in

transit service quality tend to divert travelers from automobiles to transit, thereby decreasing congestion and increasing the speed of traffic flow in the corridor. However, treating this dependence would greatly complicate the analysis. Consequently, V_A is treated as a constant in this example.

23

If equation (6.54) yields a fractional value of B with the prescribed value of t_L, then the upward rounding of B must be accompanied by an increase in t_L in order to maintain the established headway H.

24

Strictly speaking, the mileage-related costs of automobile travel also may be subsidized, as drivers typically do not pay the costs of the congestion and pollution that they cause. However, the costs of these externalities, which can be very difficult to estimate (see the discussion in section 4.8), will not be treated in this example.

25

In this example, the average CO emissions density in the corridor is directly proportional to total CO emissions in the corridor. (The density is equal to total emissions divided by the corridor's area.) Therefore, a separate computation of the CO emissions density has not been made. However, it should be borne in mind that CO concentrations can vary greatly among different locations along a street in response to variations in emissions rates along the street. Therefore, in a more sophisticated analysis of the effects of transit improvements on CO concentrations in a corridor, it would be useful to estimate CO emissions rates in the vicinities of high-CO locations in the corridor, rather than computing only total CO emissions or the average CO emissions density in the entire corridor.

26

Bus emissions rates, like automobile emissions rates, depend on speed. However, the speed dependence of bus emissions rates has a negligible effect on the results of this example and, therefore, is not included in the calculations.

27

Of course, actual implementation of the reserved lane would yield unique values of δ and of the time savings for buses, not ranges of values. The particular values of δ and of the time savings that would occur (the equilibrium values) can be computed using a mode choice model such as that in table 6.7. The equilibrium time savings is the one that causes the mode switch computed from the mode choice model and the corresponding δ value in the supply side analysis presented here to be equal. The equilibrium δ value is the one that yields the equilibrium time savings.

28

Analytical methods for estimating the effects of pricing measures are essentially the same as the methods that were discussed in connection with transit improvement measures and, therefore, will not be discussed separately here. Price changes are represented as changes in the values of the cost variables of travel demand models, such as the mode choice model described in section 6.6. The work trip results shown in table 6.26 are based on the application of work trip mode choice models similar to the one described in section 6.6. In some cases, automobile ownership levels also were modeled. The nonwork travel demand models that were used in most of the studies are considerably more complex than the work trip models are, owing to the wide range of travel decisions involved in nonwork travel (e.g., travel frequency, destination, mode, and whether to undertake multidestination travel). See reference [73] for a review of current models of nonwork travel demand. The Boston central area study also treated route choice (through a traffic flow assignment model), as it was necessary to predict the extent to which area licensing would cause through travelers to detour around the central area.

29

CO and HC emissions from transit vehicles almost certainly would be small compared to the estimated reductions in automobile emissions. However, transit NO_x emissions could counteract much of the estimated reduction in automobile NO_x emissions.

30

It is possible that the spatial redistribution of emissions that would accompany land use changes of the magnitudes discussed here could affect O_3 concentrations, apart from the effects of changes in total emissions levels. However, the information needed to assess this possibility is not currently available.

31

Of course, it is possible to predict some effects of some measures. For example, it can be predicted that zoning regulations, while they are in effect, will prevent certain types of development from occurring in certain areas. However, this does not provide a firm basis for long-range predictions of land use patterns since zoning regulations are only one of many factors that influence urban land use, and these regulations can be changed easily in response to market forces.

32

Although the discussion in this section has dealt only with regional-scale changes in urban land use patterns, it is also possible to think of land use measures that operate at the level of the individual facility or development. For example, a developer might be required to design a facility so as to minimize CO emissions from vehicles entering and leaving the facility. Such measures may be useful for controlling CO concentrations in the immediate vicinities of the affected facilities.

Chapter 7

1

The material in this chapter is more mathematical than that in the preceding chapters and may be skipped by readers not interested in dispersion modeling.

2

The material presented in this section is discussed in greater detail in references [1–4].

3

Techniques for solving equation (7.1) are not treated in this book. However, it is important to bear in mind that solving this equation and related partial differential equations that are discussed later in this chapter can present formidable computational difficulties.

4

Mathematically, specification of initial and boundary conditions is needed to obtain a unique solution to the system of partial differential equations represented by equation (7.9).

5

In making these comparisons, it is important to choose measurement locations that are consistent with models' capabilities to resolve spatial and temporal variations in concentrations. For example, the computed concentrations characterize concentrations throughout entire grid cells and, hence, cannot capture such detailed spatial variations in concentrations as an NO_2 peak immediately downwind of a high-volume roadway.

6

It is not necessary to neglect vertical winds in order to develop a model of pollutant dispersion in an air column. However, existing operational models of pollutant dispersion in air columns assume that the vertical wind velocity component is negligibly small, and this assumption is used in the discussion here.

7

The definition of S_i that is being used here is different from the one that was used in connection with equations (7.9) and (7.12), in which S_i represented the time rate of change of the concentration of pollutant i due to emissions and, hence, had units of mass per unit volume.

8

It is assumed here that when the height of the column increases, pollutants above the column are mixed into it. This assumption is likely to be true during daylight hours, when O_3 is formed, but is likely to be violated often at night.

9

It is possible to generalize equation (7.21) so that it applies to "boxes" with nonrectangular horizontal cross sections. The equation also can be generalized to permit arbitrary orientations of the wind direction with respect to the box. The size and shape of the box that is used in a box model can significantly affect computed pollutant concentrations. However, systematic procedures for choosing appropriate sizes and shapes have not yet been developed.

10

More detailed discussions of the lagrangian approach to dispersion modeling are available in references [1, 2, 32].

11

A pollutant particle is an idealized object whose motion is governed by the macroscopic movements of the atmosphere. In contrast to molecules, idealized particles do not have brownian motion. Since lagrangian models usually neglect brownian motion, these models are formulated in terms of particles rather than molecules.

12
It is possible to show that under conditions guaranteeing the validity of the mixing length approximation, equations (7.9) and (7.31) are equivalent; see references [1, 2, 32]. Note that as in equation (7.9), it is necessary to assume that ensemble average concentrations closely approximate temporal-spatial average concentrations in order to relate $\langle C \rangle$ in equation (7.31) to measured concentrations.

13
Since the theory on which gaussian models are based does not permit treatment of the chemical reactions involved in NO_2 formation, computations of annual average NO_2 concentrations usually are carried out by assuming that NO_x can be treated as an unreactive pollutant and that NO_2 concentrations are proportional to NO_x concentrations.

14
The integral in equation (7.31) must be approximated by a sum in this case, as the simulation approach provides estimates of the values of the function q at a discrete set of points.

15
The concentration measurements used in testing the simulation model and those used in constructing table 7.2 were obtained during the same series of dispersion experiments [50–52].

16
Several statistical models for estimating O_3 concentrations are described in reference [59]. Statistical models for estimating NO_2 concentrations are described in references [60, 61].

Index

Italic page numbers refer to tables and figures